Applied Combinatorics

Applied Combinatorics

Mitchel T. Keller

Washington and Lee University
Lexington, Virginia

William T. Trotter

Georgia Institute of Technology
Atlanta, Georgia

2017 Edition

Edition: 2017 Edition

Website: http://rellek.net/appcomb/

© 2006–2017   Mitchel T. Keller, William T. Trotter

This work is licensed under the Creative Commons Attribution-ShareAlike 4.0 International License. To view a copy of this license, visit http://creativecommons.org/licenses/by-sa/4.0/ or send a letter to Creative Commons, PO Box 1866, Mountain View, CA 94042, USA.

Summary of Contents

About the Authors	vii
Acknowledgements	ix
Preface	xi
Preface to 2017 Edition	xiii
Preface to 2016 Edition	xv
Prologue	1
1 An Introduction to Combinatorics	3
2 Strings, Sets, and Binomial Coefficients	17
3 Induction	39
4 Combinatorial Basics	59
5 Graph Theory	69
6 Partially Ordered Sets	113
7 Inclusion-Exclusion	141
8 Generating Functions	157
9 Recurrence Equations	183
10 Probability	213
11 Applying Probability to Combinatorics	229
12 Graph Algorithms	239

SUMMARY OF CONTENTS

13 Network Flows   259

14 Combinatorial Applications of Network Flows   279

15 Pólya's Enumeration Theorem   291

16 The Many Faces of Combinatorics   315

A Epilogue   331

B Background Material for Combinatorics   333

C List of Notation   361

Index   363

About the Authors

About William T. Trotter

William T. Trotter is a Professor in the School of Mathematics at Georgia Tech. He was first exposed to combinatorial mathematics through the 1971 Bowdoin Combinatorics Conference which featured an array of superstars of that era, including Gian Carlo Rota, Paul Erdős, Marshall Hall, Herb Ryzer, Herb Wilf, William Tutte, Ron Graham, Daniel Kleitman and Ray Fulkerson. Since that time, he has published more than 120 research papers on graph theory, discrete geometry, Ramsey theory, and extremal combinatorics. Perhaps his best known work is in the area of combinatorics and partially ordered sets, and his 1992 research monograph on this topic has been very influential. (He takes some pride in the fact that this monograph is still in print and copies are being sold in 2016.) He has more than 70 co-authors, but considers his extensive joint work with Graham Brightwell, Stefan Felsner, Peter Fishburn, Hal Kierstead and Endre Szemerédi as representing his best work. His career includes invited presentations at more than 50 international conferences and more than 30 meetings of professional societies. He was the founding editor of the *SIAM Journal on Discrete Mathematics* and has served on the Editorial Board of *Order* since the journal was launched in 1984, and his service includes an eight year stint as Editor-in-Chief. Currently, he serves on the editorial boards of three other journals in combinatorial mathematics.

Still he has his quirks. First, he insists on being called "Tom", as Thomas is his middle name, while continuing to sign as William T. Trotter. Second, he has invested time and energy serving five terms as department/school chair, one at Georgia Tech, two at Arizona State University and two at the University of South Carolina. In addition, he has served as a Vice Provost and as an Assistant Dean. Third, he is fascinated by computer operating systems and is always installing new ones. In one particular week, he put eleven different flavors of Linux on the same machine, interspersed with four complete installs of Windows 7. Incidentally, the entire process started and ended with Windows 7. Fourth, he likes to hit golf balls, not play golf, just hit balls. Without these diversions, he might even have had enough time to settle the Riemann hypothesis.

He has had eleven Ph.D. students, one of which is now his co-author on this text.

About the Authors

About Mitchel T. Keller

Mitchel T. Keller is a super-achiever (this description is written by WTT) extraordinaire from North Dakota. As a graduate student at Georgia Tech, he won a lengthy list of honors and awards, including a VIGRE Graduate Fellowship, an IMPACT Scholarship, a John R. Festa Fellowship and the 2009 Price Research Award. Mitch is a natural leader and was elected President (and Vice President) of the Georgia Tech Graduate Student Government Association, roles in which he served with distinction. Indeed, after completing his terms, his student colleagues voted to establish a continuing award for distinguished leadership, to be named the Mitchel T. Keller award, with Mitch as the first recipient. Very few graduate students win awards in the first place, but Mitch is the only one I know who has an award *named* after them.

Mitch is also a gifted teacher of mathematics, receiving the prestigious Georgia Tech 2008 Outstanding Teacher Award, a campus-wide competition. He is quick to experiment with the latest approaches to teaching mathematics, adopting what works for him while refining and polishing things along the way. He really understands the literature behind active learning and the principles of engaging students in the learning process. Mitch has even taught his more senior (some say ancient) co-author a thing or two and got him to try personal response systems in a large calculus section.

Mitch is off to a fast start in his own research career, and is already an expert in the subject of linear discrepancy. Mitch has also made substantive contributions to a topic known as Stanley depth, which is right at the boundary of combinatorial mathematics and algebraic combinatorics.

After finishing his Ph.D., Mitch received another signal honor, a Marshall Sherfield Postdoctoral Fellowship and spent two years at the London School of Economics. He is presently an Assistant Professor of Mathematics at Washington and Lee University, and a few years down the road, he'll probably be president of something.

On the personal side, Mitch is the keeper of the Mathematics Genealogy Project, and he is a great cook. His desserts are to die for.

Acknowledgements

We are grateful to our colleagues Alan Diaz, Thang Le, Noah Streib, Prasad Tetali and Carl Yerger, who have taught Applied Combinatorics from preliminary versions and have given valuable feedback. As this text is freely available on the internet, we welcome comments, criticisms, suggestions and corrections from anyone who takes a look at our work.

For the 2016 and subsequent editions, we are grateful to Robert A. Beezer, David Farmer, and Kent Morrison for organizing the American Institute of Mathematics workshop on MathBook XML (now PreTeXt) that enabled the new formats to be released. David Farmer's work on the initial conversion from LaTeX to PreTeXt. The PreTeXt Google Group was an important resource in resolving challenges along the way, and Rob Beezer is a wonderfully-responsive developer who gladly put up with any number of feature requests in order to make everything we wanted possible in the PreTeXt edition

Preface

At Georgia Tech, MATH 3012: Applied Combinatorics, is a junior-level course targeted primarily at students pursuing the B.S. in Computer Science. The purpose of the course is to give students a broad exposure to combinatorial mathematics, using applications to emphasize fundamental concepts and techniques. Applied Combinatorics is also required of students seeking the B.S. in Mathematics, and it is one of two discrete mathematics courses that computer engineering students may select to fulfill a breadth requirement. The course will also often contain a selection of other engineering and science majors who are interested in learning more mathematics. As a consequence, in a typical semester, some 250 Georgia Tech students are enrolled in Applied Combinatorics. Students enrolled in Applied Combinatorics at Georgia Tech have already completed the three semester calculus sequence—with many students bypassing one or more of the these courses on the basis of advanced placement scores. Also, the students will know some linear algebra and can at least have a reasonable discussion about vector spaces, bases and dimension.

Our approach to the course is to show students the beauty of combinatorics and how combinatorial problems naturally arise in many settings, particularly in computer science. While proofs are periodically presented in class, the course is not intended to teach students how to write proofs; there are other required courses in our curriculum that meet this need. Students may occasionally be asked to prove small facts, but these arguments are closer to the kind we expect from students in second or third semester calculus as contrasted with proofs we expect from a mathematics major in an upper-division course. Regardless, we cut very few corners, and our text can readily be used by instructors who elect to be even more rigorous in their approach.

This book arose from our feeling that a text that met our approach to Applied Combinatorics was not available. Because of the diverse set of instructors assigned to the course, the standard text was one that covered every topic imaginable (and then some), but provided little depth. We've taken a different approach, attacking the central subjects of the course description to provide exposure, but taking the time to go into greater depth in select areas to give the students a better feel for how combinatorics works. We have also included some results and topics that are not found in other texts at this level but help reveal the nature of combinatorics to students. We want students to understand that combinatorics is a subject that you must feel "in the gut", and we hope that our presentation achieves this goal. The emphasis throughout remains on

Preface

applications, including algorithms. We do not get deeply into the details of what it means for an algorithm to be "efficient", but we do include an informal discussion of the basic principles of complexity, intended to prepare students in computer science, engineering and applied mathematics for subsequent coursework.

The materials included in this book have evolved over time. Early versions of a few chapters date from 2004, but the pace quickened in 2006 when the authors team taught a large section of Applied Combinatorics. In the last five years, existing chapters have been updated and expanded, while new chapters have been added. As matters now stand, our book includes more material than we can cover in a single semester. We feel that the topics of Chapters 1–9 plus Chapters 12, 13 and 14 are the core of a one semester course in Applied Combinatorics. Additional topics can then be selected from the remaining chapters based on the interests of the instructor and students.

Mitchel T. Keller and William T. Trotter
Lexington, Virginia, and Atlanta, Georgia

Preface to 2017 Edition

Because I (MTK) didn't have the chance to teach from this book during the 2016–2017 academic year, there were few opportunities to examine some of the areas where improvements are due in the text. That said, some changes suggested in the Preface to 2016 Edition did come to fruition in this edition. In particular, the numbering of many things in Chapter 8 will not match the 2016 Edition in a number of places because of the addition of Example 8.7 to address the coefficients on $1/(1-x)^n$ in a way that doesn't require calculus. There is also one new exercise in Chapter 8, which has been placed at the end to retain consistency of numbering. Other than correcting errors, there have been no changes to the exercises, so faculty members teaching from the text may continue to assign the same exercise numbers with confidence that they are the same exercises they have been in the past.

The other notable update in this edition is the addition of a number of SageMathCells to Chapter 8 (including in the exercises), Section 9.6, and the Discussion that ends Chapter 9. The practice of vaguely referring readers to a generic computer algebra system but not providing any advice on how to use it had always been unsatisfying. I know there are places where further refinement is in order, but this edition starts a more coherent approach toward using technology for some of the unpleasant algebraic aspects of the text. Readers can edit the content of the SageMathCells in the body of the text in order to use them to tackle other problems, and those in the exercises are there for convenience more than anything and include only a bare skeleton of what might be useful for the exercise. Since SageMath is open source and can be run for free on CoCalc, this approach seems greatly preferable to targeting a commercial CAS. *For those, like me, who are coming to SageMath with experience using a commercial CAS, SageMath does not do implied multiplication in input very well. When a result comes up that seems strange, my first step is always to make sure that I'm not missing a * in my code.*

Of course, even in a text that's been in use for over a decade, there are typos. A number of small issues were resolved in this edition. The errata page `http://www.rellek.net/appcomb/errata/` lists the dozen mistakes corrected in this edition. Undoubtedly, there are other mistakes waiting to be found, and we welcome reports from readers. (Pull requests on GitHub are also welcome!)

What's next? We'd love to hear from readers with suggestions, but I anticipate that expanding Chapter 4 will be high on the list. If you're using SageMath (or Python)

Preface to 2017 Edition

alongside our text, contributions of code snippets that would be worth including are also welcomed.

Mitchel T. Keller
Lexington, Virginia

Preface to 2016 Edition

In April 2016, the American Institute of Mathematics hosted a weeklong workshop in San Jose to introduce authors of open textbooks to git and Robert A. Beezer's MathBook XML authoring language designed to seamlessly produce HTML, LaTeX, and other formats from a common XML source file. I (MTK) attended and eagerly began the conversion of existing LaTeX source for this now decade-old project into MathBook XML. David Farmer deserves an enormous amount of credit for automating much of the process through a finely-tuned script, but the code produced still required a good deal of cleanup. This edition, the first not labeled "preliminary", will hopefully become the first of many annual editions of *Applied Combinatorics* released under an open source license.

The main effort in producing this 2016 edition was to successfully convert to MathBook XML. Along the way, I attempted to correct all typographical errors we had noted in the past. There are undoubtedly more errors (typographical or otherwise) that will be corrected in future years, so please contact us via email if you spot any. The text now has an index, which may prove more helpful than searching PDF files when looking for the most essential locations of some common terms. Since MathBook XML makes it easy, we now also have a list of notation. Instructors will likely be glad to know that there were no changes to the exercises, so lists of assigned exercises from past years remain completely valid.[1] The only significant changes to the body of the text was to convert WTT's code snippets from C to Python/SageMath. This has allowed us to embed interactive SageMath cells that readers who use the HTML version of the text can run and edit. We've only scratched the surface with this powerful feature of MathBook XML, so look for more SageMath additions in future years.

The conversion to MathBook XML allows us to make a wider variety of formats available:

- HTML: With responsive design using CSS, we feel that the text now looks beautiful on personal computers, tablets, and even mobile phones. No longer will students be frantically resizing a PDF on their phone in order to try to read a passage from the text. The "knowls" offered in HTML also allow references to images, tables, and

---

[1] There are two exceptions to this. The first is that Exercise 8.8.7 has been modified to make the computations involved cleaner. We have preserved the exercise that was previously in this position as Exercise 8.8.27 and added a hint. The second is that a coefficient was changed in Exercise 9.9.15 to make the exercise feasible.

Preface to 2016 Edition

even theorems from other pages (or even a distance away on the same page) to provide a copy of the image/table/theorem right there, and another click/tap makes it disappear.

- PDF: Not much is changed here from previous years, other than the PDF is produced from the LaTeX that MathBook XML generates, and so numbering and order is consistent with the HTML version.

- Print: Campus bookstores have frequently produced printed versions of the text from the PDF provided online, but we have not previously been able to provide a printed, bound version for purchase. With the 2016 edition, we are pleased to launch a print version available through a number of online purchase channels. Campus bookstores may also acquire the book through wholesale channels for sale directly to students. Because of the CreativeCommons license under which the text is released, campuses retain the option of selling their own printed version of the text for students, although this is likely only financially advantageous to students if only a few chapters of the text are being used.

We have some ideas for what might be updated for the 2017 Edition (e.g., Chapter 4 needs to be expanded both in the body and the exercises, Chapter 8 would benefit from integration of SageMath to assist with generating function computations, and Chapter 16 is still not really finished). However, we would love to hear from those of you who are using the text, too. Are there additional topics you'd like to see added? Chapters in need of more exercises? Topics whose exposition could be improved? Please reach out to us via email, and we'll consider your suggestions.

Mitchel T. Keller
Lexington, Virginia

Contents

About the Authors   vii

Acknowledgements   ix

Preface   xi

Preface to 2017 Edition   xiii

Preface to 2016 Edition   xv

Prologue   1

1 An Introduction to Combinatorics   **3**
- 1.1 Introduction 3
- 1.2 Enumeration 4
- 1.3 Combinatorics and Graph Theory 5
- 1.4 Combinatorics and Number Theory 8
- 1.5 Combinatorics and Geometry 11
- 1.6 Combinatorics and Optimization 13
- 1.7 Sudoku Puzzles 15
- 1.8 Discussion 16

2 Strings, Sets, and Binomial Coefficients   **17**
- 2.1 Strings: A First Look 17
- 2.2 Permutations 19
- 2.3 Combinations 21
- 2.4 Combinatorial Proofs 22
- 2.5 The Ubiquitous Nature of Binomial Coefficients 25
- 2.6 The Binomial Theorem 29
- 2.7 Multinomial Coefficients 29
- 2.8 Discussion 31
- 2.9 Exercises 32

3 Induction   **39**
- 3.1 Introduction 39

Contents

- 3.2 The Positive Integers are Well Ordered 40
- 3.3 The Meaning of Statements 40
- 3.4 Binomial Coefficients Revisited 42
- 3.5 Solving Combinatorial Problems Recursively 43
- 3.6 Mathematical Induction 48
- 3.7 Inductive Definitions 49
- 3.8 Proofs by Induction 50
- 3.9 Strong Induction 53
- 3.10 Discussion 54
- 3.11 Exercises 55

4 Combinatorial Basics   59
- 4.1 The Pigeon Hole Principle 59
- 4.2 An Introduction to Complexity Theory 60
- 4.3 The Big "Oh" and Little "Oh" Notations 63
- 4.4 Exact Versus Approximate 64
- 4.5 Discussion 66
- 4.6 Exercises 67

5 Graph Theory   69
- 5.1 Basic Notation and Terminology for Graphs 69
- 5.2 Multigraphs: Loops and Multiple Edges 74
- 5.3 Eulerian and Hamiltonian Graphs 75
- 5.4 Graph Coloring 80
- 5.5 Planar Graphs 88
- 5.6 Counting Labeled Trees 96
- 5.7 A Digression into Complexity Theory 100
- 5.8 Discussion 101
- 5.9 Exercises 102

6 Partially Ordered Sets   113
- 6.1 Basic Notation and Terminology 114
- 6.2 Additional Concepts for Posets 119
- 6.3 Dilworth's Chain Covering Theorem and its Dual 122
- 6.4 Linear Extensions of Partially Ordered Sets 125
- 6.5 The Subset Lattice 126
- 6.6 Interval Orders 128
- 6.7 Finding a Representation of an Interval Order 129
- 6.8 Dilworth's Theorem for Interval Orders 131
- 6.9 Discussion 133

	6.10 Exercises	134

7 Inclusion-Exclusion — 141
- 7.1 Introduction 141
- 7.2 The Inclusion-Exclusion Formula 144
- 7.3 Enumerating Surjections 145
- 7.4 Derangements 147
- 7.5 The Euler ϕ Function 149
- 7.6 Discussion 151
- 7.7 Exercises 151

8 Generating Functions — 157
- 8.1 Basic Notation and Terminology 157
- 8.2 Another look at distributing apples or folders 160
- 8.3 Newton's Binomial Theorem 165
- 8.4 An Application of the Binomial Theorem 166
- 8.5 Partitions of an Integer 168
- 8.6 Exponential generating functions 170
- 8.7 Discussion 173
- 8.8 Exercises 174

9 Recurrence Equations — 183
- 9.1 Introduction 183
- 9.2 Linear Recurrence Equations 186
- 9.3 Advancement Operators 187
- 9.4 Solving advancement operator equations 190
- 9.5 Formalizing our approach to recurrence equations 198
- 9.6 Using generating functions to solve recurrences 202
- 9.7 Solving a nonlinear recurrence 205
- 9.8 Discussion 207
- 9.9 Exercises 210

10 Probability — 213
- 10.1 An Introduction to Probability 214
- 10.2 Conditional Probability and Independent Events 216
- 10.3 Bernoulli Trials 217
- 10.4 Discrete Random Variables 218
- 10.5 Central Tendency 220
- 10.6 Probability Spaces with Infinitely Many Outcomes 224
- 10.7 Discussion 225

10.8 Exercises 226

11 Applying Probability to Combinatorics 229
11.1 A First Taste of Ramsey Theory 229
11.2 Small Ramsey Numbers 230
11.3 Estimating Ramsey Numbers 231
11.4 Applying Probability to Ramsey Theory 232
11.5 Ramsey's Theorem 233
11.6 The Probabilistic Method 234
11.7 Discussion 235
11.8 Exercises 236

12 Graph Algorithms 239
12.1 Minimum Weight Spanning Trees 239
12.2 Digraphs 245
12.3 Dijkstra's Algorithm for Shortest Paths 245
12.4 Historical Notes 252
12.5 Exercises 253

13 Network Flows 259
13.1 Basic Notation and Terminology 259
13.2 Flows and Cuts 261
13.3 Augmenting Paths 263
13.4 The Ford-Fulkerson Labeling Algorithm 267
13.5 A Concrete Example 269
13.6 Integer Solutions of Linear Programming Problems 273
13.7 Exercises 274

14 Combinatorial Applications of Network Flows 279
14.1 Introduction 279
14.2 Matchings in Bipartite Graphs 280
14.3 Chain partitioning 284
14.4 Exercises 287

15 Pólya's Enumeration Theorem 291
15.1 Coloring the Vertices of a Square 292
15.2 Permutation Groups 294
15.3 Burnside's Lemma 297
15.4 Pólya's Theorem 299
15.5 Applications of Pólya's Enumeration Formula 303
15.6 Exercises 310

16 The Many Faces of Combinatorics — 315
- 16.1 On-line algorithms 315
- 16.2 Extremal Set Theory 318
- 16.3 Markov Chains 320
- 16.4 The Stable Matching Theorem 322
- 16.5 Zero–One Matrices 323
- 16.6 Arithmetic Combinatorics 325
- 16.7 The Lovász Local Lemma 326
- 16.8 Applying the Local Lemma 328

A Epilogue — 331

B Background Material for Combinatorics — 333
- B.1 Introduction 333
- B.2 Intersections and Unions 334
- B.3 Cartesian Products 337
- B.4 Binary Relations and Functions 337
- B.5 Finite Sets 339
- B.6 Notation from Set Theory and Logic 340
- B.7 Formal Development of Number Systems 341
- B.8 Multiplication as a Binary Operation 344
- B.9 Exponentiation 345
- B.10 Partial Orders and Total Orders 346
- B.11 A Total Order on Natural Numbers 347
- B.12 Notation for Natural Numbers 348
- B.13 Equivalence Relations 350
- B.14 The Integers as Equivalence Classes of Ordered Pairs . 350
- B.15 Properties of the Integers 351
- B.16 Obtaining the Rationals from the Integers 353
- B.17 Obtaining the Reals from the Rationals 355
- B.18 Obtaining the Complex Numbers from the Reals 356
- B.19 The Zermelo-Fraenkel Axioms of Set Theory 358

C List of Notation — 361

Index — 363

Prologue

A unique feature of this book is a recurring cast of characters: Alice, Bob, Carlos, Dave, Xing, Yolanda and Zori. They are undergraduate students at Georgia Tech, they're taking an 8:05am section of Math 3012: Applied Combinatorics, and they frequently go for coffee at the Clough Undergraduate Learning Center immediately after the class is over. They've become friends of sorts and you may find their conversations about Applied Combinatorics of interest, as they will may reveal subtleties behind topics currently being studied, reinforce connections with previously studied material or set the table for topics which will come later. Sometimes, these conversations will set aside in a clearly marked *Discussion* section, but they will also be sprinkled as brief remarks throughout the text.

In time, you will get to know these characters and will sense that, for example, when Dave comments on a topic, it will represent a perspective that Zori is unlikely to share. Some comments are right on target while others are "out in left field." Some may even be humorous, at least we hope this is the case. Regardless, our goal is not to entertain—although that is not all that bad a side benefit. Instead, we intend that our informal approach adds to the instructional value of our text.

Now it is time to meet our characters:

Alice is a computer engineering major from Philadelphia. She is ambitious, smart and intense. Alice is quick to come to conclusions, most of which are right. On occasion, Alice is not kind to Bob.

Bob is a management major from Omaha. He is a hard working and conscientious. Bob doesn't always keep pace with his friends, but anything he understands, he owns, and in the end, he gets almost everything. On the other hand, Bob has never quite understood why Alice is short with him at times.

Carlos is a really, really smart physics major from San Antonio. He has three older brothers and two sisters, one older, one younger. His high school background wasn't all that great, but Carlos is clearly a special student at Georgia Tech. He absorbs new concepts at lightning speed and sees through to the heart of almost every topic. He thinks carefully before he says something and is admirably polite.

Dave is a discrete math major from Los Angeles. Dave is a flake. He's plenty smart enough but not all that diligent. Still, he has unique insights into things and from time to time says something worth hearing—not always but sometimes. His friends say that Dave suffers from occasional brain–mouth disconnects.

Prologue

Xing is a computer science major from New York. Xing's parents immigrated from Beijing, and he was strongly supported and encouraged in his high school studies. Xing is detail oriented and not afraid to work hard.

Yolanda is a double major (computer science and chemistry) from Cumming, a small town just north of Atlanta. Yolanda is the first in her extended family to go to a college or university. She is smart and absorbs knowledge like a sponge. It's all new to her and her horizons are raised day by day.

Zori is an applied math major from Detroit. She is bottom-line focused, has little time for puzzles and always wants to see applications to justify why something is included in the course. Zori is determined, driven and impatient at times.

CHAPTER 1

An Introduction to Combinatorics

As we hope you will sense right from the beginning, we believe that combinatorial mathematics is one of the most fascinating and captivating subjects on the planet. Combinatorics is *very* concrete and has a wide range of applications, but it also has an intellectually appealing theoretical side. Our goal is to give you a taste of both. In order to begin, we want to develop, through a series of examples, a feeling for what types of problems combinatorics addresses.

1.1 Introduction

There are three principal themes to our course:

Discrete Structures Graphs, digraphs, networks, designs, posets, strings, patterns, distributions, coverings, and partitions.

Enumeration Permutations, combinations, inclusion/exclusion, generating functions, recurrence relations, and Pólya counting.

Algorithms and Optimization Sorting, eulerian circuits, hamiltonian cycles, planarity testing, graph coloring, spanning trees, shortest paths, network flows, bipartite matchings, and chain partitions.

To illustrate the accessible, concrete nature of combinatorics and to motivate topics that we will study, this preliminary chapter provides a first look at combinatorial problems, choosing examples from enumeration, graph theory, number theory, and optimization. The discussion is very informal—but this should serve to explain why we have to be more precise at later stages. We ask lots of questions, but at this stage, you'll only be able to answer a few. Later, you'll be able to answer many more ... but as promised earlier, most likely you'll never be able to answer them all. And if we're wrong in making that statement, then you're certain to become *very* famous. Also, you'll get an A++ in the course and maybe even a Ph.D. too.

Chapter 1 An Introduction to Combinatorics

1.2 Enumeration

Many basic problems in combinatorics involve counting the number of distributions of objects into cells—where we may or may not be able to distinguish between the objects and the same for the cells. Also, the cells may be arranged in patterns. Here are concrete examples.

Amanda has three children: Dawn, Keesha and Seth.

1. Amanda has ten one dollar bills and decides to give the full amount to her children. How many ways can she do this? For example, one way she might distribute the funds is to give Dawn and Keesha four dollars each with Seth receiving the balance—two dollars. Another way is to give the entire amount to Keesha, an option that probably won't make Dawn and Seth very happy. Note that hidden within this question is the assumption that Amanda does not distinguish the individual dollar bills, say by carefully examining their serial numbers. Instead, we intend that she need only decide the *amount* each of the three children is to receive.

2. The amounts of money distributed to the three children form a sequence which if written in non-increasing order has the form: a_1, a_2, a_3 with $a_1 \geq a_2 \geq a_3$ and $a_1 + a_2 + a_3 = 10$. How many such sequences are there?

3. Suppose Amanda decides to give each child at least one dollar. How does this change the answers to the first two questions?

4. Now suppose that Amanda has ten books, in fact the top 10 books from the New York Times best-seller list, and decides to give them to her children. How many ways can she do this? Again, we note that there is a hidden assumption—the ten books are all different.

5. Suppose the ten books are labeled $B_1, B_2, \ldots, B_{10}$. The sets of books given to the three children are pairwise disjoint and their union is $\{B_1, B_2, \ldots, B_{10}\}$. How many different sets of the form $\{S_1, S_2, S_3\}$ where S_1, S_2 and S_3 are pairwise disjoint and $S_1 \cup S_2 \cup S_3 = \{B_1, B_2, \ldots, B_{10}\}$?

6. Suppose Amanda decides to give each child at least one book. How does this change the answers to the preceding two questions?

7. How would we possibly answer these kinds of questions if ten was really ten thousand (OK, we're not talking about children any more!) and three was three thousand? Could you write the answer on a single page in a book?

A circular necklace with a total of six beads will be assembled using beads of three different colors. In Figure 1.1, we show four such necklaces—however, note that the first three are actually the *same* necklace. Each has three red beads, two blues and one green. On the other hand, the fourth necklace has the same number of beads of each color but it is a *different* necklace.

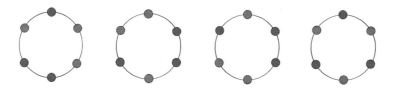

FIGURE 1.1: NECKLACES MADE WITH THREE COLORS

1. How many different necklaces of six beads can be formed using three reds, two blues and one green?

2. How many different necklaces of six beads can be formed using red, blue and green beads (not all colors have to be used)?

3. How many different necklaces of six beads can be formed using red, blue and green beads if all three colors have to be used?

4. How would we possibly answer these questions for necklaces of six thousand beads made with beads from three thousand different colors? What special software would be required to find the exact answer and how long would the computation take?

1.3 Combinatorics and Graph Theory

A **graph** G consists of a **vertex** set V and a collection E of 2-element subsets of V. Elements of E are called **edges**. In our course, we will (almost always) use the convention that $V = \{1, 2, 3, \ldots, n\}$ for some positive integer n. With this convention, graphs can be described *precisely* with a text file:

1. The first line of the file contains a single integer n, the number of vertices in the graph.

2. Each of the remaining lines of the file contains a pair of distinct integers and specifies an edge of the graph.

We illustrate this convention in Figure 1.2 with a text file and the diagram for the graph G it defines.

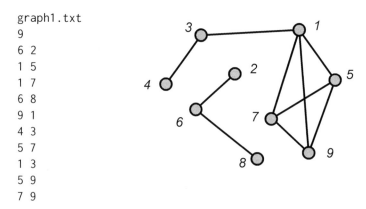

```
graph1.txt
9
6 2
1 5
1 7
6 8
9 1
4 3
5 7
1 3
5 9
7 9
```

FIGURE 1.2: A GRAPH DEFINED BY DATA

Much of the notation and terminology for graphs is quite natural. See if you can make sense out of the following statements which apply to the graph G defined above:

1. G has 9 vertices and 10 edges.

2. $\{2, 6\}$ is an edge.

3. Vertices 5 and 9 are adjacent.

4. $\{5, 4\}$ is not an edge.

5. Vertices 3 and 7 are not adjacent.

6. $P = (4, 3, 1, 7, 9, 5)$ is a path of length 5 from vertex 4 to vertex 5.

7. $C = (5, 9, 7, 1)$ is cycle of length 4.

8. G is disconnected and has two components. One of the components has vertex set $\{2, 6, 8\}$.

9. $\{1, 5, 7\}$ is a triangle.

10. $\{1, 7, 5, 9\}$ is a clique of size 4.

11. $\{4, 2, 8, 5\}$ is an independent set of size 4.

Equipped only with this little bit of background material, we are already able to pose a number of interesting and challenging problems.

Example 1.3. Consider the graph *G* shown in Figure 1.4.

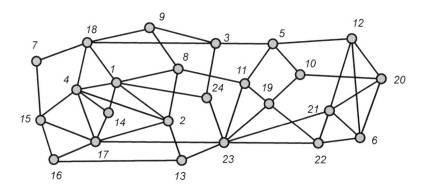

FIGURE 1.4: A CONNECTED GRAPH

1. What is the largest k for which G has a path of length k?

2. What is the largest k for which G has a cycle of length k?

3. What is the largest k for which G has a clique of size k?

4. What is the largest k for which G has an independent set of size k?

5. What is the shortest path from vertex 7 to vertex 6?

Suppose we gave the class a text data file for a graph on 1500 vertices and asked whether the graph contains a cycle of length at least 500. Raoul says yes and Carla says no. How do we decide who is right?

Suppose instead we asked whether the graph has a clique of size 500. Helene says that she doesn't think so, but isn't certain. Is it reasonable that her classmates insist that she make up her mind, one way or the other? Is determining whether this graph has a clique of size 500 harder, easier or more or less the same as determining whether it has a cycle of size 500.

We will frequently study problems in which graphs arise in a very natural manner. Here's an example.

Chapter 1 An Introduction to Combinatorics

Example 1.5. In Figure 1.6, we show the location of some radio stations in the plane, together with a scale indicating a distance of 200 miles. Radio stations that are closer than 200 miles apart must broadcast on different frequencies to avoid interference.

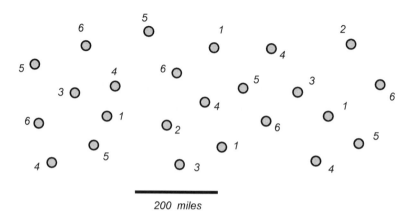

FIGURE 1.6: RADIO STATIONS

We've shown that 6 different frequencies are enough. Can you do better?
Can you find 4 stations each of which is within 200 miles of the other 3? Can you find 8 stations each of is more than 200 miles away from the other 7? Is there a natural way to define a graph associated with this problem?

Example 1.7. How big must an applied combinatorics class be so that there are either (a) six students with each pair having taken at least one other class together, or (b) six students with each pair together in a class for the first time. Is this really a hard problem or can we figure it out in just a few minutes, scribbling on a napkin?

1.4 Combinatorics and Number Theory

Broadly, number theory concerns itself with the properties of the positive integers. G.H. Hardy was a brilliant British mathematician who lived through both World Wars and conducted a large deal of number-theoretic research. He was also a pacifist who was happy that, from his perspective, his research was not "useful". He wrote in his 1940 essay *A Mathematician's Apology* "[n]o one has yet discovered any warlike purpose to be served by the theory of numbers or relativity, and it seems very unlikely that anyone will do so for many years."[1] Little did he know, the purest mathematical ideas

---

[1] G.H. Hardy, *A Mathematician's Apology*, Cambridge University Press, p. 140. (1993 printing)

of number theory would soon become indispensable for the cryptographic techniques that kept communications secure. Our subject here is not number theory, but we will see a few times where combinatorial techniques are of use in number theory.

Example 1.8. Form a sequence of positive integers using the following rules. Start with a positive integer $n > 1$. If n is odd, then the next number is $3n + 1$. If n is even, then the next number is $n/2$. Halt if you ever reach 1. For example, if we start with 28, the sequence is

$$28, 14, 7, 22, 11, 34, 17, 52, 26, 13, 40, 20, 10, 5, 16, 8, 4, 2, 1.$$

Now suppose you start with 19. Then the first few terms are

$$19, 58, 29, 88, 44, 22.$$

But now we note that the integer 22 appears in the first sequence, so the two sequences will agree from this point on. Sequences formed by this rule are called **Collatz sequences**.

Pick a number somewhere between 100 and 200 and write down the sequence you get. Regardless of your choice, you will eventually halt with a 1. However, is there some positive integer n (possibly quite large) so that if you start from n, you will never reach 1?

Example 1.9. Students in middle school are taught to add fractions by finding least common multiples. For example, the least common multiple of 15 and 12 is 60, so:

$$\frac{2}{15} + \frac{7}{12} = \frac{8}{60} + \frac{35}{60} = \frac{43}{60}.$$

How hard is it to find the least common multiple of two integers?

It's really easy if you can factor them into primes. For example, consider the problem of finding the least common multiple of 351785000 and 316752027900 if you just happen to know that

$$351785000 = 2^3 \times 5^4 \times 7 \times 19 \times 23^2 \quad \text{and}$$
$$316752027900 = 2^2 \times 3 \times 5^2 \times 7^3 \times 11 \times 23^4.$$

Then the least common multiple is

$$300914426505000 = 2^3 \times 3 \times 5^4 \times 7^3 \times 11 \times 19 \times 23^4.$$

So to find the least common multiple of two numbers, we just have to factor them into primes. That doesn't sound too hard. For starters, can you factor 1961? OK, how about 1348433? Now for a real challenge. Suppose you are told that the integer

$$c = 556849011707703570824428317333504052171636923558995115 0965$$

Chapter 1 An Introduction to Combinatorics

$$2043138898236817075547572153799$$

is the product of two primes a and b. Can you find them?

What if factoring is hard? Can you find the least common multiple of two relatively large integers, say each with about 500 digits, by another method? How should middle school students be taught to add fractions?

As an aside, we note that most calculators can't add or multiply two 20 digits numbers, much less two numbers with more than 500 digits. But it is relatively straightforward to write a computer program that will do the job for us. Also, there are some powerful mathematical software tools available. Two very well known commercial examples are *Maple*® and *Mathematica*®. In this text, we will from time to time, make use of the open source computer algebra system SageMath. We will sometimes embed interactive SageMath cells in the text, but you can also use SageMath for free online via the SageMath Cloud. For example, the SageMath cell below will produce the factorization shown above.

```
factor(300914426505000)
```

2^3 * 3 * 5^4 * 7^3 * 11 * 19 * 23^4

If you're reading this text in a web browser, go ahead and change the integer in the SageMath cell above to some other, perhaps larger, integer and click the button again to get the prime factorization of your new integer.

Now here's how we made up the challenge problem. First, we found a site on the web that lists large primes and found these two values:

$$a = 24259676230523707727576331569769824 69681 \quad \text{and}$$
$$b = 22953686867719691230002707821868552 601124472329079.$$

The SageMath code below calculates $a \times b$, and returns the result instantly.

```
a = 24259676230523707727576331569769824 69681
b = 22953686867719691230002707821868552 601124472329079
a*b
```

On the other hand, if you ask SageMath to factor c, as in the cell below, you'll likely be waiting a long time. If you get a response in more than a couple of minutes, please email us so that we can update the text with larger primes a and b!

```
factor(5568490117077035708244283173335 0405217163692\
3558995115096520431388982368170755475 72153799)
```

Questions arising in number theory can also have an enumerative flair, as the following example shows.

Example 1.10. In Figure 1.11, we show the integer partitions of 8.

8 distinct parts	4+1+1+1+1
7+1 distinct parts, odd parts	3+3+2
6+2 distinct parts	3+3+1+1 odd parts
6+1+1	3+2+2+1
5+3 distinct parts, odd parts	3+2+1+1+1
5+2+1 distinct parts	3+1+1+1+1+1 odd parts
5+1+1+1 odd parts	2+2+2+2
4+4	2+2+2+1+1
4+3+1 distinct parts	2+2+1+1+1+1
4+2+2	2+1+1+1+1+1+1
4+2+1+1	1+1+1+1+1+1+1+1 odd parts

FIGURE 1.11: THE PARTITIONS OF 8, NOTING THOSE INTO DISTINCT PARTS AND THOSE INTO ODD PARTS.

There are 22 partitions altogether, and as noted, exactly 6 of them are partitions of 8 into odd parts. Also, exactly 6 of them are partitions of 8 into distinct parts.

What would be your reaction if we asked you to find the number of integer partitions of 25892? Do you think that the number of partitions of 25892 into odd parts equals the number of partitions of 25892 into distinct parts? Is there a way to answer this question *without* actually calculating the number of partitions of each type?

1.5 Combinatorics and Geometry

There are many problems in geometry that are innately combinatorial or for which combinatorial techniques shed light on the problem.

Chapter 1 An Introduction to Combinatorics

Example 1.12. In Figure 1.13, we show a family of 4 lines in the plane. Each pair of lines intersects and no point in the plane belongs to more than two lines. These lines determine 11 regions.

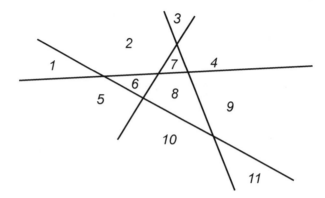

FIGURE 1.13: LINES AND REGIONS

Under these same restrictions, how many regions would a family of 8947 lines determine? Can different arrangements of lines determine different numbers of regions?

Example 1.14. Mandy says she has found a set of 882 points in the plane that determine exactly 752 lines. Tobias disputes her claim. Who is right?

Example 1.15. There are many different ways to draw a graph in the plane. Some drawings may have crossing edges while others don't. But sometimes, crossing edges must appear in any drawing. Consider the graph G shown in Figure 1.16.

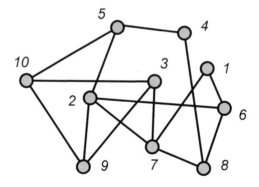

FIGURE 1.16: A GRAPH WITH CROSSING EDGES

Can you redraw G without crossing edges?

Suppose Sam and Deborah were given a homework problem asking whether a particular graph on 2843952 vertices and 9748032 edges could be drawn without edge crossings. Deborah just looked at the number of vertices and the number of edges and said that the answer is "no." Sam questions how she can be so certain—without looking more closely at the structure of the graph. Is there a way for Deborah to justify her definitive response?

1.6 Combinatorics and Optimization

You likely have already been introduced to optimization problems, as calculus students around the world are familiar with the plight of farmers trying to fence the largest area of land given a certain amount of fence or people needing to cross rivers downstream from their current location who must decide where they should cross based on the speed at which they can run and swim. However, these problems are inherently continuous. In theory, you can cross the river at any point you want, even if it were irrational. (OK, so not exactly irrational, but a good decimal approximation.) In this course, we will examine a few optimization problems that are not continuous, as only integer values for the variables will make sense. It turns out that many of these problems are very hard to solve in general.

Example 1.17. In Figure 1.18, we use letters for the labels on the vertices to help distinguish visually from the integer weights on the edges.

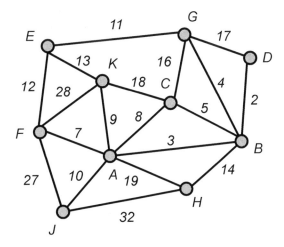

FIGURE 1.18: A LABELED GRAPH WITH WEIGHTED EDGES

Suppose the vertices are cities, the edges are highways and the weights on the edges represent **distance**.

1. What is the shortest path from vertex E to vertex B?

2. Suppose Ariel is a salesperson whose home base is city A. In what order should Ariel visit the other cities so that she goes through each of them at least once and returns home at the end—while keeping the total distance traveled to a minimum? Can Ariel accomplish such a tour visiting each city *exactly* once?

3. Sanjay is a highway inspection engineer and must traverse every highway each month. Sanjay's homebase is City E. In what order should Sanjay traverse the highways to minimize the total distance traveled? Can Sanjay make such a tour traveling along each highway exactly once?

Example 1.19. Now suppose that the vertices are locations of branch banks in Atlanta and that the weights on an edge represents the cost, in millions of dollars, of building a high capacity data link between the branch banks at it two end points. In this model, if there is no edge between two branch banks, it means that the cost of building a data link between this particular pair is prohibitively high (here we might be tempted to say the cost is infinite, but the authors don't admit to knowing the meaning of this word).

Our challenge is to decide which data links should be constructed to form a network in which any branch bank can communicate with any other branch. We assume that data can flow in either direction on a link, should it be built, and that data can be relayed through any number of data links. So to allow full communication, we should construct a **spanning tree** in this network. In Figure 1.20, we show a graph G on the left and one of its many spanning trees on the right. The weight of the spanning tree is the sum of the weights on the edges. In our model, this represents the costs, again in millions of dollars, of building the data links associated with the edges in the spanning tree. For the spanning tree shown in Figure 1.20, this total is

$$12 + 25 + 19 + 18 + 23 + 19 = 116.$$

Of all spanning trees, the bank would naturally like to find one having minimum weight.

How many spanning trees does this graph have? For a large graph, say one with 2875 vertices, does it make sense to find all spanning trees and simply take the one with minimum cost? In particular, for a positive integer n, how many trees have vertex set $\{1, 2, 3, \ldots, n\}$?

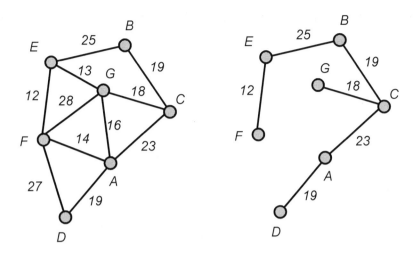

FIGURE 1.20: A WEIGHTED GRAPH AND SPANNING TREE

1.7 Sudoku Puzzles

Here's an example which has more substance than you might think at first glance. It involves Sudoku puzzles, which have become immensely popular in recent years.

Example 1.21. A **Sudoku puzzle** is a 9×9 array of cells that when completed have the integers $1, 2, \ldots, 9$ appearing exactly once in each row and each column. Also (and this is what makes the puzzles so fascinating), the numbers $1, 2, 3, \ldots, 9$ appear once in each of the nine 3×3 subsquares identified by the darkened borders. To be considered a legitimate Sudoku puzzle, there should be a *unique* solution. In Figure 1.22, we show two Sudoku puzzles. The one on the right is fairly easy, and the one on the left is far more challenging. There are many sources of Sudoku puzzles, and software that generates Sudoku puzzles and then allows you to play them with an attractive GUI is available for all operating systems we know anything about (although not recommend to play them during class!). Also, you can find Sudoku puzzles on the web at: http://www.websudoku.com. On this site, the "Evil" ones are just that.

How does Rory make up good Sudoku puzzles, ones that are difficult for Mandy to solve? How could Mandy use a computer to solve puzzles that Rory has constructed? What makes some Sudoku puzzles easy and some of them hard?

The size of a Sudoku puzzle can be expanded in an obvious way, and many newspapers include a 16×16 Sudoku puzzle in their Sunday edition (just next to a challenging crosswords puzzle). How difficult would it be to solve a 1024×1024 Sudoku puzzle, even if you had access to a powerful computer?

Chapter 1 An Introduction to Combinatorics

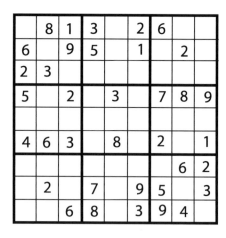

FIGURE 1.22: SUDOKU PUZZLES

1.8 Discussion

Over coffee after their first combinatorics class, Xing remarked "This doesn't seem to be going like calculus. I'm expecting the professor to teach us how to solve problems—at least some kinds of problems. Instead, a whole bunch of problems were posed and we were asked whether *we* could solve them."

Yolanda jumped in, saying "You may be judging things too quickly. I'm fascinated by these kinds of questions. They're different."

Zori grumpily laid bare her concerns: "After getting out of Georgia Tech, who's going to pay me to count necklaces, distribute library books or solve Sudoku puzzles?"

Bob politely countered, "But the problems on networks and graphs seemed to have practical applications. I heard my uncle, a very successful business guy, talk about franchising problems that sound just like those."

Alice speculated, "All those network problems sound the same to me. A fair to middling computer science major could probably write programs to solve any of them."

Dave mumbled, "Maybe not. Similar sounding problems might actually be quite different in the end. Maybe we'll learn to tell the difference."

After a bit of quiet time interrupted only by lattes disappearing, Carlos said softly, "It might not be so easy to distinguish hard problems from easy ones."

Alice followed, "Regardless, what strikes me is that we all, well almost all of us," she said, rolling her eyes at Bob, "seem to understand everything talked about in class today. It was so very concrete. I liked that."

CHAPTER 2

Strings, Sets, and Binomial Coefficients

Much of combinatorial mathematics can be reduced to the study of strings, as they form the basis of all written human communications. Also, strings are the way humans communicate with computers, as well as the way one computer communicates with another. As we shall see, sets and binomial coefficients are topics that fall under the string umbrella. So it makes sense to begin our in-depth study of combinatorics with strings.

2.1 Strings: A First Look

Let n be a positive integer. Throughout this text, we will use the shorthand notation $[n]$ to denote the n-element set $\{1, 2, \ldots, n\}$. Now let X be a set. Then a function $s \colon [n] \to X$ is also called an *X*-**string of length** n. In discussions of X-strings, it is customary to refer to the elements of X as **characters**, while the element $s(i)$ is the ith character of s. Whenever practical, we prefer to denote a string s by writing $s = "x_1 x_2 x_3 \ldots x_n"$, rather than the more cumbersome notation $s(1) = x_1, s(2) = x_2, \ldots, s(n) = x_n$.

There are a number of alternatives for the notation and terminology associated with strings. First, the characters in a string s are frequently written using subscripts as $s_1, s_2, \ldots, s_n$, so the ith-term of s can be denoted s_i rather than $s(i)$. Strings are also called **sequences**, especially when X is a set of numbers and the function s is defined by an algebraic rule. For example, the sequence of odd integers is defined by $s_i = 2i - 1$.

Alternatively, strings are called **words**, the set X is called the **alphabet** and the elements of X are called **letters**. For example, $aababbccabcbb$ is a 13-letter word on the 3-letter alphabet $\{a, b, c\}$.

In many computing languages, strings are called **arrays**. Also, when the character $s(i)$ is constrained to belong to a subset $X_i \subseteq X$, a string can be considered as an element of the cartesian product $X_1 \times X_2 \times \cdots \times X_n$, which is normally viewed as n-tuples of

Chapter 2 Strings, Sets, and Binomial Coefficients

the form $(x_1, x_2, \ldots, x_n)$ such that $x_i \in X_i$ for all $i \in [n]$.

Example 2.1. In the state of Georgia, license plates consist of four digits followed by a space followed by three capital letters. The first digit cannot be a 0. How many license plates are possible?

Solution. Let X consist of the digits $\{0, 1, 2, \ldots, 9\}$, let Y be the singleton set whose only element is a space, and let Z denote the set of capital letters. A valid license plate is just a string from

$$(X - \{0\}) \times X \times X \times X \times Y \times Z \times Z \times Z$$

so the number of different license plates is $9 \times 10^3 \times 1 \times 26^3 = 158\,184\,000$, since the size of a product of sets is the product of the sets' sizes. We can get a feel for why this is the case by focusing just on the digit part of the string here. We can think about the digits portion as being four blanks that need to be filled. The first blank has 9 options (the digits 1 through 9). If we focus on just the digit strings beginning with 1, one perspective is that they range from 1000 to 1999, so there are 1000 of them. However, we could also think about there being 10 options for the second spot, 10 options for the third spot, and 10 options for the fourth. Multiplying $10 \times 10 \times 10$ gives 1000. Since our analysis of filling the remaining digit blanks didn't depend on our choice of a 1 for the first position, we see that each of the 9 choices of initial digit gives 1 000 strings, for a total of $9\,000 = 9 \times 10^3$.

In the case that $X = \{0, 1\}$, an X-string is called a 0–1 string (also a **binary string** or **bit string**.). When $X = \{0, 1, 2\}$, an X-string is also called a **ternary** string.

Example 2.2. A machine instruction in a 32-bit operating system is just a bit string of length 32. Thus, there are 2 options for each of 32 positions to fill, making the number of such strings $2^{32} = 4\,294\,967\,296$. In general, the number of bit strings of length n is 2^n.

Example 2.3. Suppose that a website allows its users to pick their own usernames for accounts, but imposes some restrictions. The first character must be an upper-case letter in the English alphabet. The second through sixth characters can be letters (both upper-case and lower-case allowed) in the English alphabet or decimal digits (0–9). The seventh position must be '@' or '.'. The eighth through twelfth positions allow lower-case English letters, '*', '%', and '#'. The thirteenth position must be a digit. How many users can the website accept registrations from?

Solution. We can visualize the options by thinking of the 13 positions in the string as blanks that need to be filled in and putting the options for that blank above. In Table 2.4, we've used U to denote the set of upper-case letters, L for the set of lower-case letters, and D for the set of digits.

							#	#	#	#	#	
	D	D	D	D	D		%	%	%	%	%	
	L	L	L	L	L	.	*	*	*	*	*	
U	U	U	U	U	U	@	L	L	L	L	L	D
26	62	62	62	62	62	2	29	29	29	29	29	10

TABLE 2.4: STRING TEMPLATE

Below each position in the string, we've written the number of options for that position. (For example, there are 62 options for the second position, since there are 52 letters once both cases are accounted for and 10 digits. We then multiply these possibilities together, since each choice is independent of the others. Therefore, we have

$$26 \times 62^5 \times 2 \times 29^5 \times 10 = 9\,771\,287\,250\,890\,863\,360$$

total possible usernames.

2.2 Permutations

In the previous section, we considered strings in which repetition of symbols is allowed. For instance, "01110000" is a perfectly good bit string of length eight. However, in many applied settings where a string is an appropriate model, a symbol may be used in at most one position.

Example 2.5. Imagine placing the 26 letters of the English alphabet in a bag and drawing them out one at a time (without returning a letter once it's been drawn) to form a six-character string. We know there are 26^6 strings of length six that can be formed from the English alphabet. However, if we restrict the manner of string formation, not all strings are possible. The string "yellow" has six characters, but it uses the letter "l" twice and thus cannot be formed by drawing letters from a bag. However, "jacket" can be formed in this manner. Starting from a full bag, we note there are 26 choices for the first letter. Once it has been removed, there are 25 letters remaining in the bag. After drawing the second letter, there are 24 letters remaining. Continuing, we note that immediately before the sixth letter is drawn from the bag, there are 21 letters in the bag. Thus, we can form $26 \cdot 25 \cdot 24 \cdot 23 \cdot 22 \cdot 21$ six-character strings of English letters by drawing letters from a bag, a little more than half the total number of six-character strings on this alphabet.

To generalize the preceding example, we now introduce permutations. To do so, let X be a finite set and let n be a positive integer. An X-string $s = x_1 x_2 \ldots x_n$ is called

Chapter 2 Strings, Sets, and Binomial Coefficients

a **permutation** if all n characters used in s are distinct. Clearly, the existence of an X-permutation of length n requires that $|X| \geq n$.

When n is a positive integer, we define $n!$ (read "n **factorial**") by

$$n! = n \cdot (n-1) \cdot (n-2) \cdots 3 \cdot 2 \cdot 1.$$

By convention, we set $0! = 1$. As an example, $7! = 7 \cdot 6 \cdot 5 \cdot 4 \cdot 3 \cdot 2 \cdot 1 = 5040$. Now for integers m, n with $m \geq n \geq 0$ define $P(m, n)$ by

$$P(m, n) = \frac{m!}{(m-n)!} = m(m-1)\cdots(m-n+1).$$

For example, $P(9,3) = 9 \cdot 8 \cdot 7 = 504$ and $P(8,4) = 8 \cdot 7 \cdot 6 \cdot 5 = 1680$. Also, a computer algebra system will quickly report that

$$P(68, 23) = 20732231223375515741894286164203929600000.$$

Proposition 2.6. *If X is an m-element set and n is a positive integer with $m \geq n$, then the number of X-strings of length n that are permutations is $P(m, n)$.*

Proof. The proposition is true since when constructing a permutation $s = x_1 x_2, \ldots x_n$ from an m-element set, we see that there are m choices for x_1. After fixing x_1, we have that for x_2, there are $m - 1$ choices, as we can use any element of $X - \{x_1\}$. For x_3, there are $m - 2$ choices, since we can use any element in $X - \{x_1, x_2\}$. For x_n, there are $m - n + 1$ choices, because we can use any element of X except $x_1, x_2, \ldots x_{n-1}$. Noting that

$$P(m, n) = \frac{m!}{(m-n)!} = m(m-1)(m-2)\ldots(m-n+1),$$

our proof is complete. □

Note that the answer we arrived at in Example 2.5 is simply $P(26, 6)$ as we would expect in light of Proposition 2.6.

Example 2.7. It's time to elect a slate of four class officers (President, Vice President, Secretary and Treasurer) from the pool of 80 students enrolled in Applied Combinatorics. If any interested student could be elected to any position (Alice contends this is a big "if" since Bob is running), how many different slates of officers can be elected?

Solution. To count possible officer slates, work from a set X containing the names of the 80 interested students (yes, even poor Bob). A permutation of length four chosen from X is then a slate of officers by considering the first name in the permutation as the President, the second as the Vice President, the third as the Secretary, and the fourth as the Treasurer. Thus, the number of officer slates is $P(80, 4) = 37957920$.

Example 2.8. Let's return to the license plate question of Example 2.1. Suppose that Georgia required that the three letters be distinct from each other. Then, instead of having $26^3 = 17\,576$ ways to fill the last three positions on the license plate, we'd have $P(26,3) = 26 \times 25 \times 24 = 15\,600$ options, giving a total of $140\,400\,000$ license plates.

As another example, suppose that repetition of letters were allowed but the three digits in positions two through four must all be distinct from each other (but could repeat the first digit, which must still be nonzero). Then there are still 9 options for the first position and 26^3 options for the letters, but the three remaining digits can be completed in $P(10,3)$ ways. The total number of license plates would then be $9 \times P(10,3) \times 26^3$. If we want to prohibit repetition of the digit in the first position as well, we need a bit more thought. We first have 9 choices for that initial digit. Then, when filling in the next three positions with digits, we need a permutation of length 3 chosen from the remaining 9 digits. Thus, there are $9 \times P(9,3)$ ways to complete the digits portion, giving a total of $9 \times P(9,3) \times 26^3$ license plates.

2.3 Combinations

To motivate the topic of this section, we consider another variant on the officer election problem from Example 2.7. Suppose that instead of electing students to specific offices, the class is to elect an executive council of four students from the pool of 80 students. Each position on the executive council is equal, so there would be no difference between Alice winning the "first" seat on the executive council and her winning the "fourth" seat. In other words, we just want to pick four of the 80 students without any regard to order. We'll return to this question after introducing our next concept.

Let X be a finite set and let k be an integer with $0 \leq k \leq |X|$. Then a k-element subset of X is also called a **combination** of size k. When $|X| = n$, the number of k-element subsets of X is denoted $\binom{n}{k}$. Numbers of the form $\binom{n}{k}$ are called **binomial coefficients**, and many combinatorists read $\binom{n}{k}$ as "n **choose** k." When we need an in-line version, the preferred notation is $C(n,k)$. Also, the quantity $C(n,k)$ is referred to as the number of combinations of n things, taken k at a time.

Bob notes that with this notation, the number of ways a four-member executive council can be elected from the 80 interested students is $C(80,4)$. However, he's puzzled about how to compute the value of $C(80,4)$. Alice points out that it must be less than $P(80,4)$, since each executive council could be turned into $4!$ different slates of officers. Carlos agrees and says that Alice has really hit upon the key idea in finding a formula to compute $C(n,k)$ in general.

Proposition 2.9. *If n and k are integers with $0 \leq k \leq n$, then*

$$\binom{n}{k} = C(n,k) = \frac{P(n,k)}{k!} = \frac{n!}{k!(n-k)!}$$

Proof. If X is an n-element set, then $P(n,k)$ counts the number of X-permutations of length k. Each of the $C(n,k)$ k-element subsets of X can be turned into $k!$ permutations, and this accounts for each permutation exactly once. Therefore, $k!C(n,k) = P(n,k)$ and dividing by $k!$ gives the formula for the number of k-element subsets. □

Using Proposition 2.9, we can now determine that $C(80,4) = 1581580$ is the number of ways a four-member executive council could be elected from the 80 interested students.

Our argument above illustrates a common combinatorial counting strategy. We counted one thing and determined that the objects we wanted to count were *overcounted* the same number of times each, so we divided by that number ($k!$ in this case).

The following result is tantamount to saying that choosing elements to belong to a set (the executive council election winners) is the same as choosing those elements which are to be denied membership (the election losers).

Proposition 2.10. *For all integers n and k with $0 \le k \le n$,*

$$\binom{n}{k} = \binom{n}{n-k}.$$

Example 2.11. A Southern restaurant lists 21 items in the "vegetable" category of its menu. (Like any good Southern restaurant, macaroni and cheese is *one* of the vegetable options.) They sell a vegetable plate which gives the customer four different vegetables from the menu. Since there is no importance to the order the vegetables are placed on the plate, there are $C(21,4) = 5985$ different ways for a customer to order a vegetable plate at the restaurant.

Our next example introduces an important correspondence between sets and bit strings that we will repeatedly exploit in this text.

Example 2.12. Let n be a positive integer and let X be an n-element set. Then there is a natural one-to-one correspondence between subsets of X and bit strings of length n. To be precise, let $X = \{x_1, x_2, \ldots, x_n\}$. Then a subset $A \subseteq X$ corresponds to the string s where $s(i) = 1$ if and only if $i \in A$. For example, if $X = \{a,b,c,d,e,f,g,h\}$, then the subset $\{b,c,g\}$ corresponds to the bit string 01100010. There are $C(8,3) = 56$ bit strings of length eight with precisely three 1's. Thinking about this correspondence, what is the total number of subsets of an n-element set?

2.4 Combinatorial Proofs

Combinatorial arguments are among the most beautiful in all of mathematics. Oftentimes, statements that can be proved by other, more complicated methods (usually

2.4 Combinatorial Proofs

involving large amounts of tedious algebraic manipulations) have very short proofs once you can make a connection to counting. In this section, we introduce a new way of thinking about combinatorial problems with several examples. Our goal is to help you develop a "gut feeling" for combinatorial problems.

Example 2.13. Let n be a positive integer. Use Figure 2.14 to explain why

$$1 + 2 + 3 + \cdots + n = \frac{n(n+1)}{2}.$$

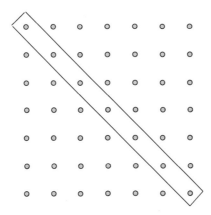

FIGURE 2.14: THE SUM OF THE FIRST n INTEGERS

Solution. Consider an $(n+1) \times (n+1)$ array of dots as depicted in Figure 2.14. There are $(n+1)^2$ dots altogether, with exactly $n+1$ on the main diagonal. The off-diagonal entries split naturally into two equal size parts, those above and those below the diagonal.

Furthermore, each of those two parts has $S(n) = 1 + 2 + 3 + \cdots + n$ dots. It follows that

$$S(n) = \frac{(n+1)^2 - (n+1)}{2} \qquad \frac{n^2 + 2n + 1 - (n+1)}{2} = \frac{n^2 + n}{2} = \frac{n(n+1)}{2}$$

and this is obvious! Now a little algebra on the right hand side of this expression produces the formula given earlier.

Example 2.15. Let n be a positive integer. Explain why

$$1 + 3 + 5 + \cdots + 2n - 1 = n^2.$$

Chapter 2 Strings, Sets, and Binomial Coefficients

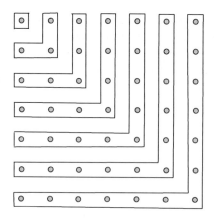

FIGURE 2.16: THE SUM OF THE FIRST n ODD INTEGERS

Solution. The left hand side is just the sum of the first n odd integers. But as suggested in Figure 2.16, this is clearly equal to n^2.

Example 2.17. Let n be a positive integer. Explain why

$$\binom{n}{0} + \binom{n}{1} + \binom{n}{2} + \cdots + \binom{n}{n} = 2^n.$$

Solution. Both sides count the number of bit strings of length n, with the left side first grouping them according to the number of 0's.

Example 2.18. Let n and k be integers with $0 \leq k < n$. Explain why

$$\binom{n}{k+1} = \binom{k}{k} + \binom{k+1}{k} + \binom{k+2}{k} + \cdots + \binom{n-1}{k}.$$

Solution. To prove this formula, we simply observe that both sides count the number of bit strings of length n that contain $k+1$ 1's with the right hand side first partitioning them according to the last occurence of a 1. (For example, if the last 1 occurs in position $k+5$, then the remaining k 1's must appear in the preceding $k+4$ positions, giving $C(k+4,k)$ strings of this type.) Note that when $k=1$ (so $k+1=2$), we have the same formula as developed earlier for the sum of the first n positive integers.

Example 2.19. Explain the identity

$$3^n = \binom{n}{0}2^0 + \binom{n}{1}2^1 + \binom{n}{2}2^2 + \cdots + \binom{n}{n}2^n.$$

Solution. Both sides count the number of $\{0, 1, 2\}$-strings of length n, the right hand side first partitioning them according to positions in the string which are not 2. (For instance, if 6 of the positions are not 2, we must first choose those 6 positions in $C(n, 6)$ ways and then there are 2^6 ways to fill in those six positions by choosing either a 0 or a 1 for each position.)

Example 2.20. Explain why, for each non-negative integer n,

$$\binom{2n}{n} = \binom{n}{0}^2 + \binom{n}{1}^2 + \binom{n}{2}^2 + \cdots + \binom{n}{n}^2.$$

Solution. Both sides count the number of bit strings of length $2n$ with half the bits being 0's, with the right side first partitioning them according to the number of 1's occurring in the first n positions of the string. Note that we are also using the trivial identity $\binom{n}{k} = \binom{n}{n-k}$.

2.5 The Ubiquitous Nature of Binomial Coefficients

In this section, we present several combinatorial problems that can be solved by appeal to binomial coefficients, even though at first glance, they do not appear to have anything to do with sets.

Example 2.21. The office assistant is distributing supplies. In how many ways can he distribute 18 identical folders among four office employees: Audrey, Bart, Cecilia and Darren, with the additional restriction that each will receive at least one folder?

Imagine the folders placed in a row. Then there are 17 gaps between them. Of these gaps, choose three and place a divider in each. Then this choice divides the folders into four non-empty sets. The first goes to Audrey, the second to Bart, etc. Thus the answer is $C(17, 3)$. In Figure 2.22, we illustrate this scheme with Audrey receiving 6 folders, Bart getting 1, Cecilia 4 and Darren 7.

FIGURE 2.22: DISTRIBUTING IDENTICAL OBJECTS INTO DISTINCT CELLS

Example 2.23. Suppose we redo the preceding problem but drop the restriction that each of the four employees gets at least one folder. Now how many ways can the distribution be made?

Solution. The solution involves a "trick" of sorts. First, we convert the problem to one that we already know how to solve. This is accomplished by *artificially* inflating everyone's allocation by one. In other words, if Bart will get 7 folders, we say that he will get 8. Also, artificially inflate the number of folders by 4, one for each of the four persons. So now imagine a row of $22 = 18 + 4$ folders. Again, choose 3 gaps. This determines a non-zero allocation for each person. The actual allocation is one less—and may be zero. So the answer is $C(21, 3)$.

Example 2.24. Again we have the same problem as before, but now we want to count the number of distributions where only Audrey and Cecilia are guaranteed to get a folder. Bart and Darren are allowed to get zero folders. Now the trick is to artificially inflate Bart and Darren's allocation, but leave the numbers for Audrey and Cecilia as is. So the answer is $C(19, 3)$.

Example 2.25. Here is a reformulation of the preceding discussion expressed in terms of integer solutions of inequalities.

We count the number of integer solutions to the inequality

$$x_1 + x_2 + x_3 + x_4 + x_5 + x_6 \leq 538$$

subject to various sets of restrictions on the values of $x_1, x_2, \ldots, x_6$. Some of these restrictions will require that the inequality actually be an equation.

The number of integer solutions is:

1. $C(537, 5)$, when all $x_i > 0$ and equality holds;

2. $C(543, 5)$, when all $x_i \geq 0$ and equality holds;

3. $C(291, 3)$, when $x_1, x_2, x_4, x_6 > 0$, $x_3 = 52$, $x_5 = 194$, and equality holds;

4. $C(537, 6)$, when all $x_i > 0$ and the inequality is strict (Imagine a new variable x_7 which is the balance. Note that x_7 must be positive.);

5. $C(543, 6)$, when all $x_i \geq 0$ and the inequality is strict (Add a new variable x_7 as above. Now it is the only one which is required to be positive.); and

6. $C(544, 6)$, when all $x_i \geq 0$.

A classical enumeration problem (with connections to several problems) involves counting lattice paths. A **lattice path** in the plane is a sequence of ordered pairs of integers:

$$(m_1, n_1), (m_2, n_2), (m_3, n_3), \ldots, (m_t, n_t)$$

so that for all $i = 1, 2, \ldots, t-1$, either

2.5 The Ubiquitous Nature of Binomial Coefficients

1. $m_{i+1} = m_i + 1$ and $n_{i+1} = n_i$, or

2. $m_{i+1} = m_i$ and $n_{i+1} = n_i + 1$.

In Figure 2.26, we show a lattice path from $(0,0)$ to $(13,8)$.

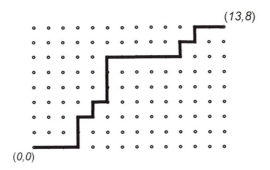

FIGURE 2.26: A Lattice Path

Example 2.27. The number of lattice paths from (m,n) to (p,q) is $C((p-m)+(q-n), p-m)$.

To see why this formula is valid, note that a lattice path is just an X-string with $X = \{H, V\}$, where H stands for *horizontal* and V stands for *vertical*. In this case, there are exactly $(p-m)+(q-n)$ moves, of which $p-m$ are horizontal.

Example 2.28. Let n be a non-negative integer. Then the number of lattice paths from $(0,0)$ to (n,n) which never go above the diagonal line $y = x$ is the **Catalan number**

$$C(n) = \frac{1}{n+1}\binom{2n}{n}.$$

To see that this formula holds, consider the family $\mathcal{P}$ of all lattice paths from $(0,0)$ to (n,n). A lattice path from $(0,0)$ to (n,n) is just a $\{H,V\}$-string of length $2n$ with exactly n H's. So $|\mathcal{P}| = \binom{2n}{n}$. We classify the paths in $\mathcal{P}$ as *good* if they never go over the diagonal; otherwise, they are *bad*. A string $s \in \mathcal{P}$ is good if the number of V's in an initial segment of s never exceeds the number of H's. For example, the string "$HHVHVVHHHVHVVV$" is a good lattice path from $(0,0)$ to $(7,7)$, while the path "$HVHVHHVVVHVHHV$" is bad. In the second case, note that after 9 moves, we have 5 V's and 4 H's.

Let $\mathcal{G}$ and $\mathcal{B}$ denote the family of all good and bad paths, respectively. Of course, our goal is to determine $|\mathcal{G}|$.

Chapter 2 Strings, Sets, and Binomial Coefficients

Consider a path $s \in \mathcal{B}$. Then there is a least integer i so that s has more V's than H's in the first i positions. By the minimality of i, it is easy to see that i must be odd (otherwise, we can back up a step), and if we set $i = 2j + 1$, then in the first $2j + 1$ positions of s, there are exactly j H's and $j + 1$ V's. The remaining $2n - 2j - 1$ positions (the "tail of s") have $n - j$ H's and $n - j - 1$ V's. We now transform s to a new string s' by replacing the H's in the tail of s by V's and the V's in the tail of s by H's and leaving the initial $2j + 1$ positions unchanged. For example, see Figure 2.29, where the path s is shown solid and s' agrees with s until it crosses the line $y = x$ and then is the dashed path. Then s' is a string of length $2n$ having $(n - j) + (j + 1) = n + 1$ V's and $(n - j - 1) + j = n - 1$ H's, so s' is a lattice path from $(0, 0)$ to $(n - 1, n + 1)$. Note that there are $\binom{2n}{n-1}$ such lattice paths.

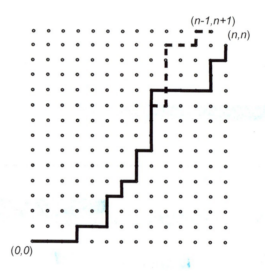

FIGURE 2.29: Transforming a Lattice Path

We can also observe that the transformation we've described is in fact a bijection between $\mathcal{B}$ and $\mathcal{P}'$, the set of lattice paths from $(0, 0)$ to $(n - 1, n + 1)$. To see that this is true, note that every path s' in $\mathcal{P}'$ must cross the line $y = x$, so there is a first time it crosses it, say in position i. Again, i must be odd, so $i = 2j + 1$ and there are j H's and $j + 1$ V's in the first i positions of s'. Therefore the tail of s' contains $n + 1 - (j + 1) = n - j$ V's and $(n - 1) - j$ H's, so interchanging H's and V's in the tail of s' creates a new string s that has n H's and n V's and thus represents a lattice path from $(0, 0)$ to (n, n), but it's still a bad lattice path, as we did not adjust the first part of the path, which results in crossing the line $y = x$ in position i. Therefore, $|\mathcal{B}| = |\mathcal{P}'|$ and thus

$$C(n) = |\mathcal{G}| = |\mathcal{P}| - |\mathcal{B}| = |\mathcal{P}| - |\mathcal{P}'| = \binom{2n}{n} - \binom{2n}{n-1} = \frac{1}{n+1}\binom{2n}{n},$$

after a bit of algebra.

It is worth observing that in Example 2.28, we made use of two common enumerative techniques: giving a bijection between two classes of objects, one of which is "easier" to count than the other, and counting the objects we do *not* wish to enumerate and deducting their number from the total.

2.6 The Binomial Theorem

Here is a truly basic result from combinatorics kindergarten.

Theorem 2.30 (Binomial Theorem). *Let x and y be real numbers with x, y and $x + y$ non-zero. Then for every non-negative integer n,*

$$(x+y)^n = \sum_{i=0}^{n} \binom{n}{i} x^{n-i} y^i.$$

Proof. View $(x + y)^n$ as a product

$$(x+y)^n = \underbrace{(x+y)(x+y)(x+y)(x+y)\ldots(x+y)(x+y)}_{n \text{ factors}}.$$

Each term of the expansion of the product results from choosing either x or y from one of these factors. If x is chosen $n - i$ times and y is chosen i times, then the resulting product is $x^{n-i}y^i$. Clearly, the number of such terms is $C(n, i)$, i.e., out of the n factors, we choose the element y from i of them, while we take x in the remaining $n - i$. □

Example 2.31. There are times when we are interested not in the full expansion of a power of a binomial, but just the coefficient on one of the terms. The Binomial Theorem gives that the coefficient of x^5y^8 in $(2x - 3y)^{13}$ is $\binom{13}{5}2^5(-3)^8$.

2.7 Multinomial Coefficients

Let X be a set of n elements. Suppose that we have two colors of paint, say red and blue, and we are going to choose a subset of k elements to be painted red with the rest

Chapter 2 Strings, Sets, and Binomial Coefficients

painted blue. Then the number of different ways this can be done is just the binomial coefficient $\binom{n}{k}$. Now suppose that we have three different colors, say red, blue, and green. We will choose k_1 to be colored red, k_2 to be colored blue, and the remaining $k_3 = n - (k_1 + k_2)$ are to be colored green. We may compute the number of ways to do this by first choosing k_1 of the n elements to paint red, then from the remaining $n - k_1$ elements choosing k_2 to paint blue, and then painting the remaining k_3 elements green. It is easy to see that the number of ways to do this is

$$\binom{n}{k_1}\binom{n-k_1}{k_2} = \frac{n!}{k_1!(n-k_1)!}\frac{(n-k_1)!}{k_2!(n-(k_1+k_2))!} = \frac{n!}{k_1!k_2!k_3!}.$$

Numbers of this form are called **multinomial coefficients**; they are an obvious generalization of the binomial coefficients. The general notation is:

$$\binom{n}{k_1, k_2, k_3, \ldots, k_r} = \frac{n!}{k_1!k_2!k_3!\ldots k_r!}.$$

For example,

$$\binom{8}{3, 2, 1, 2} = \frac{8!}{3!2!1!2!} = \frac{40320}{6 \cdot 2 \cdot 1 \cdot 2} = 1680.$$

Note that there is some "overkill" in this notation, since the value of k_r is determined by n and the values for $k_1, k_2, \ldots, k_{r-1}$. For example, with the ordinary binomial coefficients, we just write $\binom{8}{3}$ and not $\binom{8}{3,5}$.

Example 2.32. How many different rearrangements of the string:

MITCHELTKELLERANDWILLIAMTTROTTERAREGENIUSES!!

are possible if all letters and characters must be used?

Solution. To answer this question, we note that there are a total of 45 characters distributed as follows: 3 A's, 1 C, 1 D, 7 E's, 1 G, 1 H, 4 I's, 1 K, 5 L's, 2 M's, 2 N's, 1 O, 4 R's, 2 S's, 6 T's, 1 U, 1 W, and 2 !'s. So the number of rearrangements is

$$\frac{45!}{3!1!1!7!1!1!4!1!5!2!2!1!4!2!6!1!1!2!}.$$

Just as with binomial coefficients and the Binomial Theorem, the multinomial coefficients arise in the expansion of powers of a multinomial:

Theorem 2.33 (Multinomial Theorem). *Let $x_1, x_2, \ldots, x_r$ be nonzero real numbers with $\sum_{i=1}^{r} x_i \neq 0$. Then for every $n \in \mathbb{N}_0$,*

$$(x_1 + x_2 + \cdots + x_r)^n = \sum_{k_1 + k_2 + \cdots + k_r = n} \binom{n}{k_1, k_2, \ldots, k_r} x_1^{k_1} x_2^{k_2} \cdots x_r^{k_r}.$$

Example 2.34. What is the coefficient of $x^{99}y^{60}z^{14}$ in $(2x^3 + y - z^2)^{100}$? What about $x^{99}y^{61}z^{13}$?

Solution. By the Multinomial Theorem, the expansion of $(2x^3 + y - z^2)^{100}$ has terms of the form

$$\binom{100}{k_1, k_2, k_3}(2x^3)^{k_1} y^{k_2}(-z^2)^{k_3} = \binom{100}{k_1, k_2, k_3} 2^{k_1} x^{3k_1} y^{k_2}(-1)^{k_3} z^{2k_3}.$$

The $x^{99}y^{60}z^{14}$ arises when $k_1 = 33$, $k_2 = 60$, and $k_3 = 7$, so it must have coefficient

$$-\binom{100}{33, 60, 7} 2^{33}.$$

For $x^{99}y^{61}z^{13}$, the exponent on z is odd, which cannot arise in the expansion of $(2x^3 + y - z^2)^{100}$, so the coefficient is 0.

2.8 Discussion

Over coffee, Xing said that he had been experimenting with the SageMath software discussed in Chapter 1. He understood that SageMath was treating a big integer as a string. Xing enthusiastically reported that he had asked SageMath to find the sum $a + b$ of two large integers a and b, each having more than 800 digits. The software found the answer about as fast as he could hit the enter key on his netbook. "That's not so impressive," Alice interjected. "A human, even Bob, could do this in a couple of minutes using pencil and paper."

"Thanks for your kind remarks," replied Bob, with the rest of the group noting that that Alice was being pretty harsh on Bob and not for any good reason.

Dave took up Bob's case by remarking, "Very few humans, not even you Alice, would want to tackle finding the product of a and b by hand." Xing jumped back in with, "That's the point. Even a tiny netbook can find the product very, very quickly. In fact, I tried it out with two integers, each having more than one thousand digits. It found the product in about one second." Ever the skeptic, Zori said, "You mean you carefully typed in two integers of that size?" Xing quickly replied "Of course not. I just copied and pasted the data from one source to another." Yolanda said, "What a neat trick that is. Really cuts down the chance of an error."

Dave said "What about factoring? Can your netbook with its fancy software for strings factor big integers?" Xing said that he would try some sample problems and report back. Carlos said "Factoring an integer with several hundred digits is likely to be very challenging, not only for a netbook, but also for a super computer. For

Chapter 2 Strings, Sets, and Binomial Coefficients

example, suppose the given integer was either a prime or the product of two large primes. Detecting which of these two statements holds could be very difficult."

Undeterred, Dave continued, "What about exponentiation? Can your software calculate a^b when a and b are large integers?" Xing said "That shouldn't be a problem. After all, a^b is just multiplying a times itself a total of b times, and if you can do multiplication quickly, that's just a loop." Yolanda said that the way Xing was describing things, he was actually talking about a program with nested loops so it might take a long time for such a program to halt. Carlos was quiet but he thought there might be ways to speed up such computations.

By this time, Alice reinserted herself into the conversation: "Hey guys. While you were talking, I was looking for big integer topics on the web and found this problem. 'Is 838200020310007224300 a Catalan number?' How would you answer this? Do you have to use special software?"

Zori was not happy. She gloomily envisioned a future job hunt in which she was compelled to use big integer arithmetic as a job skill. Arrgghh.

2.9 Exercises

1. The Hawaiian alphabet consists of 12 letters. How many six-character strings can be made using the Hawaiian alphabet?

2. How many $2n$-digit positive integers can be formed if the digits in odd positions (counting the rightmost digit as position 1) must be odd and the digits in even positions must be even and positive?

3. Matt is designing a website authentication system. He knows passwords are most secure if they contain letters, numbers, and symbols. However, he doesn't quite understand that this additional security is defeated if he specifies in which positions each character type appears. He decides that valid passwords for his system will begin with three letters (uppercase and lowercase both allowed), followed by two digits, followed by one of 10 symbols, followed by two uppercase letters, followed by a digit, followed by one of 10 symbols. How many different passwords are there for his website system? How does this compare to the total number of strings of length 10 made from the alphabet of all uppercase and lowercase English letters, decimal digits, and 10 symbols?

4. How many ternary strings of length $2n$ are there in which the zeroes appear only in odd-numbered positions?

5. Suppose we are making license plates of the form $l_1 l_2 l_3 - d_1 d_2 d_3$ where l_1, l_2, l_3 are capital letters in the English alphabet and d_1, d_2, d_3 are decimal digits (i.e., elements of the set $\{0,1,2,3,4,5,6,7,8,9\}$) subject to the restriction that at least one digit is nonzero and at least one letter is K. How many license plates can we make?

6. Mrs. Steffen's third grade class has 30 students in it. The students are divided into three groups (numbered 1, 2, and 3), each having 10 students.

(a) The students in group 1 earned 10 extra minutes of recess by winning a class competition. Before going out for their extra recess time, they form a single file line. In how many ways can they line up?

(b) When all 30 students come in from recess together, they again form a single file line. However, this time the students are arranged so that the first student is from group 1, the second from group 2, the third from group 3, and from there on, the students continue to alternate by group in this order. In how many ways can they line up to come in from recess?

7. How many strings of the form $l_1 l_2 d_1 d_2 d_3 l_3 l_4 d_4 l_5 l_6$ are there where

- for $1 \leq i \leq 6$, l_i is an uppercase letter in the English alphabet;
- for $1 \leq i \leq 4$, d_i is a decimal digit;
- l_2 is not a vowel (i.e., $l_2 \notin \{A,E,I,O,U\}$); and
- the digits d_1, d_2, and d_3 are distinct (i.e., $d_1 \neq d_2 \neq d_3 \neq d_1$).

8. In this exercise, we consider strings made from uppercase letters in the English alphabet and decimal digits. How many strings of length 10 can be constructed in each of the following scenarios?

(a) The first and last characters of the string are letters.

(b) The first character is a vowel, the second character is a consonant, and the last character is a digit.

(c) Vowels (not necessarily distinct) appear in the third, sixth, and eighth positions and no other positions.

(d) Vowels (not necessarily distinct) appear in exactly two positions.

(e) Precisely four characters in the string are digits and no digit appears more than one time.

9. A database uses 20-character strings as record identifiers. The valid characters in these strings are upper-case letters in the English alphabet and decimal digits. (Recall there are 26 letters in the English alphabet and 10 decimal digits.) How many valid record identifiers are possible if a valid record identifier must meet *all* of the following criteria:

- Letter(s) from the set $\{A, E, I, O, U\}$ occur in *exactly* three positions of the string.
- The last three characters in the string are *distinct* decimal digits that do not appear elsewhere in the string.
- The remaining characters of the string may be filled with any of the remaining letters or decimal digits.

10. Let X be the set of the 26 lowercase English letters and 10 decimal digits. How many X-strings of length 15 satisfy *all* of the following properties (at the same time)?
- The first and last symbols of the string are distinct digits (which may appear elsewhere in the string).
- Precisely four of the symbols in the string are the letter 't'.
- Precisely three characters in the string are elements of the set $V = \{a, e, i, o, u\}$ and these characters are all distinct.

11. A donut shop sells 12 types of donuts. A manager wants to buy six donuts, one each for himself and his five employees.
  (a) Suppose that he does this by selecting a specific type of donut for each person. (He can select the same type of donut for more than one person.) In how many ways can he do this?
  (b) How many ways could he select the donuts if he wants to ensure that he chooses a different type of donut for each person?
  (c) Suppose instead that he wishes to select one donut of each of six *different* types and place them in the breakroom. In how many ways can he do this? (The order of the donuts in the box is irrelevant.)

12. The sport of korfball is played by teams of eight players. Each team has four men and four women on it. Halliday High School has seven men and 11 women interested in playing korfball. In how many ways can they form a korfball team from their 18 interested students?

13. Twenty students compete in a programming competition in which the top four students are recognized with trophies for first, second, third, and fourth places.
  (a) How many different outcomes are there for the top four places?
  (b) At the last minute, the judges decide that they will award honorable mention certificates to four individuals who did not receive trophies. In how many ways can the honorable mention recipients be selected (after the top four places have been determined)? How many total outcomes (trophies plus certificates) are there then?

14. An ice cream shop has a special on banana splits, and Xing is taking advantage of it. He's astounded at all the options he has in constructing his banana split:
   - He must choose three different flavors of ice cream to place in the asymmetric bowl the banana split is served in. The shop has 20 flavors of ice cream available.
   - Each scoop of ice cream must be topped by a sauce, chosen from six different options. Xing is free to put the same type of sauce on more than one scoop of ice cream.
   - There are 10 sprinkled toppings available, and he must choose three of them to have sprinkled over the entire banana split.

   (a) How many different ways are there for Xing to construct a banana split at this ice cream shop?
   (b) Suppose that instead of requiring that Xing choose exactly three sprinkled toppings, he is allowed to choose between zero and three sprinkled toppings. In this scenario, how many different ways are there for him to construct a banana split?

15. Suppose that a teacher wishes to distribute 25 identical pencils to Ahmed, Barbara, Casper, and Dieter such that Ahmed and Dieter receive at least one pencil each, Casper receives no more than five pencils, and Barbara receives at least four pencils. In how many ways can such a distribution be made?

16. How many integer-valued solutions are there to each of the following equations and inequalities?
   (a) $x_1 + x_2 + x_3 + x_4 + x_5 = 63$, all $x_i > 0$
   (b) $x_1 + x_2 + x_3 + x_4 + x_5 = 63$, all $x_i \geq 0$
   (c) $x_1 + x_2 + x_3 + x_4 + x_5 \leq 63$, all $x_i \geq 0$
   (d) $x_1 + x_2 + x_3 + x_4 + x_5 = 63$, all $x_i \geq 0$, $x_2 \geq 10$
   (e) $x_1 + x_2 + x_3 + x_4 + x_5 = 63$, all $x_i \geq 0$, $x_2 \leq 9$

17. How many integer solutions are there to the equation
$$x_1 + x_2 + x_3 + x_4 = 132$$
provided that $x_1 > 0$, and $x_2, x_3, x_4 \geq 0$? What if we add the restriction that $x_4 < 17$?

18. How many integer solutions are there to the inequality
$$x_1 + x_2 + x_3 + x_4 + x_5 \leq 782$$
provided that $x_1, x_2 > 0$, $x_3 \geq 0$, and $x_4, x_5 \geq 10$?

Chapter 2 Strings, Sets, and Binomial Coefficients

19. A teacher has 450 identical pieces of candy. He wants to distribute them to his class of 65 students, although he is willing to take some leftover candy home. (He does not insist on taking any candy home, however.) The student who won a contest in the last class is to receive at least 10 pieces of candy as a reward. Of the remaining students, 34 of them insist on receiving at least one piece of candy, while the remaining 30 students are willing to receive no candy.

(a) In how many ways can he distribute the candy?

(b) In how many ways can he distribute the candy if, in addition to the conditions above, one of his students is diabetic and can receive at most 7 pieces of candy? (This student is one of the 34 who insist on receiving at least one piece of candy.)

20. Give a combinatorial argument to prove the identity

$$k\binom{n}{k} = n\binom{n-1}{k-1}.$$

Hint. Think of choosing a team with a captain.

21. Let m and w be positive integers. Give a combinatorial argument to prove that for integers $k \geq 0$,

$$\sum_{j=0}^{k} \binom{m}{j}\binom{w}{k-j} = \binom{m+w}{k}.$$

22. How many lattice paths are there from $(0,0)$ to $(10,12)$?

23. How many lattice paths are there from $(3,5)$ to $(10,12)$?

24. How many lattice paths are there from $(0,0)$ to $(10,12)$ that pass through $(3,5)$?

25. How many lattice paths from $(0,0)$ to $(17,12)$ are there that pass through $(7,6)$ and $(12,9)$?

26. How many lattice paths from $(0,0)$ to $(14,73)$ are there that do *not* pass through $(6,37)$?

27. A small-town bank robber is driving his getaway car from the bank he just robbed to his hideout. The bank is at the intersection of 1st Street and 1st Avenue. He needs to return to his hideout at the intersection of 7th Street and 5th Avenue. However, one of his lookouts has reported that the town's one police officer is parked at the intersection of 4th Street and 4th Avenue. Assuming that the bank robber does not want to get arrested and drives only on streets and avenues, in how many ways can he safely return to his hideout? (Streets and avenues are uniformly spaced and numbered consecutively in this small town.)

28. The setting for this problem is the fictional town of Mascotville, which is laid out as a grid. Mascots are allowed to travel only on the streets, and not "as the yellow jacket flies." Buzz, the Georgia Tech mascot, wants to go visit his friend Thundar, the North Dakota State University mascot, who lives 6 blocks east and 7 blocks north of Buzz's hive. However, Uga VIII has recently moved into the doghouse 2 blocks east and 3 blocks north of Buzz's hive and already has a restraining order against Buzz. There's also a pair of tigers (mother and cub) from Clemson who live 1 block east and 2 blocks north of Uga VIII, and they're known for setting traps for Buzz. Buzz wants to travel from his hive to Thundar's pen every day without encountering Uga VIII or The Tiger and The Tiger Cub. However, he wants to avoid the boredom caused by using a route he's used in the past. What is the largest number of consecutive days on which Buzz can make the trip to visit Thundar without reusing a route (you may assume the routes taken by Buzz only go east and north)?

29. Determine the coefficient on $x^{15}y^{120}z^{25}$ in $(2x + 3y^2 + z)^{100}$.

30. Determine the coefficient on $x^{12}y^{24}$ in $(x^3 + 2xy^2 + y + 3)^{18}$. (Be careful, as x and y now appear in multiple terms!)

31. For each word below, determine the number of rearrangements of the word in which all letters must be used.

   (a) OVERNUMEROUSNESSES
   (b) OPHTHALMOOTORHINOLARYNGOLOGY
   (c) HONORIFICABILITUDINITATIBUS (the longest word in the English language consisting strictly of alternating consonants and vowels[1])

32. How many ways are there to paint a set of 27 elements such that 7 are painted white, 6 are painted old gold, 2 are painted blue, 7 are painted yellow, 5 are painted green, and 0 of are painted red?

33. There are many useful sets that are enumerated by the Catalan numbers. (Volume two of R.P. Stanley's *Enumerative Combinatorics* contains a famous (or perhaps infamous) exercise in 66 parts asking readers to find bijections that will show that the number of various combinatorial structures is $C(n)$, and his web page boasts an additional list of at least 100 parts.) Give bijective arguments to show that each class of objects below is enumerated by $C(n)$. (All three were selected from the list in Stanley's book.)

---

[1] http://www.rinkworks.com/words/oddities.shtml

(a) The number of ways to fully-parenthesize a product of $n+1$ factors as if the "multiplication" operation in question were not necessarily associative. For example, there is one way to parenthesize a product of two factors (a_1a_2), there are two ways to parenthesize a product of three factors $((a_1(a_2a_3))$ and $((a_1a_2)a_3))$, and there are five ways to parenthesize a product of four factors:

$$(a_1(a_2(a_3a_4))), (a_1((a_2a_3)a_4)), ((a_1a_2)(a_3a_4)), ((a_1(a_2a_3))a_4), (((a_1a_2)a_3)a_4).$$

(b) Sequences of n 1's and n -1's in which the sum of the first i terms is nonnegative for all i.

(c) Sequences $1 \leq a_1 \leq \cdots \leq a_n$ of integers with $a_i \leq i$. For example, for $n = 3$, the sequences are

$$111 \quad 112 \quad 113 \quad 122 \quad 123.$$

Hint. For part 2.9.33.c, think about drawing lattice paths on paper with grid lines and (basically) the number of boxes below a lattice path in a particular column.

CHAPTER 3

Induction

The twin concepts of recursion and induction are fundamentally important in combinatorial mathematics and computer science. In this chapter, we give a number of examples of how recursive formulas arise naturally in combinatorial problems, and we explain how they can be used to make computations. We also introduce the Principle of Mathematical Induction and give several examples of how it is applied to prove combinatorial statements. Our treatment will also include some code snippets that illustrate how functions are defined recursively in computer programs.

3.1 Introduction

A professor decides to liven up the next combinatorics class by giving a door prize. As students enter class (on time, because to be late is a bit insensitive to the rest of the class), they draw a ticket from a box. On each ticket, a positive integer has been printed. No information about the range of ticket numbers is given, although they are guaranteed to be distinct. The box of tickets was shaken robustly before the drawing, so the contents are thoroughly mixed, and the selection is done without looking inside the box.

After each student has selected a ticket, the professor announces that a cash prize of one dollar (this is a university, you know) will be awarded to the student holding the lowest numbered ticket—from among those drawn.

Must the prize be awarded? In other words, given a set of positive integers, in this case the set of ticket numbers chosen by the students, must there be a least one? More generally, is it true that in any set of positive integers, there is always a least one? What happens if there is an enrollment surge and there are infinitely many students in the class and each has a ticket?

Chapter 3 Induction

3.2 The Positive Integers are Well Ordered

Most likely, you answered the questions posed in Section 3.1 with an enthusiastic "yes", in part because you wanted the shot at the money, but more concretely because it seems so natural. But you may be surprised to learn that this is really a much more complex subject than you might think at first. In Appendix B, we discuss the development of the number systems starting from the Peano Postulates. Although we will not devote much space in this chapter to this topic, it is important to know that the positive integers come with "some assembly required." In particular, the basic operations of addition and multiplication don't come for free; instead they have to be defined.

As a by-product of this development, we get the following fundamentally important property of the set $\mathbb{N}$ of positive integers:

Principle 3.1 (Well Ordered Property of the Positive Integers). *Every non-empty set of positive integers has a least element.*

An immediate consequence of the well ordered property is that the professor will indeed have to pay someone a dollar—even if there are infinitely many students in the class.

3.3 The Meaning of Statements

Have you ever taken standardized tests where they give you the first few terms of a sequence and then ask you for the next one? Here are some sample questions. In each case, see if you can determine a reasonable answer for the next term.

1. $2, 5, 8, 11, 14, 17, 20, 23, 26, \ldots$ 29
2. $1, 1, 2, 3, 5, 8, 13, 21, 34, 55, 89, 144, 233, 377, \ldots$
3. $1, 2, 5, 14, 42, 132, 429, 1430, 4862, \ldots$
4. $2, 6, 12, 20, 30, 42, 56, 72, 90, 110, \ldots$ 132
5. $2, 3, 6, 11, 18, 27, 38, 51, \ldots$ 65

Pretty easy stuff! OK, now try the following somewhat more challenging sequence. Here, we'll give you a lot more terms and challenge you to find the next one.

$1, 2, 3, 4, 1, 2, 3, 4, 5, 1, 2, 3, 4, 5, 2, 3, 4, 5, 6, 2, 3, 4, 5, 6, 1, 2, 3, 4, 5, 2, 3, 4, 5, 6, \ldots$

Trust us when we say that we really have in mind something very concrete, and once it's explained, you'll agree that it's "obvious." But for now, it's far from it.

3.3 The Meaning of Statements

Here's another danger lurking around the corner when we encounter formulas like

$$1 + 2 + 3 + \cdots + n = \frac{n(n+1)}{2}$$

What do the dots in this statement mean? In fact, let's consider a much simpler question. What is meant by the following expression:

$$1 + 2 + 3 + \cdots + 6$$

Are we talking about the sum of the first six positive integers, or are we talking about the sum of the first 19 terms from the more complicated challenge sequence given above? You are supposed to answer that you don't know, and that's the correct answer.

The point here is that without a clarifying comment or two, the notation $1 + 2 + 3 + \cdots + 6$ isn't precisely defined. Let's see how to make things right.

First, let $f : \mathbb{N} \longrightarrow \mathbb{N}$ be a function. Set

$$\sum_{i=1}^{1} f(i) = f(1)$$

and if $n > 1$, define

$$\sum_{i=1}^{n} f(i) = f(n) + \sum_{i=1}^{n-1} f(i)$$

To see that these two statements imply that the expression $\sum_{i=1}^{n} f(i)$ is defined for all positive integers, apply the Well Ordered Property to the set of all positive integers for which the expression is not defined and use the recursive definition to define it for the least element.

So if we want to talk about the sum of the first six positive integers, then we should write:

$$\sum_{i=1}^{6} i$$

Now it is clear that we are talking about a computation that yields 21 as an answer.

A second example: previously, we defined $n!$ by writing

$$n! = n \times (n-1) \times (n-2) \times \cdots \times 3 \times 2 \times 1$$

By this point, you should realize that there's a problem here. Multiplication, like addition, is a binary operation. And what do those dots mean? Here's a way to do the job more precisely. Define $n!$ to be 1 if $n = 1$. And when $n > 1$, set $n! = n(n-1)!$.

Chapter 3 Induction

Definitions like these are called **recursive definitions**. They can be made with different starting points. For example, we could have set $n! = 1$ when $n = 0$, and when $n > 0$, set $n! = n(n-1)!$.

Here's a code snippet in SageMath, which is based on Python, so this also works as Python code.

```
def sumrecursive(n):
    if n == 1:
        return 2;
    else:
        return sumrecursive(n-1) + (n*n - 2*n + 3)
sumrecursive(3)
```

11

What is the value of sumrecursive(4)? (In order to make sure you understand how this recursive function works, calculate out sumrecursive(4) should be by hand before modifying the SageMath cell above.) Does it make sense to say that sumrecursive(n) is defined for all positive integers n? Did you recognize that this program provides a precise meaning to the expression:

$$2 + 3 + 6 + 11 + 18 + 27 + 38 + 51 + \cdots + (n^2 - 2n + 3)$$

3.4 Binomial Coefficients Revisited

The binomial coefficient $\binom{n}{k}$ was originally defined in terms of the factorial notation, and with our recursive definitions of the factorial notation, we also have a complete and legally-correct definition of binomial coefficients. The following recursive formula provides an efficient computational scheme.

Let n and k be integers with $0 \le k \le n$. If $k = 0$ or $k = n$, set $\binom{n}{k} = 1$. If $0 < k < n$, set

$$\binom{n}{k} = \binom{n-1}{k-1} + \binom{n-1}{k}.$$

This recursion has a natural combinatorial interpretation. Both sides count the number of k-element subsets of $\{1, 2, \ldots, n\}$, with the right-hand side first grouping them into those which contain the element n and then those which don't. The traditional form of displaying this recursion is shown in Table 3.2. This pattern is called "Pascal's triangle." Other than the 1s at the ends of each row, an entry of the triangle is determined by adding the entry to the left and the entry to the right in the row above.

3.5 Solving Combinatorial Problems Recursively

					1					
				1		1				
			1		2		1			
		1		3		3		1		
	1		4		6		4		1	
1		5		10		10		5		1
1	6		15		20		15		6	1
1	7	21		35		35		21	7	1
1	8	28	56		70		56	28	8	1

TABLE 3.2: PASCAL'S TRIANGLE

Xing was intrigued by the fact that he now had two fundamentally different ways to calculate binomial coefficients. One way is to write $\binom{n}{m} = P(n,m)/(n-m)!$ and just carry out the specified arithmetic. The second way is to use the recursion of Pascal's triangle, so that you are just performing additions. So he experimented by writing a computer program to calculate binomial coefficients, using a library that treats big integers as strings. Which of the two ways do you think proved to be faster when n say was between 1800 and 2000 and m was around 800?

3.5 Solving Combinatorial Problems Recursively

In this section, we present examples of combinatorial problems for which solutions can be computed recursively. In Chapter 9, we return to these problems and obtain even more compact solutions. Our first problem is one discussed in our introductory chapter.

Example 3.3. A family of n lines is drawn in the plane with (1) each pair of lines crossing and (2) no three lines crossing in the same point. Let $r(n)$ denote the number of regions into which the plane is partitioned by these lines. Evidently, $r(1) = 2, r(2) = 4$, $r(3) = 7$ and $r(4) = 11$. To determine $r(n)$ for all positive integers, it is enough to note that $r(1) = 1$, and when $n > 1$, $r(n) = n + r(n-1)$. This formula follows from the observation that if we label the lines as $L_1, L_2, \ldots, L_n$, then the $n-1$ points on line L_n where it crosses the other lines in the family divide L_n into n segments, two of which are infinite. Each of these segments is associated with a region determined by the first $n-1$ lines that has now been subdivided into two, giving us n more regions than were determined by $n-1$ lines. This situation is illustrated in Figure 3.4, where the line containing the three dots is L_4. The other lines divide it into four segments, which then divide larger regions to create regions 1 and 5, 2 and 6, 7 and 8, and 4 and 9.

Chapter 3 Induction

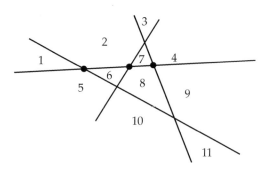

FIGURE 3.4: LINES AND REGIONS IN THE PLANE

With the recursive formula, we thus have $r(5) = 5 + 11 = 16$, $r(6) = 6 + 16 = 22$ and $r(7) = 7 + 22 = 29$. Even by hand, it wouldn't be all that much trouble to calculate $r(100)$. We could do it before lunch.

Example 3.5. A $2 \times n$ checkerboard will be tiled with rectangles of size 2×1 and 1×2. Find a recursive formula for the number $t(n)$ of tilings. Clearly, $t(1) = 1$ and $t(2) = 2$. When $n > 2$, consider the rectangle that covers the square in the upper right corner. If it is vertical, then preceding it, we have a tiling of the first $n - 1$ columns. If it is horizontal, then so is the rectangle immediately underneath it, and proceeding them is a tiling of the first $n - 2$ columns. This shows that $t(n) = t(n - 1) + t(n - 2)$. In particular, $t(3) = 1 + 2 = 3$, $t(4) = 2 + 3 = 5$ and $t(5) = 3 + 5 = 8$.

Again, if compelled, we could get $t(100)$ by hand, and a computer algebra system could get $t(1000)$.

Example 3.6. Call a ternary string *good* if it never contains a 2 followed immediately by a 0; otherwise, call it *bad*. Let $g(n)$ be the number of good strings of length n. Obviously $g(1) = 3$, since all strings of length 1 are good. Also, $g(2) = 8$ since the only bad string of length 2 is $(2, 0)$. Now consider a value of n larger than 2.

Partition the set of good strings of length n into three parts, according to the last character. Good strings ending in 1 can be preceded by any good string of length $n - 1$, so there are $g(n - 1)$ such strings. The same applies for good strings ending in 2. For good strings ending in 0, however, we have to be more careful. We can precede the 0 by a good string of length $n-1$ provided that the string does not end in 2. There are $g(n-1)$ good strings of length $n - 1$ and of these, exactly $g(n - 2)$ end in a 2. Therefore there are $g(n - 1) - g(n - 2)$ good strings of length n that end in a 0. Hence the total number of good strings of length n satisfies the recursive formula $g(n) = 3g(n - 1) - g(n - 2)$. Thus $g(3) = 3 \cdot 8 - 3 = 21$ and $g(4) = 3 \cdot 21 - 8 = 55$.

3.5 Solving Combinatorial Problems Recursively

Once more, $g(100)$ is doable by hand, while even a modest computer can be coaxed into giving us $g(5000)$.

3.5.1 Finding Greatest Common Divisors

There is more meat than you might think to the following elementary theorem, which seems to simply state a fact that you've known since second grade.

Theorem 3.7 (Division Theorem). *Let m and n be positive integers. Then there exist unique integers q and r so that*
$$m = q \cdot n + r \quad \text{and} \quad 0 \leq r < n.$$
*We call q the **quotient** and r the **remainder**.*

Proof. We settle the claim for existence. The uniqueness part is just high-school algebra. If the theorem fails to hold, then let t be the least positive integer for which there are integers m and n with $m + n = t$, but there do not exist integers q and r with $m = qn + r$ and $0 \leq r < n$.

First, we note that $n \neq 1$, for if $n = 1$, then we could take $q = m$ and $r = 0$. Also, we cannot have $m = 1$, for if $m = 1$, then we can take $q = 0$ and $r = 1$. Now the statement holds for the pair $m - 1, n$ so there are integers q and r so that
$$m - 1 = q \cdot n + r \quad \text{and} \quad 0 \leq r < n.$$
Since $r < n$, we know that $r + 1 \leq n$. If $r + 1 < n$, then
$$m = q \cdot n + (r + 1) \quad \text{and} \quad 0 \leq r + 1 < n.$$
On the other hand, if $r + 1 = n$, then
$$m = q \cdot n + (r + 1) = nq + n = (q + 1)n = (q + 1)n + 0.$$
The contradiction completes the proof. □

Recall that an integer n is a **divisor** of an integer m if there is an integer q such that $m = qn$. (We write $n \mid m$ and read "n divides m".) An integer d is a **common divisor** of integers m and n if d is a divisor of both m and n. The **greatest common divisor** of m and n, written $\gcd(m, n)$, is the largest of all the common divisors of m and n.

Here's a particularly elegant application of the preceding basic theorem:

Theorem 3.8 (Euclidean Algorithm). *Let m, n be positive integers with $m > n$ and let q and r be the unique integers for which*
$$m = q \cdot n + r \quad \text{and} \quad 0 \leq r < n.$$
If $r > 0$, then $\gcd(m, n) = \gcd(n, r)$. If $r = 0$, then n divides m, and $\gcd(m, n) = n$.

Chapter 3 Induction

Proof. Consider the expression $m = q \cdot n + r$, which is equivalent to $m - q \cdot n = r$. If a number d is a divisor of m and n, then d must also divide r. Similarly, if d is a divisor of n and r, then d must also divide m. □

Here is a code snippet that computes the greatest common divisor of m and n when m and n are positive integers with $m \geq n$. We use the familiar notation m%n to denote the remainder r in the expression $m = q \cdot n + r$, with $0 \leq r < n$.

```
def gcd(m,n):
    if m % n == 0:
        return n
    else:
        return gcd(n,m%n)
gcd(12,5)
```

1

Feel free to change the values 12 and 5 above in the SageMath cell in the HTML version of the text to calculate the greatest common divisor of some other integers. Just remember that the code assumes $m \geq n$ when you do so!

The disadvantage of this approach is the somewhat wasteful use of memory due to recursive function calls. It is not difficult to develop code for computing the greatest common divisor of m and n using only a loop, i.e., there are no recursive calls. With minimal extra work, such code can also be designed to solve the following diophantine equation problem:

Theorem 3.9. *Let m, n, and c be positive integers. Then there exist integers a and b, not necessarily non-negative, solving the linear diophantine equation $am + bn = c$ if and only if c is a multiple of the greatest common divisor of m and n.*

Let's see how the Euclidean algorithm can be used to write $\gcd(m, n)$ in the form $am + bn$ with $a, b \in \mathbb{Z}$ with the following example.

Example 3.10. Find the greatest common divisor d of 3920 and 252 and find integers a and b such that $d = 3920a + 252b$.

Solution. In solving the problem, we demonstrate how to perform the Euclidean algorithm so that we can find a and b by working backward. First, we note that

$$3920 = 15 \cdot 252 + 140.$$

Now the Euclidean algorithm tells us that $\gcd(3920, 252) = \gcd(252, 140)$, so we write

$$252 = 1 \cdot 140 + 112.$$

3.5 Solving Combinatorial Problems Recursively

Continuing, we have $140 = 1 \cdot 112 + 28$ and $112 = 4 \cdot 28 + 0$, so $d = 28$.

To find a and b, we now work backward through the equations we found earlier, "solving" them for the remainder term and then substituting. We begin with

$$28 = 140 - 1 \cdot 112.$$

But we know that $112 = 252 - 1 \cdot 140$, so

$$28 = 140 - 1(252 - 1 \cdot 140) = 2 \cdot 140 - 1 \cdot 252.$$

Finally, $140 = 3920 - 15 \cdot 252$, so now we have

$$28 = 2(3920 - 15 \cdot 252) - 1 \cdot 252 = 2 \cdot 3920 - 31 \cdot 252.$$

Therefore $a = 2$ and $b = -31$.

3.5.2 Sorting

One of the most common and most basic computing problems is sorting: Given a sequence $a_1, a_2, \ldots, a_n$ of n distinct integers, rearrange them so that they are in increasing order. We describe here an easy recursive strategy for accomplishing this task. This strategy is known as **Merge Sort**, and it is one of several optimal algorithms for sorting. Introductory computer science courses treat this topic in greater depth. In our course, we simply need some good strategy and merge sort works fine for our purposes.

To present merge sort, must first develop a strategy for solving a special case of the sorting problem. Suppose we have $s + t$ distinct integers

$$\{u_0, u_1, \ldots, u_{s-1}, v_0, v_1, \ldots, v_{t-1}\}$$

arranged as two lists with $u_0 < u_1 < \ldots < u_{s-1}$ and $v_0 < v_1 < \ldots < v_{t-1}$. How do we *merge* these two sequences into a single increasing sequence of length $s+t$. Imagine the two sequences placed on two horizontal lines, one immediately under the other. Then let u be the least integer in the first sequence and v the least integer in the second. At the moment, this implies that $u = u_0$ and $v = v_0$, but integers will be deleted from the two sequences as the process is carried out. Regardless, the meaning of u and v will be preserved. Also, set $i = 0$. Then take a_i as the minimum of u and v and delete a_i from the sequence in which it occurs. Then increase i by 1 and repeat. Here is a code snippet for accomplishing a merge operation, with u_p now written as u[p] and v_q now written as v[q].

```
u = [1,2,7,9,11,15]
v = [3,5,8,100,130,275]
a = []
```

Chapter 3 Induction

```
p = 0
q = 0
for i in range(len(u) + len(v)):
    if (p < len(u) and q < len(v)):
        a.append(min(u[p], v[q]))
        if (min(u[p], v[q])==u[p]):
            p = p+1
        else:
            q = q+1
    elif (p >= len(u)):
        a.append(v[q])
        q = q+1
    else:
        a.append(u[p])
        p = p+1
a
```

[1, 2, 3, 5, 7, 8, 9, 11, 15, 100, 130, 275]

Now that we have a good strategy for merging, it is easy to develop a recursive strategy for sorting. Given a sequence $a_1, a_2, \ldots, a_n$ of n distinct integers, we set $s = \lceil n/2 \rceil$ and $t = \lfloor n/2 \rfloor$. Then let $u_i = a_i$ for $i = 1, 2, \ldots, s$ and $v_j = a_{s+j}$, for $j = 1, 2, \ldots, t$. Sort the two subsequences and then merge them. For a concrete example, given the sequence $(2, 8, 5, 9, 3, 7, 4, 1, 6)$, we split into $(2, 8, 5, 9, 3)$ and $(7, 4, 1, 6)$. These subsequences are sorted (by a recursive call) into $(2, 3, 5, 8, 9)$ and $(1, 4, 6, 7)$, and then these two sorted sequences are merged.

For running time, if $S(n)$ is the number of operations it takes to sort a sequence of n distinct integers, then $S(2n) \leq 2S(n) + 2n$, since it clearly takes $2n$ steps to merge two sorted sequences of length n. This leads to the bound $S(n) < Cn \log n$ for some positive constant C, and in computer science courses, you will learn (here it is an exercise) that this is optimal.

3.6 Mathematical Induction

Now we move on to induction, the powerful twin of recursion.

Let n be a positive integer. Consider the following mathematical statements, each of which involve n:

1. $2n + 7 = 13$.

2. $3n - 5 = 9$.

3. $n^2 - 5n + 9 = 3$.

4. $8n - 3 < 48$.

5. $8n - 3 > 0$.

6. $(n+3)(n+2) = n^2 + 5n + 6$.

7. $n^2 - 6n + 13 \geq 0$.

Such statements are called **open statements**. Open statements can be considered as **equations**, i.e., statements that are valid for certain values of n. Statement 1 is valid only when $n = 3$. Statement 2 is never valid, i.e., it has no solutions among the positive integers. Statement 3 has exactly two solutions, and Statement 4 has six solutions. On the other hand, Statements 5, 6 and 7 are valid for all positive integers.

At this point, you are probably scratching your head, thinking that this discussion is trivial. But let's consider some statements that are a bit more complex.

1. The sum of the first n positive integers is $n(n+1)/2$.

2. The sum of the first n odd positive integers is n^2.

3. $n^n \geq n! + 4,000,000,000n2^n$ when $n \geq 14$.

How can we establish the validity of such statements, provided of course that they are actually true? The starting point for providing an answer is the following property:

Principle 3.11 (Principle of Mathematical Induction). *Let S_n be an open statement involving a positive integer n. If S_1 is true, and if for each positive integer k, assuming that the statement S_k is true implies that the statement S_{k+1} is true, then S_n is true for every positive integer n.*

With a little thought, you should see that the Principle of Mathematical Induction is logically equivalent to the Well Ordered Property of the Positive Integers. If you haven't already done so, now might be a good time to look over Appendix B on background material.

3.7 Inductive Definitions

Although it is primarily a matter of taste, recursive definitions can also be recast in an inductive setting. As a first example, set $1! = 1$ and whenever $k!$ has been defined, set $(k+1)! = (k+1)k!$.

As a second example, set

$$\sum_{i=1}^{1} f(i) = f(1) \quad \text{and} \quad \sum_{i=1}^{k+1} f(i) = \sum_{i=1}^{k} f(i) + f(k+1)$$

Chapter 3 Induction

In this second example, we are already using an abbreviated form, as we have omitted some English phrases. But the meaning should be clear.

Now let's back up and give an example which would really be part of the development of number systems. Suppose you knew everything there was to know about the *addition* of positive integers but had never heard anything about *multiplication*. Here's how this operation can be defined.

Let m be a positive integer. Then set

$$m \cdot 1 = m \quad \text{and} \quad m \cdot (k+1) = m \cdot k + m$$

You should see that this *defines* multiplication but doesn't do anything in terms of establishing such familiar properties as the commutative and associative properties. Check out some of the details in Appendix B.

3.8 Proofs by Induction

No discussion of recursion and induction would be complete without some obligatory examples of proofs using induction. We start with the "Hello World" example.

Proposition 3.12. *For every positive integer n, the sum of the first n positive integers is $n(n+1)/2$, i.e.,*

$$\sum_{i=1}^{n} i = \frac{n(n+1)}{2}.$$

For our first version of a proof of Proposition 3.12, we clearly identify the open statement S_n and describe the proof carefully in terms of S_n. As you develop more experience with writing proofs by induction, this will become less essential, as you'll see in the second version of the proof.

Proof. Let n be a positive integer, and let S_n be the open statement

$$\sum_{i=1}^{n} i = \frac{n(n+1)}{2}.$$

We will prove that S_n is true for all positive integers by induction. For the basis step, we must prove that S_1 is true. When $n = 1$, the left-hand side of S_n is just 1, while the right-hand side evaluates to $1(1+1)/2 = 1$. Therefore, S_1 is true.

Next we assume that for some positive integer k, S_k is true. That is, we assume

$$\sum_{i=1}^{k} i = \frac{k(k+1)}{2}.$$

3.8 Proofs by Induction

We now seek to prove that S_{k+1} is true, and begin by considering the left-hand side of S_{k+1}. We notice that

$$\sum_{i=1}^{k+1} i = \left(\sum_{i=1}^{k} i\right) + (k+1) = \frac{k(k+1)}{2} + (k+1),$$

since our inductive hypothesis that S_k is true gives us the simpler formula for the summation. Now continuing with a bit of algebra, we find

$$\frac{k(k+1)}{2} + (k+1) = \frac{k^2 + 3k + 2}{2} = \frac{(k+1)(k+2)}{2}.$$

Therefore, S_{k+1} is true. Since we have shown that S_1 is true and that for every positive integer k, if S_k is true, then S_{k+1} is true, we conclude that S_n is true for all positive integers n by the Principle of Mathematical Induction. □

Before looking at a refined version of this proof, let's take a moment to discuss the key steps in every proof by induction. The first step is the **basis step**, in which the open statement S_1 is shown to be true. (It's worth noting that there's nothing special about 1 here. If we want to prove only that S_n is true for all integers $n \geq 5$, then proving that S_5 is true is our basis step.) When proving the basis step, if S_n is an equation, we do not just write down S_1 and move on. We need to *prove* that S_1 is true. Notice how in the proof above, we discussed the left-hand side of S_1 and the right-hand side of S_1 and concluded that they were equal.

After the basis step comes the **inductive step**, in which we assume that S_k is true for *some* positive integer k and prove that S_{k+1} is true. When doing this, we call S_k our **inductive hypothesis**. In the inductive step, the most common mistake students make is starting with the entirety of S_{k+1} and manipulating it until they obtain a true statement. This is dangerous, as it is possible to start with something false and through valid algebraic steps, obtain a true statement. Instead, the best option is to work as with the basis step: if S_{k+1} is an equation or inequality, work on one side until you find a place to apply the inductive hypothesis and then continue until you obtain the other side. If the algebra gets tricky along the way, you can also work with the left-hand side of S_{k+1} and separately work with the right-hand side of S_{k+1}. If you're able to manipulate both sides to be in the same form, then you have shown they are equal and S_{k+1} is true.

Now let's take a look at a more refined proof of Proposition 3.12. From here on, when we give a proof by induction, we'll use this style. As you're getting started with induction proofs, you may find it useful to be more explicit about the steps as we did in the first proof above.

Chapter 3 Induction

Proof. We first prove the assertion when $n = 1$. For this value of n, the left-hand side is just 1, while the right-hand side evaluates to $1(1+1)/2 = 1$.

Now assume that for some positive integer k, the formula holds when $n = k$, i.e., assume that

$$\sum_{i=1}^{k} i = \frac{k(k+1)}{2}.$$

Then it follows that

$$\sum_{i=1}^{k+1} i = \left(\sum_{i=1}^{k} i\right) + (k+1) = \frac{k(k+1)}{2} + (k+1) = \frac{k^2 + 3k + 2}{2} = \frac{(k+1)(k+2)}{2}.$$

Thus the formula also holds when $n = k+1$. By the Principle of Mathematical Induction, it holds for all positive integers n. □

The preceding arguments are 100% correct... but some combinatorial mathematicians would argue that they may actually hide what is really going on. These folks would much prefer a combinatorial proof, as was provided in Section 2.4. Our perspective is that you should prefer to give a combinatorial proof—when you can find one. But if pressed, you should be able to give a formal proof by mathematical induction.

Here's a second example, also quite a classic. Again, recall that we gave a combinatorial proof in the last chapter. As you read the proof, make sure you can identify the open statement S_n, the basis step, and the inductive step.

Proposition 3.13. *For each positive integer n, the sum of the first n odd positive integers is n^2, i.e.,*

$$\sum_{i=1}^{n} (2i - 1) = n^2.$$

Proof. We will prove this by induction. First, note that the formula holds when $n = 1$. Now suppose that k is a positive integer and that the formula holds when $n = k$, i.e., assume

$$\sum_{i=1}^{k} (2i - 1) = k^2.$$

Then

$$\sum_{i=1}^{k+1} (2i - 1) = \left(\sum_{i=1}^{k} 2i - 1\right) + 2k + 1 = k^2 + (2k+1) = (k+1)^2.$$

Therefore, the proposition follows by the Principle of Mathematical Induction. □

3.9 Strong Induction

Here's a more general version of the first result in this section, and again we note that we gave a combinatorial proof in Section 2.4.

Proposition 3.14. *Let n and k be non-negative integers with $n \geq k$. Then*

$$\sum_{i=k}^{n} \binom{i}{k} = \binom{n+1}{k+1}.$$

Proof. Fix a non-negative integer k. We then prove the formula by induction on n. If $n = k$, note that the left hand side is just $\binom{k}{k} = 1$, while the right hand side is $\binom{k+1}{k+1}$ which is also 1. Now assume that m is a non-negative integer, with $m \geq k$, and that the formula holds when $n = m$, i.e., assume that

$$\sum_{i=k}^{m} \binom{i}{k} = \binom{m+1}{k+1}.$$

Then

$$\sum_{i=k}^{m+1} \binom{i}{k} = \sum_{i=k}^{m} \binom{i}{k} + \binom{m+1}{k}$$
$$= \binom{m+1}{k+1} + \binom{m+1}{k}$$
$$= \binom{m+2}{k+1}.$$

Therefore, the proposition follows by the Principle of Mathematical Induction. □

3.9 Strong Induction

There are occasions where the Principle of Mathematical Induction, at least as we have studied it up to this point, does not seem sufficient. Here is a concrete example. The professor asked Bob to study a function $f(n)$ defined recursively by $f(n) = 2f(n-1) - f(n-2)$ with $f(1) = 3$ and $f(2) = 5$. Specifically, the professor asked Bob to compute $f(10^{10})$, which seems like a daunting task. Over coffee, Bob scribbled on a napkin and determined that $f(3) = 7$ and $f(4) = 9$, and on the basis of these calculations alone, he thought that it might just be possible that $f(n) = 2n + 1$ for all $n \geq 1$. If this were true, he could simply report that $f(10^{10}) = 2 \cdot 10^{10} + 1 = 20000000001$.

Bob was beginning to understand proofs by induction, so he tried to prove that $f(n) = 2n + 1$ for all $n \geq 1$ by induction. For the base step, he noted that $f(1) = 3 =$

Chapter 3 Induction

$2 \cdot 1 + 1$, so all is ok to this point. For the inductive step, he assumed that $f(k) = 2k + 1$ for some $k \geq 1$ and then tried to prove that $f(k+1) = 2(k+1) + 1$. If this step could be completed, then the proof by induction would be done.

But at this point, Bob seemed to hit a barrier, because

$$f(k+1) = 2f(k) - f(k-1) = 2(2k+1) - f(k-1),$$

using the inductive hypothesis to replace $f(k)$ by $2k + 1$. However, he's was totally perplexed about what to do with the $f(k-1)$. If he knew that $f(k-1) = 2(k-1) + 1$, then the right hand side would result in $2(2k+1) - (2k-1) = 2k + 3 = 2(k+1) + 1$, which is exactly what he wants. Bob always plays by the rules, and he has to admit that he doesn't know that $f(k-1) = 2(k-1) + 1$. He only knows that $f(k) = 2k + 1$.

Bob was about to throw in the towel and ask his computer to start making the calculations recursively, when Carlos comes along and asks what he's doing. Carlos sees right away that the approach Bob was taking to prove that $f(n) = 2n + 1$ by induction won't work—but after a moment's reflection, Carlos says that there's a stronger form of an inductive proof that will do the trick. Carlos patiently explained to Bob a proposition which is called the **Strong Principle of Mathematical Induction**. To prove that an open statement S_n is valid for all $n \geq 1$, it is enough to

a Show that S_1 is valid, and

b Show that S_{k+1} is valid whenever S_m is valid for all integers m with $1 \leq m \leq k$.

The validity of this proposition is trivial since it is *stronger* than the principle of induction. What is novel here is that in order to prove a statement, it is sometimes to your advantage to prove something even stronger. Combinatorial mathematicians call this the "bootstrap" phenomenon.

Equipped with this observation, Bob saw clearly that the strong principle of induction was enough to prove that $f(n) = 2n + 1$ for all $n \geq 1$. So he could power down his computer and enjoy his coffee.

3.10 Discussion

The group was debating the value of combinatorial proofs versus formal proofs by induction. Xing said that he actually preferred to do a proof by induction, as a combinatorial proof, it could be argued, wasn't really a proof. Dave mumbled "Combinatorial proofs can always be made rigorous." They went back and forth for a while and then Alice said "But the professor never explained that weird sequence

$$1, 2, 3, 4, 1, 2, 3, 4, 5, 1, 2, 3, 4, 5, 2, 3, 4, 5, 6, 2, 3, 4, 5, 6, 1, 2, 3, 4, 5, 2, 3, 4, 5, 6, \ldots,$$

did he?"

Dave was on a roll. He asked, "Who has change for a dollar?" but nobody understood why he would derail an argument over proofs when everybody had already paid for the coffee. Alice was more to the point "You know Dave, sometimes I just don't understand why you say the things you do." Dave smiled (maybe it was more of a smirk) "It's about making change. The terms in this sequence are the fewest number of coins required to make change." Bob said "I don't get it." Dave continued "The term a_n is the fewest number of U.S. coins required to total to n cents." Now everyone groaned, everyone except Carlos, who thought that at least this time, Dave was really clever.

"Well", said Bob, "that takes care of the strange sequence, but I still don't see any difference between induction and recursion." Dave couldn't keep quiet "No one does." Xing thought differently and said "In many programming languages, you try to avoid recursion, preferring to use loops instead. Otherwise, you wind up overloading the stack. As just one example, you can compute the greatest common divisor d of m and n, as well as find a and b so that $d = am + bn$ using a loop—with very little storage. The recursive approach discussed previously, with the inherent back tracking at the end, isn't really necessary." Yolanda was impressed with Xing's extensive programming experience and knowledge, but Alice was less so.

Zori was losing her patience and was especially grumpy today "I don't see any value to any of this stuff. Who's going to pay me to find greatest common divisors?" Dave said "Nobody." Alice said, "But maybe there are some principles here that have practical application." Carlos joined in, saying "I think the basic principles behind establishing that a computer program does what you intend have a lot to do with induction and recursion." Bob said "I don't understand. When I write a program, I just pay attention to details and after just a few corrections, they always work." Alice was brutal "Maybe that's because you don't do anything complicated." Carlos was more gentle "Big software projects might have hundreds of thousands of lines of code, and pieces of the final product might be written by different groups of programmers at different moments in time. Establishing correctness can be a very difficult task." Zori's ears perked up as she thought she saw something in this last bit of conversation that might be a way to earn a salary.

3.11 Exercises

1. A database uses record identifiers that are alphanumeric strings in which the 10 decimal digits and 26 upper-case letters are valid symbols. The criteria that define a valid record identifier are recursive. A valid record identifier of length $n \geq 2$ can be constructed in the following ways:

- beginning with any upper-case letter other than D and followed by any valid

Chapter 3 Induction

   record identifier of length $n - 1$;

- beginning with $1C, 2K,$ or $7J$ and followed by any valid record identifier of length $n - 2$; or
- beginning with D and followed by any string of $n - 1$ decimal digits.

Let $r(n)$ denote the number of valid record identifiers of length n. We take $r(0) = 1$ and note that $r(1) = 26$. Find a recursion for $r(n)$ when $n \geq 2$ and use it to compute $r(5)$.

2. Consider a $1 \times n$ checkerboard. The squares of the checkerboard are to be painted white and gold, but no two consecutive squares may both be painted white. Let $p(n)$ denote the number of ways to paint the checkerboard subject to this rule. Find a recursive formula for $p(n)$ valid for $n \geq 3$.

3. Give a recursion for the number $g(n)$ of ternary strings of length n that do not contain 102 as a substring.

4. A $2 \times n$ checkerboard is to be tiled using two types of tiles. The first tile is a 1×1 square tile. The second tile is called an L-tile and is formed by removing the upper-right 1×1 square from a 2×2 tile. The L-tiles can be used in any of the four ways they can be rotated. (That is, the "missing square" can be in any of four positions.) Let $t(n)$ denote the number of tilings of the $2 \times n$ checkerboard using 1×1 tiles and L-tiles. Find a recursive formula for $t(n)$ and use it to determine $t(7)$.

5. Let S be the set of strings on the alphabet $\{0, 1, 2, 3\}$ that do not contain 12 or 20 as a substring. Give a recursion for the number $h(n)$ of strings in S of length n.

Hint. Check your recursion by manually computing $h(1), h(2), h(3),$ and $h(4)$.

6. Find $d = \gcd(5544, 910)$ as well as integers a and b such that $5544a + 910b = d$.

7. Find $\gcd(827, 249)$ as well as integers a and b such that $827a + 249b = 6$.

8. Let $a, b, m,$ and n be integers and suppose that $am + bn = 36$. What can you say about $\gcd(m, n)$?

9. (A challenging problem) For each formula, give both a proof using the Principle of Mathematical Induction and a combinatorial proof. One of the two will be easier while the other will be more challenging.

(a) $1^2 + 2^2 + 3^2 + \cdots + n^2 = \dfrac{n(n+1)(2n+1)}{6}$

(b) $\binom{n}{0}2^0 + \binom{n}{1}2^1 + \binom{n}{2}2^2 + \cdots + \binom{n}{n}2^n = 3^n$

10. Show that for all integers $n \geq 4$, $2^n < n!$.

11. Show that for all positive integers n,

$$\sum_{i=0}^{n} 2^i = 2^{n+1} - 1.$$

12. Show that for all positive integers n, $7^n - 4^n$ is divisible by 3.

13. Show that for all positive integers n, $9^n - 5^n$ is divisible by 4.

14. It turns out that if a and b are positive integers with $a > b + 1$, then there is a positive integer $M > 1$ such that $a^n - b^n$ is divisible by M for all positive integers n. Determine M in terms of a and b and prove that it is a divisor of $a^n - b^n$ for all positive integers n.

15. Use mathematical induction to prove that for all integers $n \geq 1$,

$$n^3 + (n+1)^3 + (n+2)^3$$

is divisible by 9.

16. Give a proof by induction of the Binomial Theorem (Theorem 2.30). How do you think it compares to the combinatorial argument given in Chapter 2?

17. Consider the recursion given by $f(n) = 2f(n-1) - f(n-2) + 6$ for $n \geq 2$ with $f(0) = 2$ and $f(1) = 4$. Use mathematical induction to prove that $f(n) = 3n^2 - n + 2$ for all integers $n \geq 0$.

18. Consider the recursion given by $f(n) = f(n-1) + f(n-2)$ for $n \geq 3$ with $f(1) = f(2) = 1$. Show that $f(n)$ is divisible by 3 if and only if n is divisible by 4.

19. Suppose that $x \in \mathbb{R}$ and $x > -1$. Prove that for all integers $n \geq 0$, $(1+x)^n \geq 1 + nx$.

20. Show that there is a positive constant c so that any algorithm that sorts a sequence of n positive integers must, in worst case, take $cn \log n$ steps.

Hint. Hint: There are $n!$ permutations of a set of n distinct integers. Each operation reduces the number of possibilities by a multiplicative fraction which is at most $1/2$. So if there are t operations, then $2^t \geq n!$. Now look up Stirling's approximation for $n!$ and continue from there.

CHAPTER 4

Combinatorial Basics

Dave hates doing the same thing twice. He sees himself as a free spirit and never wants to fall into a rut. Alice says that this approach to life requires one to have lots and lots of options, for if you have to do a lot of something, like get up in the morning and get dressed, then you may not be able to avoid mindless repetition, dull and boring as it may seem.

4.1 The Pigeon Hole Principle

A function $f : X \longrightarrow Y$ is said to be 1–1 (read **one-to-one**) when $f(x) \neq f(x')$ for all $x, x' \in X$ with $x \neq x'$. A 1–1 function is also called an **injection** or we say that f is **injective**. When $f : X \longrightarrow Y$ is 1–1, we note that $|X| \leq |Y|$. Conversely, we have the following self-evident statement, which is popularly called the "Pigeon Hole" principle.

Proposition 4.1 (Pigeon Hole Principle). *If $f : X \longrightarrow Y$ is a function and $|X| > |Y|$, then there exists an element $y \in Y$ and distinct elements $x, x' \in X$ so that $f(x) = f(x') = y$.*

In more casual language, if you must put $n + 1$ pigeons into n holes, then you must put two pigeons into the same hole.

Here is a classic result, whose proof follows immediately from the Pigeon Hole Principle.

Theorem 4.2 (Erdős/Szekeres). *If m and n are non-negative integers, then any sequence of $mn + 1$ distinct real numbers either has an increasing subsequence of $m + 1$ terms, or it has a decreasing subsequence of $n + 1$ terms.*

Proof. Let $\sigma = (x_1, x_2, x_3, \ldots, x_{mn+1})$ be a sequence of $mn + 1$ distinct real numbers. For each $i = 1, 2, \ldots, mn + 1$, let a_i be the maximum number of terms in a increasing subsequence of σ with x_i the first term. Also, let b_i be the maximum number of terms in a decreasing subsequence of σ with x_i the last term. If there is some i for which $a_i \geq m + 1$, then σ has an increasing subsequence of $m + 1$ terms. Conversely, if for

Chapter 4 Combinatorial Basics

some i, we have $b_i \geq n + 1$, then we conclude that σ has a decreasing subsequence of $n + 1$ terms.

It remains to consider the case where $a_i \leq m$ and $b_i \leq n$ for all $i = 1, 2, \ldots, mn + 1$. Since there are mn ordered pairs of the form (a, b) where $1 \leq a \leq m$ and $1 \leq b \leq n$, we conclude from the Pigeon Hole principle that there must be integers i_1 and i_2 with $1 \leq i_1 < i_2 \leq mn + 1$ for which $(a_{i_1}, b_{i_1}) = (a_{i_2}, b_{i_2})$. Since x_{i_1} and x_{i_2} are distinct, we either have $x_{i_1} < x_{i_2}$ or $x_{i_1} > x_{i_2}$. In the first case, any increasing subsequence with x_{i_2} as its first term can be extended by prepending x_{i_1} at the start. This shows that $a_{i_1} > a_{i_2}$. In the second case, any decreasing sequence of with x_{i_1} as its last element can be extended by adding x_{i_2} at the very end. This shows $b_{i_2} > b_{i_1}$. □

In Chapter 11, we will explore some powerful generalizations of the Pigeon Hole Principle. All these results have the flavor of the general assertion that total disarray is impossible.

4.2 An Introduction to Complexity Theory

Discussion 4.3. Bob says that he's really getting to like this combinatorial mathematics stuff. The concrete nature of the subject is appealing. But he's not sure that he understands the algorithmic component. Sometimes he sees how one might actually compute the answer to a problem—provided he had access to a powerful computer. At other times, it seems that a computational approach might be out of reach, even with the world's best and fastest computers at ready access. Carlos says it can be much worse than that. There are easily stateable problems that no one knows how to attack even if all the world's computational power is used in concert. And there's nothing on the horizon that will change that. In fact, build faster computers and you just change the threshold for what is computable. There will still be easily understood problems that will remain unresolved.

4.2.1 Three Questions

We consider three problems with a common starting point. You are given[1] a set S of $10,000$ distinct positive integers, each at most $100,000$, and then asked the following questions.

1. Is $83,172$ one of the integers in the set S?

2. Are there three integers in S whose sum is $143,297$?

---

[1]The particulars of how the set is given to you aren't important to the discussion. For example, the data could be given as a text file, with one number on each line.

4.2 An Introduction to Complexity Theory

3. Can the set S be partitioned as $S = A \cup B$ with $A \cap B = \emptyset$, so that $\sum_{a \in A} a = \sum_{b \in B} b$.

The first of the three problems sounds easy, and it is. You just consider the numbers in the set one by one and test to see if any of them is 83,172. You can stop if you ever find this number and report that the answer is yes. If you return a no answer, then you will have to have read every number in the list. Either way, you halt with a correct answer to the question having done at most 10,000 tests, and even the most modest netbook can do this in a heartbeat. And if the list is expanded to 1,000,000 integers, all at most a billion, you can still do it easily. More generally, if you're given a set S of n numbers and an integer x with the question "Is x a member of S?", you can answer this question in n steps, with each step an operation of testing a number in S to see if it is exactly equal to n. So the running time of this algorithm is proportional to n, with the constant depending on the amount of time it takes a computer to perform the basic operation of asking whether a particular integer is equal to the target value.

The second of the three problems is a bit more challenging. Now it seems that we must consider the 3-element subsets of a set of size 10,000. There are $C(10,000, 3)$ such sets. On the one hand, testing three numbers to see if their sum is 143,297 is very easy, but there are lots and lots of sets to test. Note that $C(10,000, 3) = 166,616,670,000$, and not too many computers will handle this many operations. Moreover, if the list is expanded to a million numbers, then we have more than 10^{17} triples to test, and that's off the table with today's hardware.

Nevertheless, we can consider the general case. We are given a set S of n integers and a number x. Then we are asked whether there are three integers in S whose sum is x. The algorithm we have described would have running time proportional to n^3, where the constant of proportionality depends on the time it takes to test a triple of numbers to see if there sum is x. Of course, this depends in turn on just how large the integer x and the integers in S can be.

The third of the three problems is different. First, it seems to be much harder. There are 2^{n-1} complementary pairs of subsets of a set of size n, and one of these involves the empty set and the entire set. But that leaves $2^{n-1} - 1$ pairs to test. Each of these tests is not all that tough. A netbook can easily report whether a two subsets have the same sum, even when the two sets form a partition of a set of size 10,000, but there are approximately 10^{3000} partitions to test and no piece of hardware on the planet will touch that assignment. And if we go up to a set of size 1,000,000, then the combined computing power of all the machines on earth won't get the job done.

In this setting, we have an algorithm, namely testing all partitions, but it is totally unworkable for n element sets when n is large since it has running time proportional to 2^n.

4.2.2 Certificates

Each of the three problems we have posed is in the form of a "yes/no" question. A "yes" answer to any of the three can be justified by providing a certificate that can be checked efficiently. For example, if you answer the first question with a yes, then you might provide the additional information that you will find 83,172 as the integer on line 584 in the input file. Of course, you could also provide the source code for the computer program, and let a referee run the entire procedure.

Similarly, if you answer the second question with a yes, then you could specify the three numbers and specify where in the input file they are located. An impartial referee could then verify, if it mattered, that the sum of the three integers was really 143,297 and that they were located at the specified places in the input file. Alternatively, you could again provide the source code which would require the referee to test all triples and verify that there is one that works.

Likewise, a yes for the third question admits a modest size certificate. You need only specify the elements of the subset A. The referee, who is equipped with a computer, can (a) check to see that all numbers in A belong to S; (b) form a list of the subset B consisting of those integers in S that do not belong to A; and (c) compute the sums of the integers in A and the integers in B and verify that the two sums are equal. But in this case, you would not provide source code for the algorithm, as there does not appear (at least nothing in our discussion thus far provides one) to be a reasonable strategy for deciding this problem when the problem size is large.

Now let's consider the situation with a "no" answer. When the answer to the first question is no, the certificate can again be a computer program that will enable the referee to consider all the elements of S and be satisfied that the number in question is not present. A similar remark holds for the second question, i.e., the program is the certificate.

But the situation with the third question is again very different. Now we can't say to the referee "We checked all the possibilities and none of them worked." This could not possibly be a true statement. And we have no computer program that can be run by us or by the referee. The best we could say is that we tried to find a suitable partition and were unable to do so. As a result, we don't know what the correct answer to the question actually is.

4.2.3 Operations

Many of the algorithms we develop in this book, as well as many of the computer programs that result from these algorithms involve basic steps that are called **operations**. The meaning of the word operation is intentionally left as an imprecise notion. An operation might be just comparing two integers to see if they are equal; it might be updating the value of a variable x and replacing it by $x^2 - 3x + 7$; and it might be checking whether two set sums are equal. In the third instance, we would typically limit the size of the two subsets as well as the integers in them. As a consequence, we want to be able to say that there is some constant c so that an operation can be carried out in time at most c on a computer. Different computers yield different values of c, but that is a discrepancy which we can safely ignore.

4.2.4 Input Size

Problems come in various sizes. The three problems we have discussed in this chapter have the same input size. Roughly speaking this size is 10,000 blocks, with each block able to hold an integer of size at most 100,000. In this text, we will say that the input size of this problem is $n = 10,000$, and in some sense ignoring the question of the size of the integers in the set. There are obvious limitations to this approach. We could be given a set S of size 1 and a candidate element x and be asked whether x belongs to S. Now suppose that x is a bit string the size of a typical compact disk, i.e., some 700 megabytes in length. Just reading the single entry in S to see if it's exactly x will take some time.

In a similar vein, consider the problem of determining whether a file x is located anywhere in the directory structure under y in a unix file system. If you go on the basis of name only, then this may be relatively easy. But what if you want to be sure that an exact copy of x is present? Now it is much more challenging.

4.3 The Big "Oh" and Little "Oh" Notations

Let $f : \mathbb{N} \longrightarrow \mathbb{R}$ and $g : \mathbb{N} \longrightarrow \mathbb{R}$ be functions. We write $f = O(g)$, and say f is "**Big Oh**" of g, when there is a constant c and an integer n_0 so that $f(n) \leq cg(n)$ whenever $n > n_0$. Although this notation has a long history, we can provide a quite modern justification. If f and g both describe the number of operations required for two algorithms given input size n, then the meaning of $f = O(g)$ is that f is no harder than g when the problem size is large.

We are particularly interested in comparing functions against certain natural benchmarks, e.g., $\log \log n$, $\log n$, $\sqrt{n}$, n^α where $\alpha < 1$, n, n^2, n^3, n^c where $c > 1$ is a constant, $n^{\log n}$, 2^n, $n!$, 2^{n^2}, etc.

For example, in Subsection 3.5.2 we learned that there are sorting algorithms with running time $O(n \log n)$ where n is the number of integers to be sorted. As a second example, we will learn that we can find all shortest paths in an oriented graph on n vertices with non-negative weights on edges with an algorithm having running time $O(n^2)$. At the other extreme, no one knows whether there is a constant c and an algorithm for determining whether the chromatic number of a graph is at most three which has running time $O(n^c)$.

It is important to remember that when we write $f = O(g)$, we are implying in some sense that f is no bigger than g, but it may in fact be much smaller. By contrast, there will be times when we really know that one function dominates another. And we have a second kind of notation to capture this relationship.

Let $f : \mathbb{N} \longrightarrow \mathbb{R}$ and $g : \mathbb{N} \longrightarrow \mathbb{R}$ be functions with $f(n) > 0$ and $g(n) > 0$ for all n. We write $f = o(g)$, and say that f is "**Little oh**" of g, when $\lim_{n\to\infty} f(n)/g(n) = 0$. For example $\ln n = o(n^2)$; $n^\alpha = o(n^\beta)$ whenever $0 < \alpha < \beta$; and $n^{100} = o(c^n)$ for every $c > 1$. In particular, we write $f(n) = o(1)$ when $\lim_{n\to\infty} f(n) = 0$.

4.4 Exact Versus Approximate

Many combinatorial problems admit "exact" solutions, and in these cases, we will usually try hard to find them. The Erdős/Szekeres Theorem from earlier in this chapter is a good example of an "exact" result[1]. By this statement, we mean that for each pair m and n of positive integers, there is a sequence of mn distinct real numbers that has neither an increasing subsequence of size $m+1$ nor a decreasing subsequence of size $n+1$. To see this, consider the sequence σ defined as follows: For each $i = 1, 2, \ldots, m$, let $B_i = \{j + (m-1)i : 1 \le j \le n\}$. Note that each B_i is a block of n consecutive integers. Then define a permutation σ of the first mn integers by setting $\alpha < \beta$ if there exist distinct integers i_1 and i_2 so that $\alpha \in B_{i_1}$ and $\beta \in B_{i_2}$. Also, for each $i = 1, 2, \ldots, m$, set $\alpha < \beta$ in σ when $1+(m-1)i \le \beta < \alpha \le n+(m-1)i$. Clearly, any increasing subsequence of σ contains at most one member from each block, so σ has no increasing sequence of size $m = 1$. On the other hand, any decreasing sequence in σ is contained in a single block, so σ has no decreasing sequence of size $n+1$.

As another example of an exact solution, the number of integer solutions to $x_1 + x_2 + \ldots x_r = n$ with $x_i > 0$ for $i - 1, 2, \ldots, r$ is exactly $C(n-1, r-1)$. On the other hand, nothing we have discussed thus far allows us to provide an exact solution for the number of partitions of an integer n.

4.4.1 Approximate and Asymptotic Solutions

Here's an example of a famous problem that we can only discuss in terms of approximate solutions, at least when the input size is suitably large. For an integer n, let $\pi(n)$ denote the number of primes among the first n positive integers. For example, $\pi(12) = 5$ since 2, 3, 5, 7 and 11 are primes. The exact value of $\pi(n)$ is known when

---
[1]Exact results are also called "best possible", "sharp" or "tight."

$n \leq 10^{23}$, and in fact:

$$\pi(10^{23}) = 1,925,320,391,606,803,968,923$$

On the other hand, you might ask whether $\pi(n)$ tends to infinity as n grows larger and larger. The answer is yes, and here's a simple and quite classic argument. Suppose to the contrary that there were only k primes, where k is a positive integer. Suppose these k primes are listed in increasing order as $p_1 < p_2 < \ldots < p_k$, and consider the number $n = 1 + p_1 p_2 \cdots p_k$. Then n is not divisible by any of these primes, and it is larger than p_k, which implies that n is either a prime number larger than p_k or divisible by a prime number larger than p_k.

So we know that $\lim_{n \to \infty} \pi(n) = \infty$. In a situation like this, mathematicians typically want to know more about how fast $\pi(n)$ goes to infinity. Some functions go to infinity "slowly", such as $\log n$ or $\log \log n$. Some go to infinity quickly, like 2^n, $n!$ or 2^{2^n}. Since $\pi(n) \leq n$, it can't go to infinity as fast as these last three functions, but it might go infinity like $\log n$ or maybe $\sqrt{n}$.

On the basis of computational results (done by hand, long before there were computers), Legendre conjectured in 1796 that $\pi(n)$ goes to infinity like $n/\ln n$. To be more precise, he conjectured that

$$\lim_{n \to \infty} \frac{\pi(n) \ln n}{n} = 1.$$

In 1896, exactly one hundred years after Legendre's conjecture, Hadamard and de la Vallée-Poussin independently published proofs of the conjecture, using techniques whose roots are in the Riemann's pioneering work in complex analysis. This result, now known simply as the *Prime Number Theorem*, continues to this day to be much studied topic at the boundary of analysis and number theory.

4.4.2 Polynomial Time Algorithms

Throughout this text, we will place considerable emphasis on problems for which a certificate can be found in polynomial time. This refers to problems for which there is some constant $c > 0$ so that there is an algorithm $\mathcal{A}$ for solving the problem which has running time $O(n^c)$ where n is the input size. The symbol $\mathcal{P}$ is suggestive of *polynomial*.

4.4.3 $\mathcal{P} = \mathcal{NP}$?

Perhaps the most famous question at the boundary of combinatorial mathematics, theoretical computer science and mathematical logic is the notoriously challenging question of deciding whether $\mathcal{P}$ is the same as $\mathcal{NP}$. This problem has the shorthand form: $\mathcal{P} = \mathcal{NP}$? Here, we present a brief informal discussion of this problem.

First, we have already introduced the class $\mathcal{P}$ consisting of all yes-no combinatorial problems which admit polynomial time algorithms. The first two problems discussed in this chapter belong to $\mathcal{P}$ since they can be solved with algorithms that have running time $O(n)$ and $O(n^3)$, respectively. Also, determining whether a graph is 2-colorable and whether it is connected both admit polynomial time algorithms.

We should emphasize that it may be very difficult to determine whether a problem belongs to class $\mathcal{P}$ or not. For example, we don't see how to give a fast algorithm for solving the third problem (subset sum), but that doesn't mean that there isn't one. Maybe we all need to study harder!

Setting that issue aside for the moment, the class $\mathcal{NP}$ consists of yes–no problems for which there is a certificate for a yes answer whose correctness can be verified in polynomial time. More formally, this is called the class of **nondeterministic polynomial time** problems. Our third problem definitely belongs to this class.

The famous question is to determine whether the two classes are the same. Evidently, any problem belonging to $\mathcal{P}$ also belongs to $\mathcal{NP}$, i.e, $\mathcal{P} \subseteq \mathcal{NP}$, but are they equal? It seems difficult to believe that there is a polynomial time algorithm for settling the third problem (the subset sum problem), and no one has come close to settling this issue. But if you get a good idea, be sure to discuss it with one or both authors of this text before you go public with your news. If it turns out that you are right, you are certain to treasure a photo opportunity with yours truly.

4.5 Discussion

Carlos, Dave and Yolanda were fascinated by the discussion on complexity. Zori was less enthusiastic but even she sensed that the question of which problems could be solved quickly had practical implications. She could even predict that people could earn a nice income solving problems faster and more accurately than their competition.

Bob remarked, "I'm not sure I understand what's being talked about here. I don't see why it can't be the case that all problems can be solved. Maybe we just don't know how to do it." Xing said, "Any finite problem *can* be solved. There is always a way to list all the possibilities, compare them one by one and take the best one as the answer." Alice joined in, "Well, a problem might take a long time just because it is big. For example, suppose you are given two DVD's, each completely full with the data for a large integer. How are you possibly going to multiply them together, even with a large computer and fancy software." Carlos then offered, "But I think there are really hard problems that any algorithm will take a long time to solve and not just because the input size is large. At this point, I don't know how to formulate such a problem but I suspect that they exist."

4.6 Exercises

1. Suppose you are given a list of n integers, each of size at most $100n$. How many operations would it take you to do the following tasks (in answering these questions, we are interested primarily in whether it will take $\log n, \sqrt{n}, n, n^2, n^3, 2^n, \ldots$ steps. In other words, ignore multiplicative constants.):

   (a) Determine if the number $2n + 7$ is in the list.
   (b) Determine if there are two numbers in the list whose sum is $2n + 7$.
   (c) Determine if there are two numbers in the list whose product is $2n + 7$ (This one is more subtle than it might appear! It may be to your advantage to sort the integers in the list).
   (d) Determine if there is a number i for which all the numbers in the list are between i and $i + 2n + 7$.
   (e) Determine the longest sequence of consecutive integers belonging to the list.
   (f) Determine the number of primes in the list.
   (g) Determine whether there are three integers x, y and z from the list so that $x + y = z$.
   (h) Determine whether there are three integers x, y and z from the list so that $x^2 + y^2 = z^2$.
   (i) Determine whether there are three integers x, y and z from the list so that $xy = z$.
   (j) Determine whether there are three integers x, y and z from the list so that $x^y = z$.
   (k) Determine whether there are two integers x and y from the list so that x^y is a prime.
   (l) Determine the longest arithmetic progression in the list (a sequence $(a_1, a_2, \ldots, a_t)$ is an arithmetic progression when there is a constant $d \neq 0$ so that $a_{i+1} = a_i + d$, for each $i = 1, 2, \ldots, t - 1$).
   (m) Determine the number of distinct sums that can be formed from members of the list (arbitrarily many integers from the list are allowed to be terms in the sum).
   (n) Determine the number of distinct products that can be formed from members of the list (arbitrarily many integers from the list are allowed to be factors in the product).
   (o) Determine for which integers m, the list contains at least 10% of the integers from $\{1, 2, \ldots, m\}$.

2. If you have to put $n + 1$ pigeons into n holes, you have to put two pigeons into the same hole. What happens if you have to put $mn + 1$ pigeons into n holes?

Chapter 4 Combinatorial Basics

3. Consider the set $X = \{1,2,3,4,5\}$ and suppose you have two holes. Also suppose that you have 10 pigeons: the 2-element subsets of X. Can you put these 10 pigeons into the two holes in a way that there is no 3-element subset $S = \{a,b,c\} \subset X$ for which all pigeons from S go in the same hole? Then answer the same question if $X = \{1,2,3,4,5,6\}$ with $15 = C(6,2)$ pigeons.

4. Let $n = 10,000$. Suppose a friend tells you that he has a secret family of subsets of $\{1,2,\ldots,n\}$, and if you guess it correctly, he will give you one million dollars. You think you know the family of subsets he has in mind and it contains exactly half the subsets, i.e., the family has 2^{n-1} subsets. Discuss how you can share your hunch with your friend in an effort to win the prize.

5. Let N denote the set of positive integers. When $f : N \to N$ is a function, let $E(f)$ be the function defined by $E(f)(n) = 2^{f(n)}$. What is $E^5(n^2)$?

CHAPTER 5

Graph Theory

In Example 1.5, we discussed the problem of assigning frequencies to radio stations in the situation where stations within 200 miles of each other must broadcast on distinct frequencies. Clearly we would like to use the smallest number of frequencies possible for a given layouts of transmitters, but how can we determine what that number is?

Suppose three new homes are being built and each of them must be provided with utility connections. The utilities in question are water, electricity, and natural gas. Each provider needs a direct line from their terminal to each house (the line can zig-zag all it wants, but it must go from the terminal to the house without passing through another provider's terminal or another house en route), and the three providers all wish to bury their lines exactly four feet below ground. Can they do this successfully without the lines crossing?

These are just two of many, many examples where the discrete structure known as a **graph** can serve as an enlightening mathematical model. Graphs are perhaps the most basic and widely studied combinatorial structure, and they are prominently featured in this text. Many of the concepts we will study, while presented in a more abstract mathematical sense, have their origins in applications of graphs as models for real-world problems.

5.1 Basic Notation and Terminology for Graphs

A **graph G** is a pair (V, E) where V is a set (almost always finite) and E is a set of 2-element subsets of V. Elements of V are called **vertices** and elements of E are called **edges**. We call V the **vertex set** of **G** and E is the **edge set**. For convenience, it is customary to abbreviate the edge $\{x, y\}$ as just xy. Remember though that $xy \in E$ means exactly the same as $yx \in E$. If x and y are distinct vertices from V, x and y are **adjacent** when $xy \in E$; otherwise, we say they are **non-adjacent**. We say the edge xy is **incident to** the vertices x and y.

For example, we could define a graph **G** = (V, E) with vertex set $V = \{a, b, c, d, e\}$ and edge set $E = \{\{a, b\}, \{c, d\}, \{a, d\}\}$. Notice that no edge is incident to e, which is perfectly permissible based on our definition. It is quite common to identify a graph with a visualization in which we draw a point for each vertex and a line connecting two vertices if they are adjacent. The graph **G** we've just defined is shown in Figure 5.1. It's important to remember that while a drawing of a graph is a helpful tool, it is not the same as the graph. We could draw **G** in any of several different ways without changing what it is as a graph.

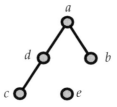

FIGURE 5.1: A GRAPH ON 5 VERTICES

As is often the case in science and mathematics, different authors use slightly different notation and terminology for graphs. As an example, some use **nodes** and **arcs** rather than vertices and edges. Others refer to vertices as **points** and in this case, they often refer to **lines** rather than edges. We will try to stick to vertices and edges but confess that we may occasionally lapse into referring to vertices as points. Also, following the patterns of many others, we will also say that adjacent vertices are **neighbors**. And we will use the more or less standard terminology that the **neighborhood** of a vertex x is the set of vertices adjacent to x. Thus, using the graph **G** we have depicted in Figure 5.1, vertices d and a are neighbors, and the neighborhood of d is $\{a,c\}$ while the neighborhood of e is the empty set. Also, the **degree** of a vertex v in a graph **G**, denoted $\deg_G(v)$, is then the number of vertices in its neighborhood, or equivalently, the number of edges incident to it. For example, we have $\deg_G(d) = \deg_G(a) = 2$, $\deg_G(c) = \deg_G(b) = 1$, and $\deg_G(e) = 0$. If the graph being discussed is clear from context, it is not uncommon to omit the subscript and simply write $\deg(v)$ for the degree of v.

When $\mathbf{G} = (V, E)$ and $\mathbf{H} = (W, F)$ are graphs, we say **H** is a **subgraph** of **G** when $W \subseteq V$ and $F \subseteq E$. We say **H** is an **induced subgraph** when $W \subseteq V$ and $F = \{xy \in E : x, y \in W\}$. In other words, an induced subgraph is defined completely by its vertex set and the original graph **G**. We say **H** is a **spanning subgraph** when $W = V$. In Figure 5.2, we show a graph, a subgraph and an induced subgraph. Neither of these subgraphs is a spanning subgraph. A graph $\mathbf{G} = (V, E)$ is called a **complete graph** when xy is an edge in **G** for every distinct pair $x, y \in V$. Conversely, **G** is an **independent graph** if $xy \notin E$, for every distinct pair $x, y \in V$. It is customary to denote a complete graph on n vertices by $\mathbf{K}_n$ and an independent graph on n vertices by $\mathbf{I}_n$. In Figure 5.3, we show

5.1 Basic Notation and Terminology for Graphs

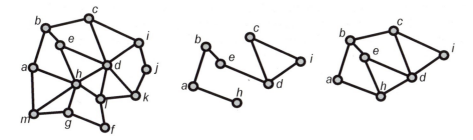

FIGURE 5.2: A Graph, a Subgraph and an Induced Subgraph

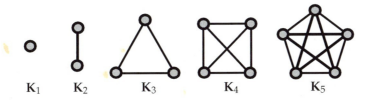

FIGURE 5.3: Small complete graphs

the complete graphs with at most 5 vertices. A sequence $(x_1, x_2, \ldots, x_n)$ of vertices in a graph $\mathbf{G} = (V, E)$ is called a **walk** when $x_i x_{i+1}$ is an edge for each $i = 1, 2, \ldots, n-1$. Note that the vertices in a walk need not be distinct. On the other hand, if the vertices are distinct, then the sequence is called a **path**, and often to emphasize where a path starts and ends, we will say that a sequence $(x_1, x_2, \ldots, x_n)$ of distinct vertices is a path from x_1 to x_n in $\mathbf{G}$. Similarly, when $n \geq 3$, a path $(x_1, x_2, \ldots, x_n)$ of n distinct vertices is called a **cycle** when $x_1 x_n$ is also an edge in $\mathbf{G}$. It is customary to denote a path on n vertices by $\mathbf{P}_n$, while $\mathbf{C}_n$ denotes a cycle on n vertices. The **length** of a path or a cycle is the number of edges it contains. Therefore, the length of $\mathbf{P}_n$ is $n-1$ and the length of $\mathbf{C}_n$ is n. In Figure 5.4, we show the paths of length at most 4, and in Figure 5.5, we show the cycles of length at most 5.

Chapter 5 Graph Theory

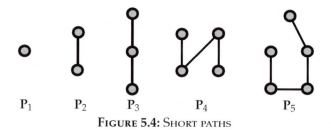

FIGURE 5.4: SHORT PATHS

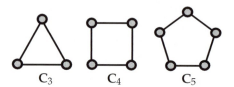

FIGURE 5.5: SMALL CYCLES

If $G = (V, E)$ and $H = (W, F)$ are graphs, we say **G** is **isomorphic** to **H** and write $G \cong H$ when there exists a bijection $f : V \xrightarrow[\text{onto}]{1\text{-}1} W$ so that x is adjacent to y in **G** if and only if $f(x)$ is adjacent to $f(y)$ in **H**. Often writers will say that **G** "contains" **H** when there is a subgraph of **G** which is isomorphic to **H**. In particular, it is customary to say that **G** contains the cycle C_n (same for P_n and K_n) when **G** contains a subgraph isomorphic to C_n. The graphs in Figure 5.6 are isomorphic. An isomorphism between these graphs is given by

$$f(a) = 5, \quad f(b) = 3, \quad f(c) = 1, \quad f(d) = 6, \quad f(e) = 2, \quad f(h) = 4.$$

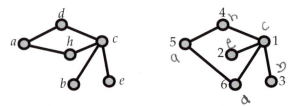

FIGURE 5.6: A PAIR OF ISOMORPHIC GRAPHS

On the other hand, the graphs shown in Figure 5.7 are *not* isomorphic, even though they have the same number of vertices and the same number of edges. Can you tell why?

5.1 Basic Notation and Terminology for Graphs

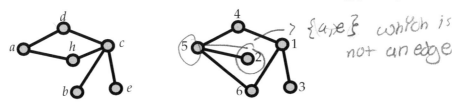

FIGURE 5.7: A PAIR OF NONISOMORPHIC GRAPHS

A graph **G** is **connected** when there is a path from x to y in **G**, for every $x, y \in V$; otherwise, we say **G** is **disconnected**. The graph of Figure 5.1 is disconnected (a sufficient justification for this is that there is no path from e to c), while those in Figure 5.6 are connected. If **G** is disconnected, we call a maximal connected subgraph of **G** a **component**. By this we mean that a subgraph **H** of **G** is a component of **G** provided that there does not exist a connected subgraph **H'** of **G** such that **H** is a subgraph of **H'**.

A graph is **acyclic** when it does not contain any cycle on three or more vertices. Acyclic graphs are also called **forests**. A connected acyclic graph is called a **tree**. When **G** = (V, E) is a connected graph, a subgraph **H** = (W, F) of **G** is called a **spanning tree** if **H** is both a spanning subgraph of **G** and a tree. In Figure 5.8, we show a graph and one of its spanning trees. We will return to the subject of spanning trees in Chapter 12.

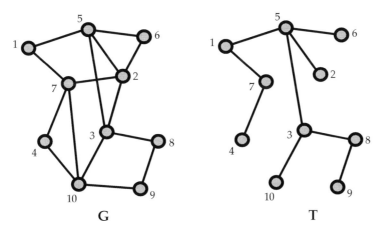

FIGURE 5.8: A GRAPH AND A SPANNING TREE

The following theorem is very elementary, and some authors refer to it as the "first theorem of graph theory". However, this basic result can be surprisingly useful.

73

Chapter 5 Graph Theory

Theorem 5.9. Let $\deg_G(v)$ denote the degree of vertex v in graph $\mathbf{G} = (V, E)$. Then

$$\sum_{v \in V} \deg_G(v) = 2|E|. \tag{5.1.1}$$

Proof. We consider how many times an edge $e = vw \in E$ contributes to each side of (5.1.1). The $\deg_G(x)$ and $\deg_G(y)$ terms on the left hand side each count e once, so e is counted twice on that side. On the right hand side, e is clearly counted twice. Therefore, we have the equality claimed. □

Corollary 5.10. *For any graph, the number of vertices of odd degree is even.*

We will return to the topic of trees later, but before moving on, let us prove one elementary proposition about trees. First, a **leaf** in a tree $\mathbf{T}$ is a vertex v with $\deg_T(v) = 1$.

Proposition 5.11. *Every tree on $n \geq 2$ vertices has at least two leaves.*

Proof. Our proof is by induction on n. For $n = 2$, there is precisely one tree, which is isomorphic to $\mathbf{K}_2$. Both vertices in this graph are leaves, so the proposition holds for $n = 2$. Now suppose that for some integer $m \geq 2$, every tree on at most m vertices has at least two leaves and let $\mathbf{T} = (V, E)$ be a tree on $m + 1$ vertices. Pick an edge $e \in E$ and form a new graph $\mathbf{T}' = (V', E')$ by deleting e from $\mathbf{T}$. That is, $V' = V$ and $E' = E - \{e\}$. Now since $\mathbf{T}'$ does not contain a path from one endpoint of e to its other endpoint, $\mathbf{T}'$ is not connected. However, deleting an edge cannot create a cycle, so $\mathbf{T}'$ is a forest. Furthermore, it has precisely two components, each of which is a tree with at most m vertices. If each component has at least two vertices, then by induction, each has at least two leaves. In the worst case scenario, two of these leaves are the endpoints of e, so at least two of the vertices are leaves in $\mathbf{T}$, too. If each component of $\mathbf{T}'$ has only one vertex, then $\mathbf{T} \cong \mathbf{K}_2$, which has two leaves. If exactly one of the components has only one vertex, then it must be a leaf in $\mathbf{T}$. Thus, applying the inductive hypothesis to the other component ensures that there is a second leaf in $\mathbf{T}$. □

5.2 Multigraphs: Loops and Multiple Edges

Consider a graph in which the vertices represent cities and the edges represent highways. Certain pairs of cities are joined by an edge while other pairs are not. The graph may or may not be connected (although a disconnected graph is likely to result in disgruntled commuters). However, certain aspects of real highway networks are not captured by this model. First, between two nearby cities, there can actually be several interconnecting highways, and traveling on one of them is fundamentally different from traveling on another. This leads to the concept of **multiple edges**, i.e., allowing for more than one edge between two adjacent vertices. Also, we could have a highway which leaves a city, goes through the nearby countryside and the returns to the same city where it originated. This leads to the concept of a **loop**, i.e., an edge with both end points being the same vertex. Also, we can allow for more than one loop with the same end point.

Accordingly, authors frequently lead off a discussion on a graph theory topic with a sentence or two like:

1. In this paper, all graphs will be **simple**, i.e., we will not allow loops or multiple edges.

2. In this paper, graphs can have loops and multiple edges.

The terminology is far from standard, but in this text, a graph will always be a **simple** graph, i.e., no loops or multiple edges. When we want to allow for loops and multiple edges, we will use the term **multigraph**. This suggests the question of what we would call a graph if it is allowed to have loops but not multiple edges, or if multiple edges are allowed but not loops. If we *really* needed to talk about such graphs, then the English language comes to our rescue, and we just state the restriction explicitly!

5.3 Eulerian and Hamiltonian Graphs

Graph theory is an area of mathematics that has found many applications in a variety of disciplines. Throughout this text, we will encounter a number of them. However, graph theory traces its origins to a problem in Königsberg, Prussia (now Kaliningrad, Russia) nearly three centuries ago. The river Pregel passes through the city, and there are two large islands in the middle of the channel. These islands were connected to the mainland by seven bridges as indicated in Figure 5.12. It is said that the citizens of Königsberg often wondered if it was possible for one to leave his home, walk through the city in such a way that he crossed each bridge precisely one time, and end up at home again. Leonhard Euler settled this problem in 1736 by using graph theory in the form of Theorem 5.13. Let **G** be a graph without isolated vertices. We say that **G** is **eulerian** provided that there is a sequence $(x_0, x_1, x_2, \ldots, x_t)$ of vertices from **G**, with repetition allowed, so that

1. $x_0 = x_t$;

2. for every $i = 0, 1, \ldots t-1$, $x_i x_{i+1}$ is an edge of **G**;

3. for every edge $e \in E$, there is a unique integer i with $0 \leq i < t$ for which $e = x_i x_{i+1}$.

Chapter 5 Graph Theory

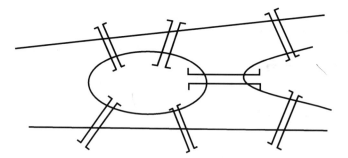

FIGURE 5.12: THE BRIDGES OF KÖNIGSBERG

When **G** is eulerian, a sequence satisfying these three conditions is called an **eulerian circuit**. A sequence of vertices $(x_0, x_1, \ldots, x_t)$ is called a **circuit** when it satisfies only the first two of these conditions. Note that a sequence consisting of a single vertex is a circuit. Before proceeding to Euler's elegant characterization of eulerian graphs, let's use SageMath to generate some graphs that are and are not eulerian.

Run the code below. It will execute until it finds a graph **G** that is eulerian. The output that will be produced is a list of the degrees of the vertices of the graph **G** followed by a drawing of **G**.

```
vertices = 13
edges = 28
g = graphs.RandomGNM(vertices,edges)
while (not g.is_eulerian() or not g.is_connected()):
    g = graphs.RandomGNM(vertices,edges)
print g.degree_sequence()
g.show()
```

We encourage you to evaluate the run the code above multiple times, even changing the number of vertices and edges. If it seems to be running a log time, it may be that you have made the number of edges too small, so try increasing it a bit. Do you notice anything about the degrees of the vertices in the graphs produced?

Now let's try to find a graph **H** that is *not* eulerian. Again, the output is the list of degrees of **H** followed by a drawing of **H**.

```
vertices = 15
edges = 25
g = graphs.RandomGNM(vertices,edges)
while (g.is_eulerian() or not g.is_connected()):
    g = graphs.RandomGNM(vertices,edges)
print g.degree_sequence()
```

5.3 Eulerian and Hamiltonian Graphs

```
g.show()
```

One thing you probably noticed in running this second block of code is that it tended to come back much faster than the first. That would suggest that the non-eulerian graphs outnumber the eulerian graphs. Did you notice anything different about the degrees of the vertices in these graphs compared to the ones that were eulerian?

The following elementary theorem completely characterizes eulerian graphs. Its proof gives an algorithm that is easily implemented.

Theorem 5.13. *A graph* **G** *is eulerian if and only if it is connected and every vertex has even degree.*

Proof. Clearly, an eulerian graph must be connected. Also, if $(x_0, x_1, \ldots, x_t)$ is an eulerian circuit in **G**, then for each $i = 0, 1, \ldots, t-1$, we can view the edge $x_i x_{i+1}$ as exiting x_i and entering x_{i+1}. The degree of every vertex must be even, since for each vertex x, the number of edges exiting x equals the number of edges entering x. Furthermore, each edge incident with x either exits from x or enters x.

We now describe a deterministic process that will either (a) find an eulerian circuit, (b) show that the graph is disconnected, or (c) find a vertex of odd degree. The description is simplified by assuming that the vertices in **G** have been labelled with the positive integers $1, 2, \ldots, n$, where n is the number of vertices in **G**. Furthermore, we take $x_0 = 1$.

We launch our algorithm with a trivial circuit C consisting of the vertex $x_0 = (1)$. Thereafter suppose that we have a partial circuit C defined by $(x_0, x_1, \ldots, x_t)$ with $x_0 = x_t = 1$. The edges of the form $x_i x_{i+1}$ have been **traversed**, while the remaining edges in **G** (if any) have not. If the third condition for an euler circuit is satisfied, we are done, so we assume it does not hold.

We then choose the least integer i for which there is an edge incident with x_i that has not already been traversed. If there is no such integer, since there are edges that have not yet been traversed, then we have discovered that the graph is disconnected. So we may assume that the integer i exists. Set $u_0 = x_i$. We define a sequence $(u_0, u_1, \ldots, u_s)$ recursively. If $j \geq 0$, set

$$N_j = \{y : u_j y \text{ is an edge in } \mathbf{G} \text{ and has not yet been traversed.}\}$$

If $N_j \neq \emptyset$, we take u_{j+1} as the least positive integer in N_j. If $N_j = \emptyset$, then $j \geq 1$ and we take $s = j$ and halt this subroutine.

When the subroutine halts, we consider two cases. If $u_0 \neq u_s$, then u_0 and u_s are vertices of odd degree in **G**. So we are left to consider the case where $u_0 = u_s = x_i$. In this case, we simply expand our original sequence $(x_0, x_1, \ldots, x_t)$ by replacing the integer x_i by the sequence $(u_0, u_1, \ldots, u_s)$. □

Chapter 5 Graph Theory

As an example, consider the graph **G** shown in Figure 5.14. Evidently, this graph is connected and all vertices have even degree. Here is the sequence of circuits starting with the trivial circuit C consisting only of the vertex 1.

$$
\begin{aligned}
C &= (1) \\
&= (1, 2, 4, 3, 1) \quad \text{start next from 2} \\
&= (1, 2, 5, 8, 2, 4, 3, 1) \quad \text{start next from 4} \\
&= (1, 2, 5, 8, 2, 4, 6, 7, 4, 9, 6, 10, 4, 3, 1) \quad \text{start next from 7} \\
&= (1, 2, 5, 8, 2, 4, 6, 7, 9, 11, 7, 4, 9, 6, 10, 4, 3, 1) \quad \text{Done!!}
\end{aligned}
$$

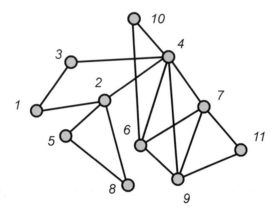

FIGURE 5.14: An Eulerian Graph

You should note that Theorem 5.13 holds for loopless graphs in which multiple edges are allowed. Euler used his theorem to show that the multigraph of Königsberg shown in Figure 5.15, in which each land mass is a vertex and each bridge is an edge, is *not* eulerian, and thus the citizens could not find the route they desired. (Note that in Figure 5.15 there are multiple edges between the same pair of vertices.)

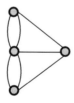

FIGURE 5.15: The multigraph of Königsberg's bridges

5.3 Eulerian and Hamiltonian Graphs

A graph $\mathbf{G} = (V, E)$ is said to be **hamiltonian** if there exists a sequence $(x_1, x_2, \ldots, x_n)$ so that

1. every vertex of **G** appears exactly once in the sequence;

2. $x_1 x_n$ is an edge of **G**; and

3. for each $i = 1, 2, \ldots, n-1$, $x_i x_{i+1}$ is an edge in **G**.

Such a sequence of vertices is called a **hamiltonian cycle**.

The first graph shown in Figure 5.16 both eulerian and hamiltonian. The second is hamiltonian but not eulerian.

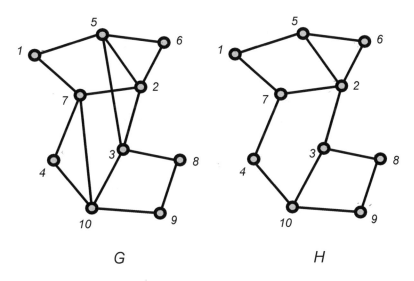

Figure 5.16: Eulerian and Hamiltonian Graphs

In Figure 5.17, we show a famous graph known as the Petersen graph. It is not hamiltonian.

79

Chapter 5 Graph Theory

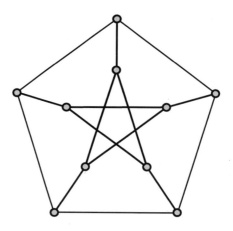

FIGURE 5.17: THE PETERSEN GRAPH

Unlike the situation with eulerian circuits, there is no known method for quickly determining whether a graph is hamiltonian. However, there are a number of interesting conditions which are sufficient. Here is one quite well known example, due to Dirac.

Theorem 5.18. *If G is a graph on n vertices and each vertex in G has at least $\lceil \frac{n}{2} \rceil$ neighbors, then G is hamiltonian.*

Proof. Suppose the theorem fails and let n be the least positive integer for which there exists a graph **G** on n vertices so that each vertex in **G** has at least $\lceil n/2 \rceil$ neighbors, yet there is no hamiltonian cycle in **G**. Clearly, $n \geq 4$.

Now let t be the largest integer for which **G** has a path $P = (x_1, x_2, \ldots, x_t)$ on t vertices. Clearly all neighbors of both x_1 and x_t appear on this path. By the pigeon hole principle, there is some integer i with $1 \leq i < t$ so that $x_1 x_{i+1}$ and $x_i x_t$ are edges in **G**. However, this implies that

$$C = (x_1, x_2, x_3, \ldots, x_i, x_t, x_{t-1}, x_{t-2}, \ldots, x_{i+1})$$

is a cycle of length t in **G**. In turn, this requires $\lceil n/2 \rceil < t < n$. But if y is any vertex not on the cycle, then y must have a neighbor on C, which implies that **G** has a path on $t + 1$ vertices. The contradiction completes the proof. □

5.4 Graph Coloring

Let's return now to the subject of Example 1.5, assigning frequencies to radio stations so that they don't interfere. The first thing that we will need to do is to turn the map

5.4 Graph Coloring

of radio stations into a suitable graph, which should be pretty natural at this juncture. We define a graph $G = (V, E)$ in which V is the set of radio stations and $xy \in E$ if and only if radio station x and radio station y are within 200 miles of each other. With this as our model, then we need to assign different frequencies to two stations if their corresponding vertices are joined by an edge. This leads us to our next topic, coloring graphs.

When $G = (V, E)$ is a graph and C is a set of elements called **colors**, a **proper coloring** of G is a function $\phi : V \to C$ such that if $\phi(x) \neq \phi(y)$ whenever xy is an edge in G. The least t for which G has a proper coloring using a set C of t colors is called the **chromatic number** of G and is denoted $\chi(G)$. In Figure 5.19, we show a proper coloring of a graph using 5 colors. Now we can see that our radio frequency assignment problem is the much-studied question of finding the chromatic number of an appropriate graph.

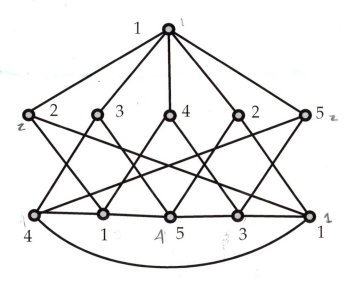

FIGURE 5.19: A PROPER COLORING USING 5 COLORS

Discussion 5.20. Everyone agrees that the graph G in Figure 5.19 has chromatic number at most 5. However, there's a bit of debate going on about if $\chi(G) = 5$. Bob figures the authors would not have used five colors if they didn't need to. Carlos says he's glad they're having the discussion, since all having a proper coloring does is provide them with an upper bound on $\chi(G)$. Bob sees that the graph has a vertex of degree 5 and claims that must mean $\chi(G) = 5$. Alice groans and draws a graph with 101 vertices, one of which has degree 100, but with chromatic number 2. Bob is shocked, but agrees with her. Xing wonders if the fact that the graph does not contain a K_3 has any

Chapter 5 Graph Theory

bearing on the chromatic number. Dave's in a hurry to get to the gym, but on his way out the door he says they can get a proper 4-coloring pretty easily, so $\chi(\mathbf{G}) \leq 4$. The rest decide it's time to keep reading.

- What graph did Alice draw that shocked Bob?
- What changes did Dave make to the coloring in Figure 5.19 to get a proper coloring using four colors?

5.4.1 Bipartite Graphs

A graph $\mathbf{G} = (V, E)$ with $\chi(\mathbf{G}) \leq 2$ is called a **2-colorable graph**. A couple of minutes of reflection should convince you that for $n \geq 2$, the cycle $\mathbf{C}_{2n}$ with $2n$ vertices is 2-colorable. On the other hand, $\mathbf{C}_3 \cong \mathbf{K}_3$ is clearly not 2-colorable. Furthermore, no odd cycle $\mathbf{C}_{2n+1}$ for $n \geq 1$ is 2-colorable. It turns out that the property of containing an odd cycle is the only impediment to being 2-colorable, which means that recognizing 2-colorable graphs is easy, as the following theorem shows.

Theorem 5.21. *A graph is 2-colorable if and only if it does not contain an odd cycle.*

Proof. Let $\mathbf{G} = (V, E)$ be a 2-colorable graph whose coloring function partitions V as $A \cup B$. Since there are no edges between vertices on the same side of the partition, any cycle in $\mathbf{G}$ must alternate vertices between A and B. In order to complete the cycle, therefore, the number of vertices in the cycle from A must be the same as the number from B, implying that the cycle has even length.

Now suppose that $\mathbf{G}$ does not contain an odd cycle. Note that we may assume that $\mathbf{G}$ is connected, as each component may be colored individually. The **distance** $d(u, v)$ between vertices $u, v \in V$ is the length of a shortest path from u to v, and of course $d(u, u) = 0$. Fix a vertex $v_0 \in V$ and define

$$A = \{v \in V : d(u_0, v) \text{ is even}\} \quad \text{and} \quad B = \{v \in V : d(v_0, v) \text{ is odd}\}.$$

We claim that coloring the vertices of A with color 1 and the vertices of B with color 2 is a proper coloring. suppose not. Then without loss of generality, there are vertices $x, y \in A$ such that $xy \in E$. Since $x, y \in A$, $d(v_0, x)$ and $d(v_0, y)$ are both even. Let

$$v_0, x_1, x_2, \ldots, x_n = x$$

and

$$v_0, y_1, y_2, \ldots, y_m = y$$

be shortest paths from v_0 to x and y, respectively. If $x_i \neq y_j$ for all $1 \leq i \leq n$ and $1 \leq j \leq m$, then since m and n are both even,

$$v_0, x_1, x_2, \ldots, x_n = x, y = y_m, y_{m-1}, \ldots, y_2, y_1, v_0$$

is an odd cycle in **G**, which is a contradiction. Thus, there must be i, j such that $x_i = y_j$, and we may take i, j as large as possible. (That is, after $x_i = y_j$, the two paths do not intersect again.) Thus,

$$x_i, x_{i+1}, \ldots, x_n = x, y = y_m, y_{m-1}, \ldots, y_j = x_i$$

is a cycle in **G**. How many vertices are there in this cycle? A quick count shows that it has

$$n - (i - 1) + m - (j - 1) - 1 = n + m - (i + j) + 1$$

vertices. We know that n and m are even, and notice that i and j are either both even or both odd, since $x_i = y_j$ and the odd-subscripted vertices of our path belong to B while those with even subscripts belong to A. Thus, $i + j$ is even, so $n + m - (i + j) + 1$ is odd, giving a contradiction. □

A graph **G** is called a **bipartite graph** when there is a partition of the vertex V into two sets A and B so that the subgraphs induced by A and B are independent graphs, i.e., no edge of **G** has both of its endpoints in A or in B. Evidently, bipartite graphs are 2-colorable. On the other hand, when a 2-colorable graph is disconnected, there is more than one way to define a suitable partition of the vertex set into two independent sets.

Bipartite graphs are commonly used as models when there are two distinct types of objects being modeled and connections are only allowed between two objects of different types. For example, on one side, list candidates who attend a career fair and on the other side list the available positions. The edges might naturally correspond to candidate/position pairs which link a person to a responsibility they are capable of handling.

As a second example, a bipartite graph could be used to visualize the languages spoken by a group of students. The vertices on one side would be the students with the languages listed on the other side. We would then have an edge xy when student x spoke language y. A concrete example of this graph for our favorite group of students is shown in Figure 5.22, although Alice isn't so certain there should be an edge connecting Dave and English. One special class of bipartite graphs that bears mention is the class of **complete bipartite graphs**. The complete bipartite graph $K_{m,n}$ has vertex set $V = V_1 \cup V_2$ with $|V_1| = m$ and $|V_2| = n$. It has an edge xy if and only if $x \in V_1$ and $y \in V_2$. The complete bipartite graph $K_{3,3}$ is shown in Figure 5.23.

Chapter 5 Graph Theory

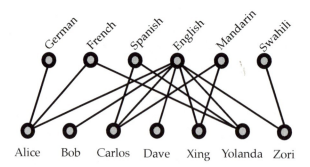

FIGURE 5.22: A BIPARTITE GRAPH

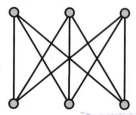

FIGURE 5.23: THE COMPLETE BIPARTITE GRAPH $K_{3,3}$

5.4.2 Cliques and Chromatic Number

A **clique** in a graph $G = (V, E)$ is a set $K \subseteq V$ such that the subgraph induced by K is isomorphic to the complete graph $K_{|K|}$. Equivalently, we can say that every pair of vertices in K are adjacent. The **maximum clique size** or **clique number** of a graph G, denoted $\omega(G)$, is the largest t for which there exists a clique K with $|K| = t$. For example, the graph in Figure 5.14 has clique number 4 while the graph in Figure 5.19 has maximum clique size 2.

For every graph G, it is obvious that $\chi(G) \geq \omega(G)$. On the other hand, the inequality may be far from tight. Before proving showing how bad it can be, we need to introduce a more general version of the Pigeon Hole Principle. Consider a function $f: X \to Y$ with $|X| = 2|Y| + 1$. Since $|X| > |Y|$, the Pigeon Hole Principle as stated in Proposition 4.1 only tells us that there are distinct $x, x' \in X$ with $f(x) = f(x')$. However, we can say more here. Suppose that each element of Y has at most two elements of X mapped to it. Then adding up the number of elements of X based on how many are mapped to each element of Y would only allow X to have (at most) $2|Y|$ elements. Thus, there must be $y \in Y$ so that there are three distinct elements $x, x', x'' \in X$ with $f(x) = f(x') = f(x'') = y$. This argument generalizes to give the following version of

5.4 Graph Coloring

the Pigeon Hole Principle:

Proposition 5.24 (Generalized Pigeon Hole Principle). *If $f: X \to Y$ is a function and $|X| \geq (m-1)|Y| + 1$, then there exists an element $y \in Y$ and distinct elements $x_1, \ldots, x_m \in X$ so that $f(x_i) = y$ for $i = 1, \ldots, m$.*

We are now prepared to present the following proposition showing that clique number and chromatic number need not be close at all. We give two proofs. The first is the work of J. Kelly and L. Kelly, while the second is due to J. Mycielski.

Proposition 5.25. *For every $t \geq 3$, there exists a graph G_t so that $\chi(G_t) = t$ and $\omega(G_t) = 2$.*

Proof. We proceed by induction on t. For $t = 3$, we take G_3 to be the cycle C_5 on five vertices. Now assume that for some $t \geq 3$, we have determined the graph G_t. Suppose that G_t has n_t vertices. Label the vertices of G_t as $x_1, x_2, \ldots, x_{n_t}$. Construct G_{t+1} as follows. Begin with an independent set I of cardinality $t(n_t - 1) + 1$. For every subset S of I with $|S| = n_t$, label the elements of S as $y_1, y_2, \ldots, y_{n_t}$. For this particular n_t-element subset attach a copy of G_t with y_i adjacent to x_i for $i = 1, 2, \ldots, n_t$. Vertices in copies of G_t for distinct n_t-element subsets of I are nonadjacent, and a vertex in I has at most one neighbor in a particular copy of G_t.

To see that $\omega(G_{t+1}) = 2$, it will suffice to argue that G_{t+1} contains no triangle (K_3). Since G_t is triangle-free, any triangle in G_{t+1} must contain a vertex of I. Since none of the vertices of I are adjacent, any triangle in G_{t+1} contains only one point of I. Since each vertex of I is adjacent to at most one vertex of any fixed copy of G_t, if $y \in I$ is part of a triangle, the other two vertices must come from distinct copies of G_t. However, vertices in different copies of G_t are not adjacent, so $\omega(G_{t+1}) = 2$. Notice that $\chi(G_{t+1}) \geq t$ since G_{t+1} contains G_t. On the other hand, $\chi(G_{t+1}) \leq t + 1$ since we may use t colors on the copies of G_t and a new color on the independent set I. To see that $\chi(G_{t+1}) = t+1$, observe that if we use only t colors, then by the generalized Pigeon Hole Principle, there is an n_t-element subset of I in which all vertices have the same color. Then this color cannot be used in the copy of G_t which is attached to that n_t-element subset. □

Proof. We again start with G_3 as the cycle C_5. As before we assume that we have constructed for some $t \geq 3$ a graph G_t with $\omega(G_t) = 2$ and $\chi(G_t) = t$. Again, label the vertices of G_t as $x_1, x_2, \ldots, x_{n_t}$. To construct G_{t+1}, we now start with an independent set I, but now I has only n_t points, which we label as $y_1, y_2, \ldots, y_{n_t}$. We then add a copy of G_t with y_i adjacent to x_j if and only if x_i is adjacent to x_j. Finally, attach a new vertex z adjacent to all vertices in I.

Clearly, $\omega(G_{t+1}) = 2$. Also, $\chi(G_{t+1}) \geq t$, since it contains G_t as a subgraph. Furthermore, $\chi(G_{t+1}) \leq t+1$, since we can color G_t with colors from $\{1, 2, \ldots, t\}$, use color $t+1$ on the independent set I, and then assign color 1 to the new vertex z. We claim that in fact $\chi(G_{t+1}) = t + 1$. Suppose not. Then we must have $\chi(G_{t+1}) = t$. Let ϕ be a proper coloring of G_{t+1}. Without loss of generality, ϕ uses the colors in $\{1, 2, \ldots, t\}$ and ϕ assigns color t to z. Then consider the nonempty set S of vertices in the copy of G_t to which ϕ assigns color t. For each x_i in S, change the color on x_i so that it matches the color assigned to y_i by ϕ, which cannot be t, as z is colored t. What results is a proper coloring of the copy of G_t with only $t-1$ colors since x_i and y_i are adjacent to the same vertices of the copy of G_t. The contradiction shows that $\chi(G_{t+1}) = t+1$, as claimed. □

Since a 3-clique looks like a triangle, Proposition 5.25 is often stated as "There exist triangle-free graphs with large chromatic number." As an illustration of the construction in the proof of Mycielski, we again refer to Figure 5.19. The graph shown is G_4. We will return to the topic of graphs with large chromatic number in Section 11.6 where we show that are there graphs with large chromatic number which lack not only cliques of more than two vertices but also *cycles* of fewer than g vertices for *any* value of g. In other words, there is a graph G with $\chi(G) = 10^6$ but no cycle with fewer than 10^{10} vertices!

5.4.3 Can We Determine Chromatic Number?

Suppose you are given a graph G. It's starting to look like it is not easy to find an algorithm that answers the question "Is $\chi(G) \leq t$?" It's easy to verify a certificate (a proper coloring using at most t colors), but how could you even find a proper coloring, not to mention one with the fewest number of colors? Similarly for the question "Is $\omega(G) \geq k$?", it is easy to verify a certificate. However, finding a maximum clique appears to be a very hard problem. Of course, since the gap between $\chi(G)$ and $\omega(G)$ can be arbitrarily large, being able to find one value would not (generally) help in finding the value of the other. No polynomial-time algorithm is known for either of these problems, and many believe that no such algorithm exists. In this subsection, we look at one approach to finding chromatic number and see a case where it does work efficiently.

A very naïve algorithmic way to approach graph coloring is the First Fit, or "greedy", algorithm. For this algorithm, fix an ordering of the vertex set $V = \{v_1, v_2, \ldots v_n\}$. We define the coloring function ϕ one vertex at a time in increasing order of subscript. We begin with $\phi(v_1) = 1$ and then we define $\phi(v_{i+1})$ (assuming vertices $v_1, v_2, \ldots, v_i$ have been colored) to be the least positive integer color that has not already been used on any of its neighbors in the set $\{v_1, \ldots v_i\}$. Figure 5.26 shows two different orderings of the same graph. Exercise 5.9.24 demonstrates that the ordering of V is vital to the ability of the First Fit algorithm to color G using $\chi(G)$ colors. In general, finding an optimal ordering is just as difficult as coloring G. Thus, this very simple algorithm does not work well in general. However, for some classes of graphs, there is a "natural" ordering that leads to optimal performance of First Fit. Here is one such example—one that we

5.4 Graph Coloring

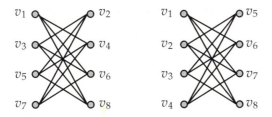

FIGURE 5.26: TWO ORDERINGS OF THE VERTICES OF A BIPARTITE GRAPH.

will study again in the next chapter in a different context.

Given an indexed family of sets $\mathcal{F} = \{S_\alpha : \alpha \in V\}$, we associate with $\mathcal{F}$ a graph **G** defined as follows. The vertex set of **G** is the set V and vertices x and y in V are adjacent in **G** if and only if $S_x \cap S_y \neq \emptyset$. We call **G** an **intersection graph**. It is easy to see that every graph is an intersection graph (*Why?*), so it makes sense to restrict the sets which belong to $\mathcal{F}$. For example, we call **G** an **interval graph** if it is the intersection graph of a family of closed intervals of the real line $\mathbb{R}$. For example, in Figure 5.27, we show a collection of six intervals of the real line on the left. On the right, we show the corresponding interval graph having an edge between vertices x and y if and only if intervals x and y overlap.

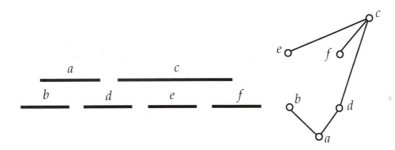

FIGURE 5.27: A COLLECTION OF INTERVALS AND ITS INTERVAL GRAPH

Theorem 5.28. *If* $\mathbf{G} = (V, E)$ *is an interval graph, then* $\chi(\mathbf{G}) = \omega(\mathbf{G})$.

Proof. For each $v \in V$, let $I(v) = [a_v, b_v]$ be a closed interval of the real line so that uv is an edge in **G** if and only if $I(u) \cap I(v) \neq \emptyset$. Order the vertex set V as $\{v_1, v_2, \ldots v_n\}$ such that $a_1 \leq a_2 \leq \cdots \leq a_n$. (Ties may be broken arbitrarily.) Apply the First Fit coloring algorithm to **G** with this ordering on V. Now when the First Fit coloring algorithm colors v_i, all of its neighbors have left end point at most a_i. Since they are neighbors of v_i, however, we know that their right endpoints are all at least a_i. Thus,

87

Chapter 5 Graph Theory

v_i and its previously-colored neighbors form a clique. Hence, v_i is adjacent to at most $\omega(G) - 1$ other vertices that have already been colored, so when the algorithm colors v_i, there will be a color from $\{1, 2, \ldots, \omega(G)\}$ not already in use on its neighbors. The algorithm will assign v_i the smallest such color. Thus, we never need to use more than $\omega(G)$ colors, so $\chi(G) = \omega(G)$. □

A graph G is said to be **perfect** if $\chi(H) = \omega(H)$ for every induced subgraph H. Since an induced subgraph of an interval graph is an interval graph, Theorem 5.28 shows interval graphs are perfect. The study of perfect graphs originated in connection with the theory of communications networks and has proved to be a major area of research in graph theory for many years now.

5.5 Planar Graphs

Let's return to the problem of providing lines for water, electricity, and natural gas to three homes which we discussed in the introduction to this chapter. How can we model this problem using a graph? The best way is to have a vertex for each utility and a vertex for each of the three homes. Then what we're asking is if we can draw the graph that has an edge from each utility to each home so that none of the edges cross. This graph is shown in Figure 5.29. You should recognize it as the complete bipartite graph $K_{3,3}$ we introduced earlier in the chapter.

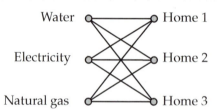

FIGURE 5.29: A GRAPH OF CONNECTING HOMES TO UTILITIES

While this example of utility lines might seem a bit contrived, since there's really no good reason that the providers can't bury their lines at different depths, the question of whether a graph can be drawn in the plane such that edges intersect only at vertices is a long-studied question in mathematics that does have useful applications. One area where it arises is in the design of microchips and circuit boards. In those contexts, the material is so thin that the option of placing connections at different depths either does not exist or is severely restricted. There is much deep mathematics that underlies this area, and this section is intended to introduce a few of the key concepts.

5.5 Planar Graphs

By a **drawing** of a graph, we mean a way of associating its vertices with points in the Cartesian plane $\mathbb{R}^2$ and its edges with simple polygonal arcs whose endpoints are the points associated to the vertices that are the endpoints of the edge. You can think of a polygonal arc as just a finite sequence of line segments such that the endpoint of one line segment is the starting point of the next line segment, and a simple polygonal arc is one that does not cross itself. (Our choice of polygonal arcs rather than arbitrary curves actually doesn't cause an impediment, since by taking very, very, very short line segments we can approximate any curve.) A **planar drawing** of a graph is one in which the polygonal arcs corresponding to two edges intersect only at a point corresponding to a vertex to which they are both incident. A graph is **planar** if it has a planar drawing. A **face** of a planar drawing of a graph is a region bounded by edges and vertices and not containing any other vertices or edges.

Figure 5.30 shows a planar drawing of a graph with 6 vertices and 9 edges. Notice how one of the edges is drawn as a true polygonal arc rather than a straight line segment. This drawing determines 5 regions, since we also count the unbounded region that surrounds the drawing.

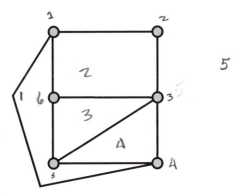

FIGURE 5.30: A PLANAR DRAWING OF A GRAPH

Figure 5.31 shows a planar drawing of the complete graph K_4. There are 4 vertices, 6 edges, and 4 faces in the drawing.

Chapter 5 Graph Theory

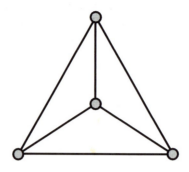

FIGURE 5.31: A PLANAR DRAWING OF K_4

What happens if we compute the number of vertices minus the number of edges plus the number of faces for these drawings? We have

$$6 - 9 + 5 = 2$$
$$4 - 6 + 4 = 2$$

While it might seem like a coincidence that this computation results in 2 for these planar drawings, there's a more general principle at work here, and in fact it holds for *any* planar drawing of *any* planar graph.

In fact, the number 2 here actually results from a fundamental property of the plane, and there are a corresponding theorems for other surfaces. However, we only need the result as stated above.

Theorem 5.32 (Euler's Formula). *Let G be a connected planar graph with n vertices and m edges. Every planar drawing of G has f faces, where f satisfies*

$$n - m + f = 2.$$

Proof. Our proof is by induction on the number m of edges. If $m = 0$, then since **G** is connected, our graph has a single vertex, and so there is one face. Thus $n - m + f = 1 - 0 + 1 = 2$ as needed. Now suppose that we have proven Euler's formula for all graphs with less than m edges and let **G** have m edges. Pick an edge e of **G**. What happens if we form a new graph **G'** by deleting e from **G**? If **G'** is connected, our inductive hypothesis applies. Say that **G'** has n' vertices, m' edges, and f' faces. Then by induction, these numbers satisfy

$$n' - m' + f' = 2.$$

Since we only deleted one edge, $n' = n$ and $m' = m - 1$. What did the removal of e do to the number of faces? In $\mathbf{G'}$ there's a new face that was formerly two faces divided by e in $\mathbf{G}$. Thus, $f' = f - 1$. Substituting these into $n' - m' + f' = 2$, we have

$$n - (m - 1) + (f - 1) = 2 \iff n - m + f = 2.$$

Thus, if $\mathbf{G'}$ is connected, we are done. If $\mathbf{G'}$ is disconnected, however, we cannot apply the inductive assumption to $\mathbf{G'}$ directly. Fortunately, since we removed only one edge, $\mathbf{G'}$ has two components, which we can view as two connected graphs $\mathbf{G'_1}$ and $\mathbf{G'_2}$. Each of these has fewer than m edges, so we may apply the inductive hypothesis to them. For $i = 1, 2$, let n'_i be the number of vertices of $\mathbf{G'_i}$, m'_i the number of edges of $\mathbf{G'_i}$, and f'_i the number of faces of $\mathbf{G'_i}$. Then by induction we have

$$n'_1 - m'_1 + f'_1 = 2 \quad \text{and} \quad n'_2 - m'_2 + f'_2 = 2.$$

Adding these together, we have

$$(n'_1 + n'_2) - (m'_1 + m'_2) + (f'_1 + f'_2) = 4.$$

But now $n = n'_1 + n'_2$, and $m'_1 + m'_2 = m - 1$, so the equality becomes

$$n - (m - 1) + (f'_1 + f'_2) = 4 \iff n - m + (f'_1 + f'_2) = 3.$$

The only thing we have yet to figure out is how $f'_1 + f'_2$ relates to f, and we have to hope that it will allow us to knock the 3 down to a 2. Every face of $\mathbf{G'_1}$ and $\mathbf{G'_2}$ is a face of $\mathbf{G}$, since the fact that removing e disconnects $\mathbf{G}$ means that e must be part of the boundary of the unbounded face. Further, the unbounded face is counted twice in the sum $f'_1 + f'_2$, so $f = f'_1 + f'_2 - 1$. This gives exactly what we need to complete the proof. □

Taken by itself, Euler's formula doesn't seem that useful, since it requires counting the number of faces in a planar embedding. However, we can use this formula to get a quick way to determine that a graph is not planar. Consider a drawing without edge crossings of a graph on n vertices and m edges, with $n \geq 3$. We consider pairs (e, F) where e is an edge of $\mathbf{G}$ and F is a face that has e as part of its boundary. How many such pairs are there? Let's call the number of pairs p. Each edge can bound either one or two faces, so we have that $p \leq 2m$. We can also bound p by counting the number of pairs in which a face F appears. Each face is bounded by at least 3 edges, so it appears in at least 3 pairs, and so $p \geq 3f$. Thus $3f \leq 2m$ or $f \leq 2m/3$. Now, utilizing Euler's formula, we have

$$m = n + f - 2 \leq n + \frac{2m}{3} - 2 \iff \frac{m}{3} \leq n - 2.$$

Thus, we've proven the following theorem.

Read over in more depth.

Chapter 5 Graph Theory

Theorem 5.33. *A planar graph on n vertices has at most $3n - 6$ edges when $n \geq 3$.*

The contrapositive of this theorem, namely that an n-vertex graph with more than $3n - 6$ edges is not planar, is usually the most useful formulation of this result. For instance, we've seen (Figure 5.31) that $\mathbf{K}_4$ is planar. What about $\mathbf{K}_5$? It has 5 vertices and $C(5, 2) = 10 > 9 = 3 \cdot 5 - 6$ edges, so it is not planar, and thus for $n \geq 5$, $\mathbf{K}_n$ is not planar, since it contains $\mathbf{K}_5$. It's important to note that Theorem 5.33 is not the be-all, end-all of determining if a graph is planar. To see this, let's return to the subject of drawing $\mathbf{K}_{3,3}$ in the plane. This graph has 6 vertices and 9 edges, so it passes the test of Theorem 5.33. However, if you spend a couple minutes trying to find a way to draw $\mathbf{K}_{3,3}$ in the plane without any crossing edges, you'll pretty quickly begin to believe that it can't be done—and you'd be right!

To see why $\mathbf{K}_{3,3}$ is not planar, we'll have to return to Euler's formula, and we again work with edge-face pairs. For $\mathbf{K}_{3,3}$, we see that every edge would have to be part of the boundary of two faces, and faces are bounded by cycles. Also, since the graph is bipartite, there are no odd cycles. Thus, counting edge-face pairs from the edge perspective, we see that there are $2m = 18$ pairs. If we let f_k be the number of faces bounded by a cycle of length k, then $f = f_4 + f_6$. Thus, counting edge-face pairs from the face perspective, there are $4f_4 + 6f_6$ pairs. From Euler's formula, we see that the number of faces f must be 5, so then $4f_4 + 6f_6 \geq 20$. But from our count of edge-face pairs, we have $2m = 4f_4 + 6f_6$, giving $18 \geq 20$, which is clearly absurd. Thus, $\mathbf{K}_{3,3}$ is not planar.

At this point, you're probably asking yourself "So what?" We've invested a fair amount of effort to establish that $\mathbf{K}_5$ and $\mathbf{K}_{3,3}$ are nonplanar. Clearly any graph that contains them is also nonplanar, but there are a lot of graphs, so you might think that we could be at this forever. Fortunately, we won't be, since at its core, planarity really comes down to just these two graphs, as we shall soon see.

If $\mathbf{G} = (V, E)$ is a graph and $uv \in E$, then we may form a new graph $\mathbf{G}'$ called an **elementary subdivision** of $\mathbf{G}$ by adding a new vertex v' and replacing the edge uv by edges uv' and $v'v$. In other words, $\mathbf{G}'$ has vertex set $V' = V \cup \{v'\}$ and edge set $E' = (E - \{uv\}) \cup \{uv', v'v\}$. Two graphs $\mathbf{G}_1$ and $\mathbf{G}_2$ are **homeomorphic** if they can be obtained from the same graph by a (potentially trivial) sequence of elementary subdivisions.

The purpose of discussing homeomorphic graphs is that two homeomorphic graphs have the same properties when it comes to being drawn in the plane. To see this, think about what happens to $\mathbf{K}_5$ if we form an elementary subdivision of it via any one of its edges. Clearly it remains nonplanar. In fact, if you take any nonplanar graph and form the elementary subdivision using any one of its edges, the resulting graph is nonplanar. The following very deep theorem was proved by the Polish mathematician Kazimierz Kuratowski in 1930. Its proof is beyond the scope of this text.

5.5 Planar Graphs

Theorem 5.34 (Kuratowski's Theorem). *A graph is planar if and only if it does not contain a subgraph homeomorphic to either K_5 or $K_{3,3}$.*

Kuratowski's Theorem gives a useful way for checking if a graph is planar. Although it's not always easy to find a subgraph homeomorphic to K_5 or $K_{3,3}$ by hand, there are efficient algorithms for planarity testing that make use of this characterization. To see this theorem at work, let's consider the Petersen graph shown in Figure 5.17. The Petersen graph has 10 vertices and 15 edges, so it passes the test of Theorem 5.33, and our argument using Euler's formula to prove that $K_{3,3}$ is nonplanar was complex enough, we probably don't want to try it for the Petersen graph. To use Kuratowski's Theorem here, we need to decide if we would rather find a subgraph homeomorphic to K_5 or to $K_{3,3}$. Although the Petersen graph looks very similar to K_5, it's actually simultaneously *too* similar and too different for us to be able to find a subgraph homeomorphic to K_5, since each vertex has degree 3. Thus, we set out to find a subgraph of the Petersen graph homeomorphic to $K_{3,3}$. To do so, note that $K_{3,3}$ contains a cycle of length 6 and three edges that are in place between vertices opposite each other on the cycle. We identify a six-cycle in the Petersen graph and draw it as a hexagon and place the remaining four vertices inside the cycle. Such a drawing is shown in Figure 5.35. The subgraph homeomorphic to $K_{3,3}$ is found by deleting the black vertex, as then the white vertices have degree two, and we can replace each of them and their two incident edges (shown in bold) by a single edge.

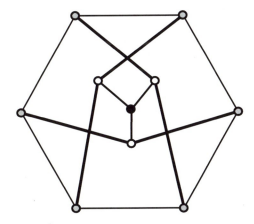

FIGURE 5.35: A MORE ILLUSTRATIVE DRAWING OF THE PETERSEN GRAPH

We close this section with a problem that brings the current section together with the topic of graph coloring. In 1852 Francis Guthrie, an Englishman who was at the time studying to be lawyer but subsequently became a professor of mathematics in South Africa, was trying to color a map of the counties of England so that any two counties that shared a boundary segment (meaning they touched in more than a single point) were colored with different colors. He noticed that he only needed four colors to do this, and was unable to draw any sort of map that would require five colors. (He was able to find a map that required four colors, an example of which is shown in Figure 5.36.)

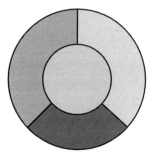

Figure 5.36: A map that requires four colors

Could it possibly be true that *every* map could be colored with only four colors? He asked his brother Frederick Guthrie, who was a mathematics student at University College, London, about the problem, and Frederick eventually communicated the problem to Augustus de Morgan (of de Morgan's laws fame), one of his teachers. It was in this way that one of the most famous (or infamous) problems, known for a century as the Four Color Problem and now the Four Color Theorem, in graph theory was born. De Morgan was very interested in the Four Color Problem, and communicated it to Sir William Rowan Hamilton, a prominent Irish mathematician and the one for whom hamiltonian cycles are named, but Hamilton did not find the problem interesting. Hamilton is one of the few people who considered the Four Color Problem but did not become captivated by it.

We'll continue our discussion of the history of the Four Color Theorem in a moment, but first, we must consider how we can turn the problem of coloring a map into a graph theory question. Well, it seems natural that each region should be assigned a corresponding vertex. We want to force regions that share a boundary to have different colors, so this suggests that we should place an edge between two vertices if and only if their corresponding regions have a common boundary. (As an example, the map in Figure 5.36 corresponds to the graph K_4.) It is not difficult to see that this produces a planar graph, since we may draw the edges through the common boundary segment. Furthermore, with a little bit of thought, you should see that given a planar drawing of a graph, you can create a map in which each vertex leads to a region and edges lead to common boundary segments. Thus, the Four Color Problem could be stated as "Does every planar graph have chromatic number at most four?"

Interest in the Four Color Problem languished until 1877, when the British mathematician Arthur Cayley wrote a letter to the Royal Society asking if the problem had been resolved. This brought the problem to the attention of many more people, and the first "proof" of the Four Color Theorem, due to Alfred Bray Kempe, was completed in 1878 and published a year later. It took 11 years before Percy John Heawood found a flaw in the proof but was able to salvage enough of it to show that every planar graph has chromatic number at most five. In 1880, Peter Guthrie Tait, a British physicist best known for his book *Treatise on Natural Philosophy* with Sir William Thomson (Lord Kelvin), made an announcement that suggested he had a proof of the Four Color Theorem utilizing hamiltonian cycles in certain planar graphs. However, consistent with the way Tait approached some conjectures in the mathematical theory of knots, it appears that he subsequently realized around 1883 that he could not prove that the hamiltonian cycles he was using actually existed and so Tait likely only believed he had a proof of the Four Color Theorem for a short time, if at all. However, it would take until 1946 to find a counterexample to the conjecture Tait had used in his attempt to prove the Four Color Theorem.

In the first half of the twentieth century, some incremental progress toward resolving the Four Color Problem was made, but few prominent mathematicians took a serious interest in it. The final push to prove the Four Color Theorem came with about at the same time that the first electronic computers were coming into widespread use in industry and research. In 1976, two mathematicians at the University of Illinois announced their computer-assisted proof of the Four Color Theorem. The proof by Kenneth Appel and Wolfgang Haken led the University of Illinois to add the phrase "FOUR COLORS SUFFICE" to its postage meter's imprint.[1]

Theorem 5.37 (Four Color Theorem). *Every planar graph has chromatic number at most four.*

Appel and Haken's proof of the Four Color Theorem was at a minimum unsatisfactory for many mathematicians, and to some it simply wasn't a proof. These mathematicians felt that the using a computer to check various cases was simply too uncertain; how could you be certain that the code that checked the 1,482 "unavoidable configurations" didn't contain any logic errors? In fact, there were several mistakes found in the cases analyzed, but none were found to be fatal flaws. In 1989, Appel and Haken published a 741-page tome entitled *Every Planar Map is Four Colorable* which provided corrections to all known flaws in their original argument. This still didn't satisfy many,

---

[1] A photograph of an envelope with such a meter mark on it can be found in the book *The Four-Color Theorem: History, Topological Foundations, and Idea of Proof* by Rudolf and Gerda Fritsch. (Springer, 1998)

and in the early 1990's a team consisting of Neil Robertson from The Ohio State University; Daniel P. Sanders, a graduate student at the Georgia Institute of Technology; Paul Seymour of Bellcore; and Robin Thomas from Georgia Tech announced a new proof of the Four Color Theorem. However, it still required the use of computers. The proof did gain more widespread acceptance than that of Appel and Haken, in part because the new proof used fewer than half (633) of the number of configurations the Appel-Haken proof used and the computer code was provided online for anyone to verify. While still unsatisfactory to many, the proof by Robertson, et al. was generally accepted, and today the issue of the Four Color Theorem has largely been put to rest. However, many still wonder if anyone will ever find a proof of this simple statement that does not require the assistance of a computer.

5.6 Counting Labeled Trees

How many trees are there with vertex set $[n] = \{1, 2, \ldots, n\}$? Let T_n be this number. For $n = 1$, there is clearly only one tree. Also, for $n = 2$, there is only one tree, which is isomorphic to K_2. In determining T_3, we finally have some work to do; however, there's not much, since all trees on 3 vertices are isomorphic to P_3. Thus, there are $T_3 = 3$ **labeled trees** on 3 vertices, corresponding to which vertex is the one of degree 2. When $n = 4$, we can begin by counting the number of nonisomorphic trees and consider two cases depending on whether the tree has a vertex of degree 3. If there is a vertex of degree 3, the tree is isomorphic to $K_{1,3}$ or it does not have a vertex of degree three, in which case it is isomorphic to P_4, since there must be precisely two vertices of degree 2 in such a graph. There are four labelings by [4] for $K_{1,3}$ (choose the vertex of degree three). How many labelings by [4] are there for P_4? There are $C(4, 2)$ ways to choose the labels i, j given to the vertices of degree 2 and two ways to select one of the remaining labels to be made adjacent to i. Thus, there are 12 ways to label P_4 by [4] and so $T_4 = 16$.

To this point, it looks like maybe there's a pattern forming. Perhaps it is the case that for all $n \geq 1$, $T_n = n^{n-2}$. This is in fact the case, but let's see how it works out for $n = 5$ before proving the result in general. What are the nonisomorphic trees on five vertices? Well, there's $K_{1,4}$ and P_5 for sure, and there's also the third tree shown in Figure 5.38. After thinking for a minute or two, you should be able to convince yourself that this is all of the possibilities. How many labelings by [5] does each of these have? There are 5 for $K_{1,4}$ since there are 5 ways to choose the vertex of degree 4. For P_5, there are 5 ways to choose the middle vertex of the path, $C(4, 2) = 6$ ways to label the two remaining vertices of degree 2 once the middle vertex is labeled, and then 2 ways to label the vertices of degree 1. This gives 60 labelings. For the last tree, there are 5 ways to label the vertex of degree 3, $C(4, 2) = 6$ ways to label the two leaves adjacent to the vertex of degree 3, and 2 ways to label the remaining two vertices, giving 60 labelings. Therefore, $T_5 = 125 = 5^3 = 5^{5-2}$.

FIGURE 5.38: THE NONISOMORPHIC TREES ON $n = 5$ VERTICES

It turns out that we are in fact on the right track, and we will now set out to prove the following:

Theorem 5.39 (Cayley's Formula). *The number T_n of labeled trees on n vertices is n^{n-2}.*

This result is usually referred to as Cayley's Formula, although equivalent results were proven earlier by James J. Sylvester (1857) and Carl W. Borchardt (1860). The reason that Cayley's name is most often affixed to this result is that he was the first to state and prove it in graph theoretic terminology (in 1889). (Although one could argue that Cayley really only proved it for $n = 6$ and then claimed that it could easily be extended for all other values of n, and whether such an extension can actually happen is open to some debate.) Cayley's Formula has many different proofs, most of which are quite elegant. If you're interested in presentations of several proofs, we encourage you to read the chapter on Cayley's Formula in *Proofs from THE BOOK* by Aigner, Ziegler, and Hofmann, which contains four different proofs, all using different proof techniques. Here we give a fifth proof, due to Prüfer and published in 1918. Interestingly, even though Prüfer's proof came after much of the terminology of graph theory was established, he seemed unaware of it and worked in the context of permutations and his own terminology, even though his approach clearly includes the ideas of graph theory. We will use a recursive technique in order to find a bijection between the set of labeled trees on n vertices and a natural set of size n^{n-2}, the set of strings of length $n - 2$ where the symbols in the string come from $[n]$.

We define a recursive algorithm that takes a tree **T** on $k \geq 2$ vertices labeled by elements of a set S of positive integers of size k and returns a string of length $k - 2$ whose symbols are elements of S. (The set S will usually be $[k]$, but in order to define a recursive procedure, we need to allow that it be an arbitrary set of k positive integers.) This string is called the **Prüfer code** of the tree **T**. Let prüfer(**T**) denote the Prüfer code of the tree **T**, and if v is a leaf of **T**, let **T**$-v$ denote the tree obtained from **T** by removing v (i.e., the subgraph induced by all the other vertices). We can then define prüfer(**T**) recursively by the following procedure.

1. If **T** ≅ **K**$_2$, return the empty string.

2. Else, let v be the leaf of **T** with the smallest label and let u be its unique neighbor. Let i be the label of u. Return $(i, \text{prüfer}(T - v))$.

Example 5.40. Before using Prüfer codes to prove Cayley's Formula, let's take a moment to make sure we understand how they are computed given a tree. Consider the 9-vertex tree **T** in Figure 5.41.

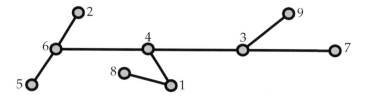

FIGURE 5.41: A LABELED 9-VERTEX TREE

How do we compute prüfer(**T**)? Since **T** has more than two vertices, we use the second step and find that v is the vertex with label 2 and u is the vertex with label 6, so prüfer(**T**) = $(6, \text{prüfer}(T - v))$. The graph **T** − v is shown in Figure 5.42.

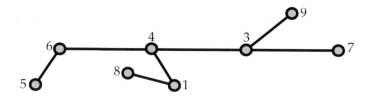

FIGURE 5.42: THE TREE **T** − v

The recursive call prüfer(**T** − v) returns $(6, \text{prüfer}(T - v - v'))$, where v' is the vertex labeled 5. Continuing recursively, the next vertex deleted is 6, which appends a 4 to the string. Then 7 is deleted, appending 3. Next 8 is deleted, appending 1. This is followed by the deletion of 1, appending 4. Finally 4 is deleted, appending 3, and the final recursive call has the subtree isomorphic to **K**$_2$ with vertices labeled 3 and 9, and an empty string is returned. Thus, prüfer(**T**) = 6643143.

We're now prepared to give a proof of Cayley's Formula.

Proof. It is clear that prüfer(**T**) takes an n-vertex labeled tree with labels from $[n]$ and returns a string of length $n - 2$ whose symbols are elements of $[n]$. What we have yet

5.6 Counting Labeled Trees

to do is determine a way to take such a string and construct an n-vertex labeled tree from it. If we can find such a construction, we will have a bijection between the set $\mathcal{T}_n$ of labeled trees on n vertices and the set of strings of length $n-2$ whose symbols come from $[n]$, which will imply that $T_n = n^{n-2}$.

First, let's look at how prüfer(T) behaves. What numbers actually appear in the Prüfer code? The numbers that appear in the Prüfer code are the labels of the *non-leaf* vertices of **T**. The label of a leaf simply cannot appear, since we always record the label of the *neighbor* of the leaf we are deleting, and the only way we would delete the neighbor of a leaf is if that neighbor were also a leaf, which can only happen $\mathbf{T} \cong \mathbf{K}_2$, in which case prüfer(**T**) simply returns the empty string. Thus if $I \subset [n]$ is the set of symbols that appear in prüfer(**T**), the labels of the leaves of **T** are precisely the elements of $[n] - I$.

With the knowledge of which labels belong to the leaves of **T** in hand, we are ready to use induction to complete the proof. Our goal is to show that if given a string $\mathbf{s} = s_1 s_2 \cdots s_{n-2}$ whose symbols come from a set S of n elements, there is a unique tree **T** with prüfer(**T**) = **s**. If $n = 2$, the only such string is the empty string, so 1 and 2 both label leaves and we can construct only $\mathbf{K}_2$. Now suppose we have the result for some $m \geq 2$, and we try to prove it for $m + 1$. We have a string $\mathbf{s} = s_1 s_2 \cdots s_{m-1}$ with symbols from $[m+1]$. Let I be the set of symbols appearing in **s** and let k be the least element of $[m+1] - I$. By the previous paragraph, we know that k is the label of a leaf of **T** and that its unique neighbor is the vertex labeled s_1. The string $\mathbf{s}' = s_2 s_3 \cdots s_{m-1}$ has length $m-2$ and since k does not appear in **s**, its symbols come from $S = [m+1] - \{k\}$, which has size m. Thus, by induction, there is a unique tree **T**' whose Prüfer code is $\mathbf{s}'$. We form **T** from **T**' by attaching a leaf with label k to the vertex of **T**' with label s_1 and have a tree of the desired type. □

Example 5.43. We close this section with an example of how to take a Prüfer code and use it to construct a labeled tree. Consider the string $\mathbf{s} = 75531$ as a Prüfer code. Then the tree **T** corresponding to **s** has 7 vertices, and its leaves are labeled 2, 4, and 6. The inductive step in our proof attaches the vertex labeled 2 to the vertex labeled 7 in the tree **T**' with Prüfer code 5531 and vertex labels $\{1, 3, 4, 5, 6, 7\}$, since 2 is used to label the last vertex added. What are the leaves of **T**'? The symbols in $\{4, 6, 7\}$ do not appear in 5531, so they must be the labels of leaves, and the construction says that we would attach the vertex labeled 4 to the vertex labeled 5 in the tree we get by induction. In Table 5.44, we show how this recursive process continues.

Chapter 5 Graph Theory

Prüfer code	Label set	Edge added
75531	{1,2,3,4,5,6,7}	2–7
5531	{1,3,4,5,6,7}	4–5
531	{1,3,5,6,7}	6–5
31	{1,3,5,7}	5–3
1	{1,3,7}	3–1
(empty string)	{1,7}	1–7

TABLE 5.44: TURNING THE PRÜFER CODE 75531 INTO A LABELED TREE

We form each row from the row above it by removing the first label used on the edge added from the label set and removing the first symbol from the Prüfer code. Once the Prüfer code becomes the empty string, we know that the two remaining labels must be the labels we place on the ends of K_2 to start building T. We then work back up the edge added column, adding a new vertex and the edge indicated. The tree we construct in this manner is shown in Figure 5.45.

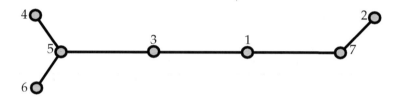

FIGURE 5.45: THE LABELED TREE WITH PRÜFER CODE 75531

5.7 A Digression into Complexity Theory

We have already introduced in Chapter 4 a few notions about efficient algorithms. We also discussed the difficulty of determining a graph's chromatic number and clique number earlier in this chapter. We conclude with a brief discussion of some issues involving computational complexity for other problems discussed in this chapter.

Let's begin with some problems for which there are polynomial-time algorithms. Suppose you are given a graph on n vertices and asked whether or not the graph is connected. Here a positive answer can be justified by providing a spanning tree. On the other hand, a negative answer can be justified by providing a partition of the vertex sets $V = V_1 \cup V_2$ with V_1 and V_2 non-empty subsets and having no edges with one endpoint in V_1 and the other in V_2. In Chapter 12 we will discuss two efficient algorithms that find spanning trees in connected graphs. They can easily be modified to produce a partition showing the graph is disconnected.

If you are asked whether a connected graph is eulerian, then a positive answer can be justified by producing the appropriate sequence. We gave an algorithm to do this earlier in the chapter. A negative answer can be justified by producing a vertex of odd degree, and our algorithm will identify such a vertex if it exists. (Depending on the data structures used to represent the graph, it may be most efficient to simply look for vertices of odd degree without using the algorithm to find an eulerian circuit.)

On the surface, the problem of determining if a graph is hamiltonian looks similar to that of determining if the graph is eulerian. Both call for a sequence of vertices in which each pair of consecutive vertices is joined by an edge. Of course, each problem has an additional requirement on yes certificates. However, justifying a negative answer to the question of whether a graph is hamiltonian is not straightforward. Theorem 5.18 only gives a way to confirm that a graph *is* hamiltonian; there are many nonhamiltonian graphs that do not satisfy its hypothesis. At this time, no one knows how to efficiently justify a negative answer—at least not in the general case.

5.8 Discussion

Over coffee, today's conversation was enthusiastic and heated at times. Zori got things off with a blast "I don't think graphs are of any use at all..." but she wasn't even able to finish the sentence before Yolanda uncharacteristically interrupted her with "You're off base on this one. I see lots of ways graphs can be used to model real world problems. The professor actually showed us examples back in our first class. But now that we're talking in more depth about graphs, things are even clearer." Bob added, "These eulerian and hamiltonian cycle problems are certain to have applications in network routing problems." Xing reinforced Bob with "Absolutely. There are important questions in network integrity and information exchange that are very much the same as these basic problems." Alice piled on "Even the notion of chromatic number clearly has practical applications." By this time, Zori realized her position was indefensible but she was reluctant to admit it. She offered only a "Whatever."

Things quieted down a bit and Dave said "Finding a hamiltonian cycle can't be all that hard, if someone guarantees that there is one. This extra information must be of value in the search." Xing added "Maybe so. It seems natural that it should be easier to find something if you know it's there." Alice asked "Does the same thing hold for chromatic number?" Bob didn't understand her question "Huh?" Alice continued, this time being careful not to even look Bob's way "I mean if someone tells you that a graph is 3-colorable, does that help you to find a coloring using only three colors?" Dave said "Seems reasonable to me."

Chapter 5 Graph Theory

After a brief pause, Carlos offered "I don't think this extra knowledge is of any help. I think these problems are pretty hard, regardless." They went back and forth for a while, but in the end, the only thing that was completely clear is that graphs and their properties had captured their attention, at least for now.

5.9 Exercises

1. The questions in this exercise pertain to the graph **G** shown in Figure 5.46.

    (a) What is the degree of vertex 8?

    (b) What is the degree of vertex 10?

    (c) How many vertices of degree 2 are there in **G**? List them.

    (d) Find a cycle of length 8 in **G**.

    (e) What is the length of a shortest path from 3 to 4?

    (f) What is the length of a shortest path from 8 to 7?

    (g) Find a path of length 5 from vertex 4 to vertex 6.

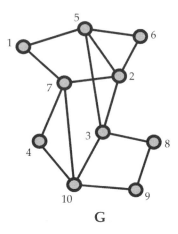

FIGURE 5.46: A GRAPH

2. Draw a graph with 8 vertices, all of odd degree, that does not contain a path of length 3 or explain why such a graph does not exist.

3. Draw a graph with 6 vertices having degrees 5, 4, 4, 2, 1, and 1 or explain why such a graph does not exist.

4. For the next Olympic Winter Games, the organizers wish to expand the number of teams competing in curling. They wish to have 14 teams enter, divided into two pools of seven teams each. Right now, they're thinking of requiring that in preliminary play each team will play seven games against distinct opponents. Five of the opponents will come from their own pool and two of the opponents will come from the other pool. They're having trouble setting up such a schedule, so they've come to you. By using an appropriate graph-theoretic model, either argue that they cannot use their current plan or devise a way for them to do so.

5.9 Exercises

5. For this exercise, consider the graph **G** in Figure 5.47.

   (a) Let $V_1 = \{g, j, c, h, e, f\}$ and $E_1 = \{ge, jg, ch, ef\}$. Is (V_1, E_1) a subgraph of **G**?

   (b) Let $V_2 = \{g, j, c, h, e, f\}$ and $E_2 = \{ge, jg, ch, ef, cj\}$. Is (V_2, E_2) a subgraph of **G**?

   (c) Let $V_3 = \{a, d, c, h, b\}$ and $E_3 = \{ch, ac, ad, bc\}$. Is (V_3, E_3) an induced subgraph of **G**?

   (d) Draw the subgraph of **G** induced by $\{g, j, d, a, c, i\}$.

   (e) Draw the subgraph of **G** induced by $\{c, h, f, i, j\}$.

   (f) Draw a subgraph of **G** having vertex set $\{e, f, b, c, h, j\}$ that is *not* an induced subgraph.

   (g) Draw a spanning subgraph of **G** with exactly 10 edges.

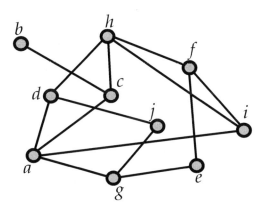

FIGURE 5.47: A GRAPH G

6. Prove that every tree on n vertices has exactly $n - 1$ edges.

7. Figure 5.48 contains four graphs on six vertices. Determine which (if any) pairs of graphs are isomorphic. For pairs that are isomorphic, give an isomorphism between the two graphs. For pairs that are not isomorphic, explain why.

103

Chapter 5 Graph Theory

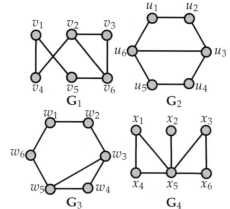

FIGURE 5.48: ARE THESE GRAPHS ISOMORPHIC?

8. Find an eulerian circuit in the graph **G** in Figure 5.49 or explain why one does not exist.

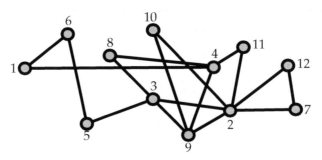

FIGURE 5.49: A GRAPH **G**

9. Consider the graph **G** in Figure 5.50. Determine if the graph is eulerian. If it is, find an eulerian circuit. If it is not, explain why it is not. Determine if the graph is hamiltonian. If it is, find a hamiltonian cycle. If it is not, explain why it is not.

104

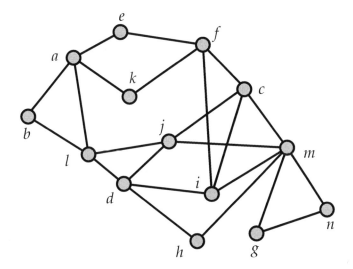

FIGURE 5.50: A GRAPH G

10. Explain why the graph **G** in Figure 5.51 does not have an eulerian circuit, but show that by adding a single edge, you can make it eulerian.

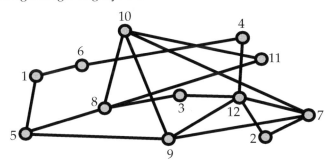

FIGURE 5.51: A GRAPH G

11. An **eulerian trail** is defined in the same manner as an eulerian circuit (see Section 5.3) except that we drop the condition that $x_0 = x_t$. Prove that a graph has an eulerian trail if and only if it is connected and has at most two vertices of odd degree.

12. Alice and Bob are discussing a graph that has 17 vertices and 129 edges. Bob argues that the graph is hamiltonian, while Alice says that he's wrong. Without knowing

Chapter 5 Graph Theory

anything more about the graph, must one of them be right? If so, who and why, and if not, why not?

13. Find the chromatic number of the graph **G** in Figure 5.52 and a coloring using $\chi(\mathbf{G})$ colors.

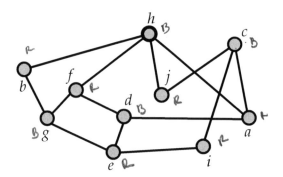

FIGURE 5.52: A GRAPH **G** TO COLOR

14. Find the chromatic number of the graph **G** in Figure 5.53 and a coloring using $\chi(\mathbf{G})$ colors.

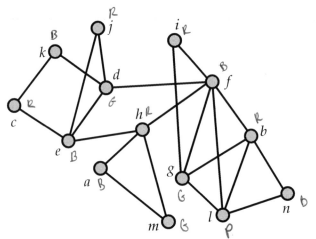

FIGURE 5.53: A GRAPH **G** TO COLOR

15. A pharmaceutical manufacturer is building a new warehouse to store its supply of

10 chemicals it uses in production. However, some of the chemicals cannot be stored in the same room due to undesirable reactions that will occur. The matrix below has a 1 in position (i, j) if and only if chemical i and chemical j cannot be stored in the same room. Develop an appropriate graph theoretic model and determine the smallest number of rooms into which they can divide their warehouse so that they can safely store all 10 chemicals in the warehouse.

$$\begin{bmatrix} 0 & 1 & 0 & 1 & 1 & 0 & 1 & 0 & 0 & 0 \\ 1 & 0 & 0 & 1 & 1 & 0 & 0 & 0 & 0 & 1 \\ 0 & 0 & 0 & 0 & 0 & 1 & 0 & 1 & 1 & 0 \\ 1 & 1 & 0 & 0 & 1 & 0 & 0 & 0 & 0 & 0 \\ 1 & 1 & 0 & 1 & 0 & 0 & 0 & 0 & 1 & 0 \\ 0 & 0 & 1 & 0 & 0 & 0 & 1 & 0 & 0 & 1 \\ 1 & 0 & 0 & 0 & 0 & 1 & 0 & 1 & 0 & 0 \\ 0 & 0 & 1 & 0 & 0 & 0 & 1 & 0 & 0 & 0 \\ 0 & 0 & 1 & 0 & 1 & 0 & 0 & 0 & 0 & 0 \\ 0 & 1 & 0 & 0 & 0 & 1 & 0 & 0 & 0 & 0 \end{bmatrix}$$

16. A school is preparing the schedule of classes for the next academic year. They are concerned about scheduling calculus, physics, English, statistics, economics, chemistry, and German classes, planning to offer a single section of each one. Below are the lists of courses that each of six students must take in order to successfully graduate. Determine the smallest number of class periods that can be used to schedule these courses if each student can take at most one course per class period. Explain why fewer class periods cannot be used.

Student	Courses
1	Chemistry, Physics, Economics
2	English, German, Statistics
3	Statistics, Calculus, German
4	Chemistry, Physics
5	English, Chemistry
6	Chemistry, Economics

17. All trees with more than one vertex have the same chromatic number. What is it, and why?

18. Find a proper $(t + 1)$-coloring of the graph G_{t+1} in Mycielski's proof of Proposition 5.25. This establishes that $\chi(G_{t+1}) \leq t + 1$.

Chapter 5 Graph Theory

19. How many vertices does the graph G_4 from the Kelly and Kelly proof of Proposition 5.25 have?

20. Construct and draw the graph G_5 from Mycielski's proof of Proposition 5.25.

21. Find a recursive formula for the number of vertices n_t in the graph G_t from the Kelly and Kelly proof of Proposition 5.25.

22. Let b_t be the number of vertices in the graph G_t from the Mycielski's proof of Proposition 5.25. Find a recursive formula for b_t.

23. The **girth** of a graph **G** is the number of vertices in a shortest cycle of **G**. Find the girth of the graph G_t in the Kelly and Kelly proof of Proposition 5.25 and prove that your answer is correct. As a challenge, see if you can modify the construction of G_t to increase the girth. If so, how far are you able to increase it?

24. Use the First Fit algorithm to color the graph in Figure 5.26 using the two different orderings of the vertex set shown there.

25. Draw the interval graph corresponding to the intervals in Figure 5.54.

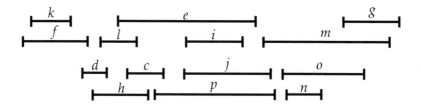

FIGURE 5.54: A COLLECTION OF INTERVALS

26. Use the First Fit coloring algorithm to find the chromatic number of the interval graph whose interval representation is shown in Figure 5.54 as well as a proper coloring using as few colors as possible.

27.

(a) From Exercise 5.9.24 you know that choosing a bad ordering of the vertices of a graph can lead to the First Fit coloring algorithm producing a coloring that is far from optimal. However, you can use this algorithm to prove a bound on the chromatic number. Show that if every vertex of **G** has degree at most D, then $\chi(G) \leq D + 1$.

(b) Give an example of a bipartite graph with $D = 1000$ to show that this bound need not be tight.

28. Is the graph in Figure 5.53 planar? If it is, find a drawing without edges crossings. If it is not give a reason why it is not.

29. Is the graph in Figure 5.55 planar? If it is, find a drawing without edge crossings. If it is not give a reason why it is not.

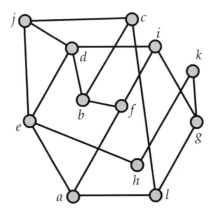

Figure 5.55: Is this graph planar?

30. Find a planar drawing of the graph $K_5 - e$, by which we mean the graph formed from the complete graph on 5 vertices by deleting any edge.

31. Exhibit a planar drawing of an eulerian planar graph with 10 vertices and 21 edges.

32. Show that every planar graph has a vertex that is incident to at most five edges.

33. Let $G = (V, E)$ be a graph with $V = \{v_1, v_2, \ldots, v_n\}$. Its *degree sequence* is the list of the degrees of its vertices, arranged in nonincreasing order. That is, the degree sequence of G is $(\deg_G(v_1), \deg_G(v_2), \ldots, \deg_G(v_n))$ with the vertices arranged such that $\deg_G(v_1) \geq \deg_G(v_2) \geq \cdots \geq \deg_G(v_n)$. Below are five sequences of integers (along with n, the number of integers in the sequence). Identify

- the *one* sequence that **cannot be the degree sequence of any graph**;
- the *two* sequences that could be the degree sequence of a **planar** graph;
- the *one* sequence that could be the degree sequence of a **tree**;
- the *one* sequence that is the degree sequence of an **eulerian** graph; and
- the *one* sequence that is the degree sequence of a graph that must be **hamiltonian**.

Explain your answers. (Note that one sequence will get two labels from above.)

Chapter 5 Graph Theory

(a) $n = 10$: $(4, 4, 2, 2, 1, 1, 1, 1, 1, 1)$
(b) $n = 9$: $(8, 8, 8, 6, 4, 4, 4, 4, 4)$
(c) $n = 7$: $(5, 4, 4, 3, 2, 1, 0)$
(d) $n = 10$: $(7, 7, 6, 6, 6, 6, 5, 5, 5, 5)$
(e) $n = 6$: $(5, 4, 3, 2, 2, 2)$

34. Below are three sequences of length 10. One of the sequences cannot be the degree sequence (see Exercise 5.9.33) of any graph. Identify it and say why. For each of the other two, say *why* (if you have enough information) a *connected* graph with that degree sequence

- is definitely hamiltonian/cannot be hamiltonian;
- is definitely eulerian/cannot be eulerian;
- is definitely a tree/cannot be a tree; and
- is definitely planar/cannot be planar.

(If you do not have enough information to make a determination for a sequence without having specific graph(s) with that degree sequence, write "not enough information" for that property.)

(a) $(6, 6, 4, 4, 4, 4, 2, 2, 2, 2)$
(b) $(7, 7, 7, 7, 6, 6, 6, 2, 1, 1)$
(c) $(8, 6, 4, 4, 4, 3, 2, 2, 1, 1)$

35. For the two degree sequences in Exercise 5.9.34 that correspond to graphs, there were some properties for which the degree sequence was not sufficient information to determine if the graph had that property. For each of those situations, see if you can draw both a graph that has the property and a graph that does not have the property.

36. Draw the 16 labeled trees on 4 vertices.

37. Determine prüfer(T) for the tree T in Figure 5.56.

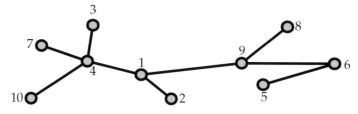

FIGURE 5.56: A 10-VERTEX TREE

38. Determine prüfer(**T**) for the tree **T** in Figure 5.57.

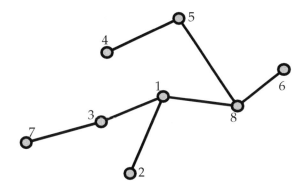

FIGURE 5.57: A 10-vertex tree

39. Determine prüfer(**T**) for the tree **T** Figure 5.58.

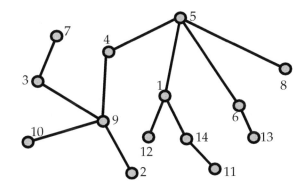

FIGURE 5.58: A 14-vertex tree

40. Construct the labeled tree **T** with Prüfer code 96113473.

41. Construct the labeled tree **T** with Prüfer code 23134.

42. Construct the labeled tree **T** with Prüfer code (using commas to separate symbols in the string, since we have labels greater than 9) 10, 1, 7, 4, 3, 4, 10, 2, 2, 8.

43. (Challenge problem) When $\mathbf{G} = (V, E)$ is a graph, let $\Delta(\mathbf{G})$ denote the maximum degree in **G**. Prove *Brooks' Theorem*: If **G** is connected and $\Delta(\mathbf{G}) = k$, then $\chi(\mathbf{G}) \leq k + 1$.

Furthermore, equality holds if and only if (a) $k = 2$ and **G** is an odd cycle, or (b) $k \neq 2$ and $\mathbf{G} = \mathbf{K}_{k+1}$.

Hint. Hint: It's clear that $\chi(\mathbf{G}) \leq k + 1$ (in fact, this was already assigned as an exercise). Assume that $\chi(\mathbf{G}) = k + 1$ but that neither conclusion (a) or (b) holds. Take a spanning tree of **G** and an appropriate ordering of the vertices, with two leaves of the tree coming first. Then show that a First Fit coloring of the graph will only use k colors.

CHAPTER 6

Partially Ordered Sets

Discussion 6.1. Alice was surfing the web and found a site listing top movies, grouped by categories (comedy, drama, family, etc) as well as by the decade in which they were released. Alice was intrigued by the critic's choices and his rankings, especially for the top seven dramas from the 1990's. Alice agreed with the critic's choices as a group but not the specific rankings. She wrote the critic's rankings on the board and just to the right, she gave her own rankings, all the time insisting that she was certainly correct in her opinions.

	Movie Critic's Ranking	Alice's Ranking
1	Saving Private Ryan	Life is Beautiful
2	Life is Beautiful	Saving Private Ryan
3	Forrest Gump	Good Will Hunting
4	Braveheart	Titanic
5	Good Will Hunting	Braveheart
6	Titanic	Forrest Gump
7	Jurassic Park	Jurassic Park

Dave studied the two rankings and listened carefully to Alice's rationale (which he felt was a bit over the top), but eventually, he held up the following diagram and offered it as a statement of those comparisons on which both Alice and the movie critic were in agreement. Do you see how Dave made up this diagram? Add your own rankings of these seven films and then draw the diagram that Dave would produce as a statement about the comparisons on which you, Alice and the movie critic were in agreement.

More generally, when humans are asked to express preferences among a set of options, they often report that establishing a totally ranked list is difficult if not impossible. Instead, they prefer to report a partial order—where comparisons are made between certain pairs of options but not between others. In this chapter, we make these observations more concrete by introducing the concept of a partially ordered set. Elementary examples include (1) a family of sets which is partially ordered by inclusion

Chapter 6 Partially Ordered Sets

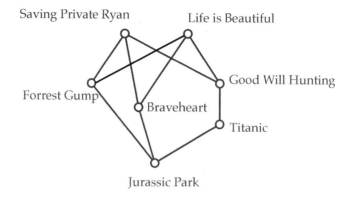

FIGURE 6.2: TOP MOVIES FROM THE 90'S

and (2) a set of positive integers which is partially ordered by division. From an applications standpoint, a complex construction job typically involves a large number of projects for which there is a notion of precedence between some but not all pairs. Also, computer file systems are modeled by trees which become partially ordered sets whenever links are added.

6.1 Basic Notation and Terminology

A **partially ordered set** or **poset P** is a pair (X, P) where X is a set and P is a reflexive, antisymmetric, and transitive binary relation on X. (Refer to Section B.10 for a refresher of what these properties are if you need to.) We call X the **ground set** while P is a **partial order** on X. Elements of the ground set X are also called **points**, and the poset P is **finite** if its ground set X is a finite set.

Example 6.3. Let $X = \{a, b, c, d, e, f\}$. Consider the following binary relations on X.

$$R_1 = \{(a,a),(b,b),(c,c),(d,d),(e,e),(f,f),(a,b),(a,c),(e,f)\}$$
$$R_2 = \{(a,a),(b,b),(c,c),(d,d),(e,e),(f,f),(d,b),(d,e),(b,a),(e,a),$$
$$(d,a),(c,f)\}$$
$$R_3 = \{(a,a),(b,b),(c,c),(d,d),(e,e),(f,f),(a,c),(a,e),(a,f),(b,c),$$
$$(b,d),(b,e),(b,f),(d,e),(d,f),(e,f)\}$$
$$R_4 = \{(a,a),(b,b),(c,c),(d,d),(e,e),(f,f),(d,b),(b,a),(e,a),(c,f)\}$$
$$R_5 = \{(a,a),(c,c),(d,d),(e,e),(a,e),(c,a),(c,e),(d,e)\}$$
$$R_6 = \{(a,a),(b,b),(c,c),(d,d),(e,e),(f,f),(d,f),(b,e),(c,a),(e,b)\}$$

6.1 Basic Notation and Terminology

Which of the binary relations are partial orders on X? For those that are not partial orders on X, which property or properties are violated?

Solution. A bit of checking confirms that R_1, R_2 and R_3 are partial orders on X, so $\mathbf{P}_1 = (X, R_1)$, $\mathbf{P}_2 = (X, R_2)$ and $\mathbf{P}_3 = (X, R_3)$ are posets. Several of the other examples we will discuss in this chapter will use the poset $\mathbf{P}_3 = (X, R_3)$.

On the other hand, R_4, R_5 and R_6 are not partial orders on X. Note that R_4 is not transitive, as it contains (d, b) and (b, a) but not (d, a). The relation R_5 is not reflexive, since it doesn't contain (b, b). (Also, it also doesn't contain (f, f), but one shortcoming is enough.) Note that R_5 is a partial order on $\{a, b, d, e\}$. The relation R_6 is not antisymmetric, as it contains both (b, e) and (e, b).

When $\mathbf{P} = (X, P)$ is a poset, it is common to write $x \leq y$ in P or $y \geq x$ in P as substitutes for $(x, y) \in P$. Of course, the notations $x < y$ in P and $y > x$ in P mean $x \leq y$ in P and $x \neq y$. When the poset $\mathbf{P}$ remains fixed throughout a discussion, we will sometimes abbreviate $x \leq y$ in P by just writing $x \leq y$, etc. When x and y are distinct points from X, we say x is **covered** by y in $\mathbf{P}$[1] when $x < y$ in P, and there is no point $z \in X$ for which $x < z$ and $z < y$ in P. For example, in the poset $\mathbf{P}_3 = (X, R_3)$ from Example 6.3, d is covered by e and c covers b. However, a is not covered by f, since $a < e < f$ in R_3. We can then associate with the poset $\mathbf{P}$ a **cover graph** $\mathbf{G}$ whose vertex set is the ground set X of $\mathbf{P}$ with xy an edge in $\mathbf{G}$ if and only if one of x and y covers the other in $\mathbf{P}$. Again, for the poset $\mathbf{P}_3$ from Example 6.3, we show the cover graph on the left side of Figure 6.4. Actually, on the right side of this figure is just another drawing of this same graph.

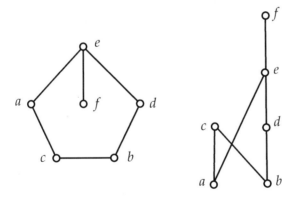

FIGURE 6.4: COVER GRAPH

---

[1]Reflecting the vagaries of the English language, mathematicians use the phrases: (1) x is covered by y in P; (2) y covers x in P; and (3) (x, y) is a cover in P interchangeably.

Chapter 6 Partially Ordered Sets

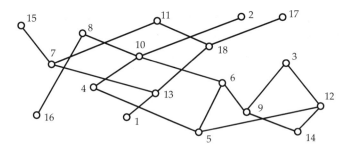

FIGURE 6.5: A Poset on 17 Points

It is convenient to illustrate a poset with a suitably drawn diagram of the cover graph in the Euclidean plane. We choose a standard horizontal/vertical coordinate system in the plane and require that the vertical coordinate of the point corresponding to y be larger than the vertical coordinate of the point corresponding to x whenever y covers x in P. Each edge in the cover graph is represented by a straight line segment which contains no point corresponding to any element in the poset other than those associated with its two end points. Such diagrams are called **Hasse diagrams** (**poset diagrams, order diagrams**, or just **diagrams**). Now it should be clear that the drawing on the right side of Figure 6.4 is a diagram of the poset $\mathbf{P}_3$ from Example 6.3, while the diagram on the left is not.

For posets of moderate size, diagrams are frequently used to define a poset—rather than the explicit binary relation notation illustrated in Example 6.3. In Figure 6.5, we illustrate a poset $\mathbf{P} = (X, P)$ with ground set $X = [17] = \{1, 2, \ldots, 17\}$. It would take several lines of text to write out the binary relation P, and somehow the diagram serves to give us a more tactile sense of the properties of the poset.

Discussion 6.6. Alice and Bob are talking about how you communicate with a computer in working with posets. Bob says that computers have incredible graphics capabilities these days and that you just give the computer a pdf scan of a diagram. Alice says that she doubts that anybody really does that. Carlos says that there are several effective strategies. One way is to label the points with positive integers from $[n]$ where n is the number of points in the ground set and then define a 0–1 $n \times n$ matrix A with entry $a(i, j) = 1$ when $i \leq j$ in P and $a(i, j) = 0$ otherwise. Alternatively, you can just provide for each element i in the ground set a vector $U(x)$ listing all elements which are greater than x in P. This vector can be what computer scientists call a **linked list**.

A partial order P is called a **total order** (also, a **linear order**) if for all $x, y \in X$, either $x \leq y$ in P or $y \leq x$ in P. For small finite sets, we can specify a linear order by listing the elements from least to greatest. For example, $L = [b, c, d, a, f, g, e]$ is the linear order on the ground set $\{a,b,c,d,e,f,g\}$ with $b<c<d<a<f<g<e$ in L.

The set of real numbers comes equipped with a natural total order. For example, $1 < 7/5 < \sqrt{2} < \pi$ in this order. But in this chapter, we will be interested primarily with partial orders that are *not* linear orders. Also, we note that special care must be taken when discussing partial orders on ground sets whose elements are real numbers. For the poset shown in Figure 6.5, note that 14 is less than 8, while 3 and 6 are incomparable. Best not to tell your parents that you've learned that under certain circumstances, 14 can be less than 8 and that you may be able to say which of 3 and 6 is larger than the other. The subtlety may be lost in the heated discussion certain to follow.

Example 6.7. There are several quite natural ways to construct posets.

1. A family $\mathcal{F}$ of sets is partially ordered by inclusion, i.e., set $A \leq B$ if and only if A is a subset of B.

2. A set X of positive integers is partially ordered by division—without remainder, i.e., set $m \leq n$ if and only if $n \equiv 0 \pmod{m}$.

3. A set X of t-tuples of real numbers is partially ordered by the rule
$$(a_1, a_2, \ldots, a_t) \leq (b_1, b_2, \ldots, b_t)$$
if and only if $a_i \leq b_i$ in the natural order on $\mathbb{R}$ for $i = 1, 2, \ldots, t$.

4. When $L_1, L_2, \ldots, L_k$ are linear orders on the same set X, we can define a partial order P on X by setting $x \leq y$ in P if and only if $x \leq y$ in L_i for all $i = 1, 2, \ldots, k$.

We illustrate the first three constructions with the posets shown in Figure 6.8. As is now clear, in the discussion at the very beginning of this chapter, Dave drew a diagram for the poset determined by the intersection of the linear orders given by Alice and the movie critic.

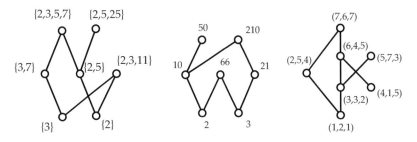

FIGURE 6.8: CONSTRUCTING POSETS

Chapter 6 Partially Ordered Sets

Distinct points x and y in a poset $\mathbf{P} = (X, P)$ are **comparable** if either $x < y$ in P or $x > y$ in P; otherwise x and y are **incomparable**. If x and y are incomparable in $\mathbf{P}$, we sometimes write $x \| y$ in $\mathbf{P}$. With a poset $\mathbf{P} = (X, P)$, we associate a **comparability graph** $\mathbf{G}_1 = (X, E_1)$ and an **incomparability graph** $\mathbf{G}_2 = (X, E_2)$. The edges in the comparability graph $\mathbf{G}_1$ consist of the comparable pairs and the edges in the incomparability graph are the incomparable pairs. We illustrate these definitions in Figure 6.9 where we show the comparability graph and the incomparability graph of the poset $\mathbf{P}_3$.

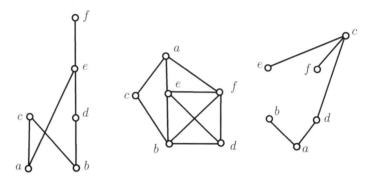

FIGURE 6.9: COMPARABILITY AND INCOMPARABILITY GRAPHS

When $\mathbf{P} = (X, P)$ is a poset and $Y \subseteq X$, the binary relation $Q = P \cap (Y \times Y)$ is a partial order on Y, and we call the poset (Y, Q) a **subposet** of $\mathbf{P}$. In Figure 6.10, we show a subposet of the poset first presented in Figure 6.5.

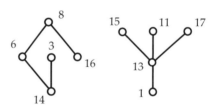

FIGURE 6.10: A SUBPOSET

When $\mathbf{P} = (X, P)$ is a poset and C is a subset of X, we say that C is a **chain** if every distinct pair of points from C is comparable in P. When P is a linear order, the entire ground set X is a chain. Dually, if A is a subset of X, we say that A is an **antichain** if every distinct pair of points from A is incomparable in P. Note that a one-element subset is both a chain and an antichain. Also, we consider the empty set as both a chain and an antichain.

The **height** of a poset $\mathbf{P} = (X, P)$, denoted height($\mathbf{P}$), is the largest h for which there exists a chain of h points in $\mathbf{P}$. Dually, the **width** of a poset $\mathbf{P} = (X, P)$, denoted width($\mathbf{P}$), is the largest w for which there exists an antichain of w points in $\mathbf{P}$.

Discussion 6.11. Given a poset $\mathbf{P} = (X, P)$, how hard is to determine its height and width? Bob says that it is very easy. For example, to find the width of a poset, just list all the subsets of X. Delete those which are not antichains. The answer is the size of the largest subset that remains. He is quick to assert that the same approach will work to find the height. Alice groans at Bob's naivety and suggests that he should read further in this chapter.

6.2 Additional Concepts for Posets

We say (X, P) and (Y, Q) are **isomorphic**, and write $(X, P) \cong (Y, Q)$ if there exists a bijection (1–1 and onto map) $f : X \to Y$ so that $x_1 \leq x_2$ in P if and only if $f(x_1) \leq f(x_2)$ in Q. In this definition, the map f is called an **isomorphism** from $\mathbf{P}$ to $\mathbf{Q}$. In Figure 6.8, the first two posets are isomorphic.

Discussion 6.12. Bob sees a pattern linking the first two posets shown in Figure 6.8 and asserts that any poset of one of these two types is isomorphic to a poset of the other type. Alice admits that Bob is right—but even more is true. The four constructions given in Example 6.7 are universal in the sense that *every* poset is isomorphic to a poset of each of the four types. Do you see why? If you get stuck answering this, we will revisit the question at the end of the chapter, and we will give you a hint.

An isomorphism from $\mathbf{P}$ to $\mathbf{P}$ is called an **automorphism** of $\mathbf{P}$. An isomorphism from $\mathbf{P}$ to a subposet of $\mathbf{Q}$ is called an **embedding** of $\mathbf{P}$ in $\mathbf{Q}$. In most settings, we will not distinguish between isomorphic posets, and we will say that a poset $\mathbf{P} = (X, P)$ is **contained** in $\mathbf{Q} = (Y, Q)$ (also $\mathbf{Q}$ **contains** $\mathbf{P}$) when there is an embedding of $\mathbf{P}$ in $\mathbf{Q}$. Also, we will say that $\mathbf{P}$ **excludes** $\mathbf{Q}$ when no subposet of $\mathbf{P}$ is isomorphic to $\mathbf{Q}$, and we will frequently say $\mathbf{P} = \mathbf{Q}$ when $\mathbf{P}$ and $\mathbf{Q}$ are isomorphic.

With the notion of isomorphism, we are lead naturally to the notion of an "unlabeled" posets, and in Figure 6.13, we show a diagram for such a poset.

Chapter 6 Partially Ordered Sets

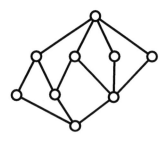

Figure 6.13: An Unlabeled Partially Ordered Set

Discussion 6.14. How hard is it to tell whether two posets are isomorphic? Bob thinks it's not too difficult. Bob says that if you give him a bijection between the ground sets, then he can quickly determine whether you have established that the two posets are isomorphic. Alice senses that Bob is confusing the issue of testing whether two posets are isomorphic with simply verifying that a particular bijection can be certified to be an isomorphism. The first problem seems much harder to her. Carlos says that he thinks it's actually very hard and that in fact, no one knows whether there is a good algorithm.

Note that the poset shown in Figure 6.13 has the property that there is only one maximal point. Such a point is sometimes called a **one**, denoted not surprisingly as 1. Also, there is only one minimal point, and it is called a **zero**, denoted 0.

The **dual of a partial order** P on a set X is denoted by P^d and is defined by $P^d = \{(y, x) : (x, y) \in P\}$. The **dual of a poset** $\mathbf{P} = (X, P)$ is denoted by $\mathbf{P}^d$ and is defined by $\mathbf{P}^d = (X, P^d)$. A poset $\mathbf{P}$ is **self-dual** if $\mathbf{P} = \mathbf{P}^d$.

A poset $\mathbf{P} = (X, P)$ is **connected** if its comparability graph is connected, i.e., for every $x, y \in X$ with $x \neq y$, there is a finite sequence $x = x_0, x_1, \ldots, x_n = y$ of points from X so that x_i is comparable to x_{i+1} in P for $i = 0, 1, 2, \ldots, n - 1$. A subposet $(Y, P(Y))$ of (X, P) is called a **component** of $\mathbf{P}$ if $(Y, P(Y))$ is connected and there is no subset $Z \subseteq X$ containing Y as a proper subset for which $(Z, P(Z))$ is connected. A one-point component is **trivial** (also, a **loose point** or **isolated point**); components of two or more points are **nontrivial**. Note that a loose point is both a minimal element and a maximal element. Returning to the poset shown in Figure 6.5, we see that it has two components.

It is natural to say that a graph **G** is a **comparability graph** when there is a poset $\mathbf{P} = (X, P)$ whose comparability graph is isomorphic to **G**. For example, we show in Figure 6.15 a graph on 6 vertices which is not a comparability graph. (We leave the task of establishing this claim as an exercise.)

6.2 Additional Concepts for Posets

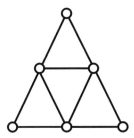

FIGURE 6.15: A Graph Which is Not a Comparability Graph

Similarly, we say that a graph **G** is a **cover graph** when there exists a poset **P** = (X, P) whose cover graph is isomorphic to **G**. Not every graph is a cover graph. In particular, any graph which contains a triangle is not a cover graph. In the exercises at the end of the chapter, you will be asked to construct triangle-free graphs which are not cover graphs—with some hints given as to how to proceed.

Discussion 6.16. Bob is quite taken with graphs associated with posets. He makes the following claims.

1. Only linear orders have paths as cover graphs.

2. A poset and its dual have the same cover graph and the same comparability graph.

3. Any two posets with the same cover graph have the same height and the same width.

4. Any two posets with the same comparability graph have the same height and the same width.

Alice shrugs and says that Bob is right half the time. Which two assertions are correct?

Undeterred, Bob notes that the comparability graph shown in Figure 6.9 is also an incomparability graph (for another poset). He goes on to posit that this is always true, i.e., whenever **G** is the comparability graph of a poset **P**, there is another poset **Q** for which **G** is the incomparability graph of **Q**. Alice says that Bob is right on the first count but she is not so sure about the second. Dave mumbles that they should take a look at the comparability graph of the third poset in Figure 6.8. This graph is not an incomparability graph. But in his typical befuddled manner, Dave doesn't offer any justification for this statement. Can you help Alice and Bob to see why Dave is correct?

Bob is on a roll and he goes on to suggest that it is relatively easy to determine whether a graph is a comparability graph (he read it on the web), but he has a sense

Chapter 6 Partially Ordered Sets

that determining whether a graph is a cover graph might be difficult. Do you think he is right—on either count?

6.3 Dilworth's Chain Covering Theorem and its Dual

In this section, we prove the following theorem of R.P. Dilworth, which is truly one of the classic results of combinatorial mathematics.

Theorem 6.17 (Dilworth's Theorem). *If $\mathbf{P} = (X, P)$ is a poset and* width$(P) = w$, *then there exists a partition* $X = C_1 \cup C_2 \cup \cdots \cup C_w$, *where C_i is a chain for $i = 1, 2, \ldots, w$. Furthermore, there is no chain partition into fewer chains.*

Before proceeding with the proof of Dilworth's theorem in Subsection 6.3.1, we pause to discuss the dual version for partitions into antichains, as it is even easier to prove.

Theorem 6.18 (Dual of Dilworth's Theorem). *If $\mathbf{P} = (X, P)$ is a poset and* height$(P) = h$, *then there exists a partition* $X = A_1 \cup A_2 \cup \cdots \cup A_h$, *where A_i is an antichain for $i = 1, 2, \ldots, h$. Furthermore, there is no partition using fewer antichains.*

Proof. For each $x \in X$, let height(x) be the largest integer t for which there exists a chain
$$x_1 < x_2 < \ldots < x_t$$
with $x = x_t$. Evidently, height$(x) \leq h$ for all $x \in X$. Then for each $i = 1, 2, \ldots, h$, let $A_i = \{x \in X : \text{height}(x) = i\}$. It is easy to see that each A_i is an antichain, as if $x, y \in A_i$ are such that $x < y$, then there is a chain $x_1 < x_2 < \cdots < x_i = x < x_{i+i} = y$, so height$(y) \geq i + 1$. Since height$(P) = h$, there is a maximum chain $C = \{x_1, x_2, \ldots, x_h\}$. If it were possible to partition $\mathbf{P}$ into $t < h$ antichains, then by the Pigeon Hole Principle, one of the antichains would contain two points from C, but this is not possible. □

When $\mathbf{P} = (X, P)$ is a poset, a point $x \in X$ with height$(x) = 1$ is called a **minimal point** of $\mathbf{P}$. We denote the set of all minimal points of a poset $\mathbf{P} = (X, P)$ by min(X, P).[1]

The argument given for the proof of Theorem 6.18 yields an efficient algorithm, one that is defined recursively. Set $\mathbf{P}_0 = \mathbf{P}$. If $\mathbf{P}_i$ has been defined and $\mathbf{P}_i \neq \emptyset$, let $A_i = \min(\mathbf{P}_i)$ and then let $\mathbf{P}_{i+1}$ denote the subposet remaining when A_i is removed from $\mathbf{P}_i$.

In Figure 6.19, we illustrate the antichain partition provided by this algorithm for the 17 point poset from Figure 6.5. The darkened points form a chain of size 5.

---

[1]Since we use the notation $\mathbf{P} = (X, P)$ for a poset, the set of minimal elements can be denoted by min$(\mathbf{P})$ or min(X, P). This convention will be used for all set valued and integer valued functions of posets.

6.3 Dilworth's Chain Covering Theorem and its Dual

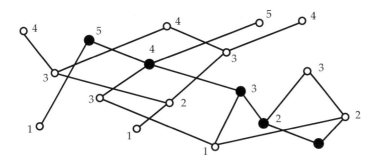

FIGURE 6.19: A Poset of Height 5

Discussion 6.20. Alice claims that it is very easy to find the set of minimal elements of a poset. Do you agree?

Dually, we can speak of the set max(**P**) of **maximal points** of **P**. We can also partition **P** into height(**P**) antichains by recursively removing the set of maximal points.

We pause to remark that when $\mathbf{P} = (X, P)$ is a poset, the set of all chains of **P** is itself partially ordered by inclusion. So it is natural to say that a chain C is **maximal** when there is no chain C' containing C as a proper subset. Also, a chain C is **maximum** when there is no chain C' with $|C| < |C'|$. Of course, a maximum chain is maximal, but maximal chains need not be maximum.

Maximal antichains and **maximum antichains** are defined analogously.

With this terminology, the thrust of Theorem 6.18 is that it is easy to find the height h of a poset as well as a maximum chain C consisting of h points from **P**. Of course, we also get a handy partition of the poset into h antichains.

6.3.1 Proof of Dilworth's Theorem

The argument for Dilworth's theorem is simplified by the following notation. When $\mathbf{P} = (X, P)$ is a poset and $x \in X$, we let $D(x) = \{y \in X : y < x \text{ in } P\}$; $D[x] = \{y \in X : y \leq x \text{ in } P\}$; $U(x) = \{y \in X : y > x \text{ in } P\}$; $U[x] = \{y \in X : y \geq x\}$; and $I(x) = \{y \in X - \{x\} : x \| y \text{ in } P\}$. When $S \subseteq X$, we let $D(S) = \{y \in X : y < x \text{ in } P,$ for some $x \in S\}$ and $D[S] = S \cup D(S)$. The subsets $U(S)$ and $U[S]$ are defined dually. We call $D(x)$, $D[x]$, $D(s)$, and $D[S]$ **down sets**, while $U(x)$, $U[x]$, $U(s)$, and $U[S]$ are **up sets**. Note that when A is a maximal antichain in **P**, the ground set X can be partitioned into pairwise disjoint sets as $X = A \cup D(A) \cup U(A)$.

We are now ready for the proof. Let $\mathbf{P} = (X, P)$ be a poset and let w denote the width of **P**. As in Theorem 6.18, the Pigeon Hole Principle implies that we require at least w chains in any chain partition of **P**. To prove that w suffice, we proceed by

Chapter 6 Partially Ordered Sets

induction on $|X|$, the result being trivial if $|X| = 1$. Assume validity for all posets with $|X| \le k$ and suppose that $\mathbf{P} = (X, P)$ is a poset with $|X| = k + 1$. Without loss of generality, $w > 1$; otherwise, the trivial partition $X = C_1$ satisfies the conclusion of the theorem. Furthermore, we observe that if C is a (nonempty) chain in (X, P), then we may assume that the subposet $(X - C, P(X - C))$ also has width w. To see this, observe that the theorem holds for the subposet, so that if $\text{width}(X - C, P(X - C)) = w' < w$, then we can partition $X - C$ as $X - C = C_1 \cup C_2 \cup \cdots \cup C_{w'}$, so that $X = C \cup C_1 \cup \cdots \cup C_{w'}$ is a partition into $w' + 1$ chains. Since $w' < w$, we know $w' + 1 \le w$, so we have a partition of X into at most w chains. Since any partition of X into chains must use at least w chains, this is exactly the partition we seek.

Choose a maximal point x and a minimal point y with $y \le x$ in P. Then let C be the chain containing only the points x and y. Note that C contains either one or two elements depending on whether x and y are distinct.

Let $Y = X - C$ and $Q = P(Y)$ and let A be a w-element antichain in the subposet (Y, Q). In the partition $X = A \cup D(A) \cup U(A)$, the fact that y is a minimal point while A is a maximal antichain imply that $y \in D(A)$. Similarly, $x \in U(A)$. In particular, this shows that x and y are distinct.

Label the elements of A as $\{a_1, a_2, \ldots, a_w\}$. Note that $U[A] \ne X$ since $y \notin U[A]$, and $D[A] \ne X$ since $x \notin D[A]$. Therefore, we may apply the inductive hypothesis to the subposets of $\mathbf{P}$ determined by $D[A]$ and $U[A]$, respectively, and partition each of these two subposets into w chains:

$$U[A] = C_1 \cup C_2 \cup \cdots \cup C_w \quad \text{and} \quad D[A] = D_1 \cup D_2 \cup \cdots \cup D_w.$$

Without loss of generality, we may assume these chains have been labeled so that $a_i \in C_i \cap D_i$ for each $i = 1, 2, \ldots, w$. However, this implies that

$$X = (C_1 \cup D_1) \cup (C_2 \cup D_2) \cup \cdots \cup (C_w \cup D_w)$$

is the desired partition which in turn completes the proof.

In Figure 6.21, we illustrate Dilworth's chain covering theorem for the poset first introduced in Figure 6.5. The darkened points form a 7-element antichain, while the labels provide a partition into 7 chains.

6.4 Linear Extensions of Partially Ordered Sets

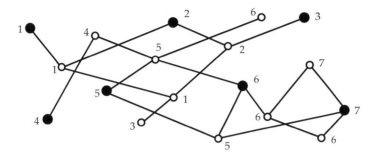

FIGURE 6.21: A POSET OF WIDTH 7

Discussion 6.22. The ever alert Alice notes that the proof given above for Dilworth's theorem does not seem to provide an efficient algorithm for finding the width w of a poset, much less a partition of the poset into w chains. Bob has yet to figure out why listing all the subsets of X is a bad idea. Carlos is sitting quietly listening to their bickering, but finally, he says that a skilled programmer can devise an algorithm from the proof. Students are encouraged to discuss this dilemma—but rest assured that we will return to this issue later in the text.

6.4 Linear Extensions of Partially Ordered Sets

Let $\mathbf{P} = (X, P)$ be a partially ordered set. A linear order L on X is called a **linear extension** (also, a **topological sort**) of P, if $x < y$ in L whenever $x < y$ in P. For example, the table displayed in Figure 6.23 shows that our familiar example $\mathbf{P}_3$ has 11 linear extensions.

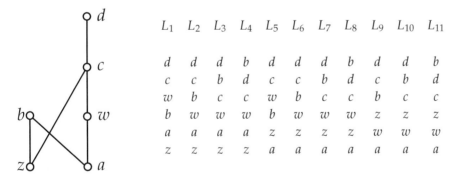

FIGURE 6.23: A POSET AND ITS LINEAR EXTENSIONS

Chapter 6 Partially Ordered Sets

Discussion 6.24. Bob says that he is not convinced that every finite poset has a linear extension. Alice says that it is easy to show that they do. Is she right?

Carlos says that there are subtleties to this question when the ground set X is infinite. You might want to do a web search on the name Szpilrajn and read about his contribution to this issue.

The classical sorting problem studied in all elementary computer science courses is to determine an unknown linear order L of a set X by asking a series of questions of the form: Is $x < y$ in L? All the well known sorting algorithms (bubble sort, merge sort, quick sort, etc.) proceed in this manner.

Here is an important special case: determine an unknown linear extension L of a poset **P** by asking a series of questions of the form: Is $x < y$ in L?

Discussion 6.25. Given the poset $\mathbf{P} = (X, P)$ shown in Figure 6.5 and the problem of determining an unknown linear extension of P, how should Alice decide which question (of the form: Is $x < y$ in L?) to ask?

How would you like to be assigned to count the number of linear extensions of this poset? In general, how hard is it to determine the number of linear extensions of a poset? Could you (and your computer) do this count for a poset on 100,000 points?

6.5 The Subset Lattice

When X is a finite set, the family of all subsets of X, partially ordered by inclusion, forms a **subset lattice**[1]. We illustrate this in Figure 6.26 where we show the lattice of all subsets of $\{1, 2, 3, 4\}$. In this figure, note that we are representing sets by bit strings, and we have further abbreviated the notation by writing strings without commas and parentheses. For a positive integer t, we let $\mathbf{2}^t$ denote the subset lattice consisting of all subsets of $\{1, 2, \ldots, t\}$ ordered by inclusion. Some elementary properties of this poset are:

1. The height is $t + 1$ and all maximal chains have exactly $t + 1$ points.

2. The size of the poset $\mathbf{2}^t$ is 2^t and the elements are partitioned into ranks (antichains) $A_0, A_1, \ldots, A_t$ with $|A_i| = \binom{t}{i}$ for each $i = 0, 1, \ldots, t$.

3. The maximum size of a rank in the subset lattice occurs in the middle, i.e. if $s = \lfloor t/2 \rfloor$, then the largest binomial coefficient in the sequence $\binom{t}{0}, \binom{t}{1}, \binom{t}{2}, \ldots, \binom{t}{t}$ is $\binom{t}{s}$. Note that when t is odd, there are two ranks of maximum size, but when t is even, there is only one.

---

[1] A **lattice** is a special type of poset. You do not have to concern yourself with the definition and can safely replace "lattice" with "poset" as you read this chapter.

6.5 The Subset Lattice

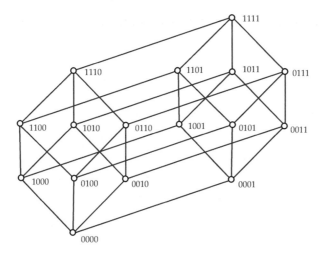

FIGURE 6.26: A SUBSET LATTICE

6.5.1 Sperner's Theorem

For the width of the subset lattice, we have the following classic result of Sperner.

Theorem 6.27 (Sperner's Theorem). *For each $t \geq 1$, the width of the subset lattice 2^t is the maximum size of a rank, i.e.,*

$$\text{width}(2^t) = \binom{t}{\lfloor \frac{t}{2} \rfloor}$$

Proof. The width of the poset 2^t is at least $C(t, \lfloor \frac{t}{2} \rfloor)$ since the set of all $\lfloor \frac{t}{2} \rfloor$-element subsets of $\{1, 2, \ldots, t\}$ is an antichain. We now show that the width of 2^t is at most $C(t, \lfloor \frac{t}{2} \rfloor)$.

Let w be the width of 2^t and let $\{S_1, S_2, \ldots, S_w\}$ be an antichain of size w in this poset, i.e., each S_i is a subset of $\{1, 2, \ldots, t\}$ and if $1 \leq i < j \leq w$, then $S_i \not\subseteq S_j$ and $S_j \not\subseteq S_i$.

For each i, consider the set $\mathcal{S}_i$ of all maximal chains which pass through S_i. It is easy to see that if $|S_i| = k_i$, then $|\mathcal{S}_i| = k_i!(t - k_i)!$. This follows from the observation that to form such a maximum chain beginning with S_i as an intermediate point, you delete the elements of S_i one at a time to form the sets of the lower part of the chain. Also, to form the upper part of the chain, you add the elements not in S_i one at a time.

Note further that if $1 \leq i < j \leq w$, then $\mathcal{S}_i \cap \mathcal{S}_j = \emptyset$, for if there was a maximum chain belonging to both $\mathcal{S}_i$ and $\mathcal{S}_j$, then it would imply that one of S_i and S_j is a subset of the other.

Chapter 6 Partially Ordered Sets

Altogether, there are exactly $t!$ maximum chains in $\mathbf{2}^t$. This implies that

$$\sum_{i=1}^{w} k_i!(t-k_i)! \leq t!.$$

This implies that

$$\sum_{i=1}^{w} \frac{k_i!(t-k_i)!}{t!} = \sum_{i=1}^{w} \frac{1}{\binom{t}{k_i}} \leq 1.$$

It follows that

$$\sum_{i=1}^{i=w} \frac{1}{\binom{t}{\lceil \frac{t}{2} \rceil}} \leq 1$$

Thus

$$w \leq \binom{t}{\lceil \frac{t}{2} \rceil}.$$

$\square$

6.6 Interval Orders

When we discussed Dilworth's Theorem, we commented that the algorithmic aspects would be deferred until later in the text. But there is one important class of orders for which the full solution is easy to obtain.

A poset $\mathbf{P} = (X, P)$ is called an **interval order** if there exists a function I assigning to each element $x \in X$ a closed interval $I(x) = [a_x, b_x]$ of the real line $\mathbb{R}$ so that for all x, $y \in X$, $x < y$ in P if and only if $b_x < a_y$ in $\mathbb{R}$. We call I an **interval representation** of $\mathbf{P}$, or just a **representation** for short. For brevity, whenever we say that I is a representation of an interval order $\mathbf{P} = (X, P)$, we will use the alternate notation $[a_x, b_x]$ for the closed interval $I(x)$. Also, we let $|I(x)|$ denote the **length** of the interval, i.e., $|I(x)| = b_x - a_x$. Returning to the poset $\mathbf{P}_3$, the representation shown in Figure 6.28 shows that it is an interval order. Note that end points of intervals used in a representation need not be distinct. In fact, distinct points x and y from X may satisfy $I(x) = I(y)$. We even allow degenerate intervals, i.e., those of the form $[a, a]$. On the other hand, a representation is said to be **distinguishing** if all intervals are non-degenerate and all end points are distinct. It is relatively easy to see that every interval order has a distinguishing representation.

As we shall soon see, interval orders can be characterized succinctly in terms of forbidden subposets. Before stating this characterization, we need to introduce a bit more notation. By $\mathbf{n}$ (for $n \geq 1$ an integer), we mean the chain with n points. More precisely, we take the ground set to be $\{0, 1, \ldots, n-1\}$ with $i < j$ in $\mathbf{n}$ if and only if $i < j$ in $\mathbb{Z}$. If $\mathbf{P} = (X, P)$ and $\mathbf{Q} = (Y, Q)$ are posets with X and Y disjoint, then $\mathbf{P} + \mathbf{Q}$ is the poset

6.7 Finding a Representation of an Interval Order

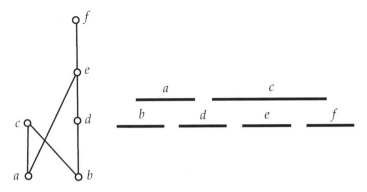

FIGURE 6.28: AN INTERVAL ORDER AND ITS REPRESENTATION

$\mathbf{R} = (X \cup Y, R)$ where the partial order is given by $z \le w$ in R if and only if (a) $z, w \in X$ and $z \le w$ in P or (b) $z, w \in Y$ and $z \le w$ in Q. Thus, $\mathbf{n} + \mathbf{m}$ consists of a chain with n points and a chain with m points and no comparabilities between them. In particular, $\mathbf{2} + \mathbf{2}$ can be viewed as a four-point poset with ground set $\{a, b, c, d\}$ and $a < b$ and $c < d$ as the only relations (other than those required to make the relation reflexive).

Theorem 6.29 (Fishburn's Theorem). *Let $\mathbf{P} = (X, P)$ be a poset. Then $\mathbf{P}$ is an interval order if and only if it excludes $\mathbf{2} + \mathbf{2}$.*

Proof. We show only that an interval order cannot contain a subposet isomorphic to $\mathbf{2} + \mathbf{2}$, deferring the proof in the other direction to the next section. Now suppose that $\mathbf{P} = (X, P)$ is a poset, $\{x, y, z, w\} \subseteq X$ and the subposet determined by these four points is isomorphic to $\mathbf{2} + \mathbf{2}$. We show that $\mathbf{P}$ is not an interval order. Suppose to the contrary that I is an interval representation of $\mathbf{P}$. Without loss of generality, we may assume that $x < y$ and $z < w$ in P. Thus $x \| w$ and $z \| y$ in P. Then $b_x < a_y$ and $b_z < a_w$ in $\mathbb{R}$ so that $a_w \le b_x < a_y \le b_z$, which is a contradiction. □

6.7 Finding a Representation of an Interval Order

In this section, we develop an algorithm for finding an interval representation of an interval order. In fact, this algorithm can be applied to any poset. Either it will find an interval representation or it will find a subposet isomorphic to $\mathbf{2} + \mathbf{2}$. As a consequence, we establish the other half of Fishburn's Theorem.

When $\mathbf{P} = (X, P)$ is an interval order and n is a positive integer, there may be many different ways to represent $\mathbf{P}$ using intervals with integer end points in $[n]$. But there is certainly a least n for which a representation can be found, and here we see that the representation is unique. The discussion will again make use of the notation for down sets and up sets that we introduced prior to the proof of Dilworth's Theorem. As a reminder, we repeat it here. For a poset $\mathbf{P} = (X, P)$ and a subset $S \subset X$, let $D(S) = \{y \in X : \text{there exists some } x \in S \text{ with } y < x \text{ in } P\}$. Also, let $D[S] = D(S) \cup S$. When $|S| = 1$, say $S = \{x\}$, we write $D(x)$ and $D[x]$ rather than $D(\{x\})$ and $D[\{x\}]$. Dually, for a subset $S \subseteq X$, we define $U(S) = \{y \in X : \text{there exists some } x \in X \text{ with } y > x \text{ in } P\}$. As before, set $U[S] = U(S) \cup S$. And when $S = \{x\}$, we just write $U(x)$ for $\{y \in X : x < y \text{ in } P\}$.

Let $\mathbf{P} = (X, P)$ be a poset. We start our procedure by finding the following subsets of the ground set: $\mathcal{D} = \{D(x) : x \in X\}$. We then distinguish two cases. In the first case, there are distinct elements x and y for which $D(x) \not\subseteq D(y)$ and $D(y) \not\subseteq D(x)$. In this case, we choose an element $z \in D(x) - D(y)$ and an element $w \in D(y) - D(x)$. It follows that the four elements in $\{x, y, z, w\}$ form a subposet of $\mathbf{P}$ which is isomorphic to $\mathbf{2 + 2}$.

Our second case is that either $D(x) \subseteq D(y)$ or $D(y) \subseteq D(x)$ for all $x, y \in X$. In this case, we will show that $\mathbf{P}$ is an interval order. Now find the family: $\mathcal{U} = \{U(x) : x \in X\}$. In this case, it is easy to see that we will always have either $U(x) \subseteq U(y)$ or $U(y) \subseteq U(x)$ for all $x, y \in X$.

Let $d = |\mathcal{D}|$. In the exercises, we will provide (actually in doing your homework, *you* will provide) the details for backing up the following statement: $|\mathcal{U}| = |\mathcal{D}|$, so for now we assume that this statement is valid. Label the sets in $\mathcal{D}$ and $\mathcal{U}$ respectively as $D_1, D_2, \ldots, D_d$ and $U_1, U_2, \ldots, U_d$ so that

$$\emptyset = D_1 \subset D_2 \subset D_3 \subset \cdots \subset D_d \quad \text{and}$$

$$U_1 \supset U_2 \cdots \supset U_{d-2} \supset U_{d-1} \supset \cdots \supset U_d = \emptyset.$$

We form an interval representation I of $\mathbf{P}$ by the following rule: For each $x \in X$, set $I(x) = [i, j]$, where $D(x) = D_i$ and $U(x) = U_j$. It is not immediately clear that this rule is legal, i.e., it might happen that applying the rule results in values of i and j for which $j < i$. But again, as a result of the exercises, we will see that this never happens. This collection of exercises is summarized in the following theorem.

Theorem 6.30. *If $\mathbf{P}$ is a poset excluding $\mathbf{2 + 2}$, then the following statements hold.*

1. *The number of down sets equals the number of up sets. That is, $|\mathcal{D}| = |\mathcal{U}|$.*

2. *For each $x \in X$, if $I(x) = [i, j]$, then $i \leq j$ in $\mathbb{R}$.*

3. *For each $x, y \in X$, if $I(x) = [i, j]$ and $I(y) = [k, l]$, then $x < y$ in P if and only if $j < k$ in $\mathbb{R}$.*

4. *The integer d is the least positive integer for which $\mathbf{P}$ has an interval representation using integer end points from $[d]$. This representation is unique.*

6.8 Dilworth's Theorem for Interval Orders

Consider the poset shown in Figure 6.31.

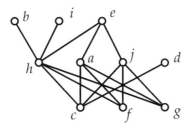

FIGURE 6.31: AN INTERVAL ORDER ON 10 POINTS

Then $d = 5$ with $D_1 = \emptyset$, $D_2 = \{c\}$, $D_3 = \{c, f, g\}$, $D_4 = \{c, f, g, h\}$, and $D_5 = \{a, c, f, g, h, j\}$. Also $U_1 = \{a, b, d, e, h, i, j\}$, $U_2 = \{a, b, e, h, i, j\}$, $U_3 = \{b, e, i\}$, $U_4 = \{e\}$, and $U_5 = \emptyset$. So

$I(a) = [3, 4]$ $I(b) = [4, 5]$ $I(c) = [1, 1]$ $I(d) = [2, 5]$ $I(e) = [5, 5]$
$I(f) = [1, 2]$ $I(g) = [1, 2]$ $I(h) = [3, 3]$ $I(i) = [4, 5]$ $I(j) = [3, 4]$

To illustrate the situation where this process can be used to determine when a poset is not an interval order, consider again the poset shown in Figure 6.31. Erase the line joining points c and d. For the resulting poset, you will then find that $D(j) = \{f, g\}$ and $D(d) = \{c\}$. Therefore, the four points c, d, f and j form a copy of **2 + 2** in this modified poset.

6.8 Dilworth's Theorem for Interval Orders

As remarked previously, we do not yet have an efficient process for determining the width of a poset and a minimum partition into chains. For interval orders, there is indeed a simple way to find both. The explanation is just to establish a connection with coloring of interval graphs as discussed in Chapter 5.

Let $\mathbf{P} = (X, P)$ be an interval order and let $\{[a_x, b_x] : x \in X\}$ be intervals of the real line so that $x < y$ in $\mathbf{P}$ if and only $b_x < a_y$. Then let $\mathbf{G}$ be the interval graph determined by this family of intervals. Note that if x and y are distinct elements of X, then x and y are incomparable in $\mathbf{P}$ if and only if xy is an edge in $\mathbf{G}$. In other words, $\mathbf{G}$ is just the incomparability graph of $\mathbf{P}$.

Recall from Chapter 5 that interval graphs are perfect, i.e., $\chi(\mathbf{G}) = \omega(\mathbf{G})$ for every interval graph $\mathbf{G}$. Furthermore, you can find an optimal coloring of an interval graph by applying first fit to the vertices in a linear order that respects left end points. Such a coloring concurrently determines a partition of $\mathbf{P}$ into chains.

Chapter 6 Partially Ordered Sets

In fact, if you want to skip the part about interval representations, take any linear ordering of the elements as $x_1, x_2, \ldots, x_n$ so that $i < j$ whenever $D(x)$ is a proper subset of $D(y)$. Then apply First Fit with respect to chains. For example, using the 10 point interval order illustrated in Figure 6.31, here is such a labeling:

$$x_1 = g \qquad x_2 = f \qquad x_3 = c \qquad x_4 = d \qquad x_5 = h$$
$$x_6 = a \qquad x_7 = j \qquad x_8 = b \qquad x_9 = i \qquad x_{10} = e$$

Now apply the First Fit algorithm to the points of **P**, in this order, to assign them to chains $C_1, C_2, \ldots$. In other words, assign x_1 to chain C_1. Thereafter if you have assigned points $x_1, x_2, \ldots, x_i$ to chains, then assign x_{i+1} to chain C_j where j is the least positive integer for which x_{i+1} is comparable to x_k whenever $1 \le k \le i$ and x_k has already been assigned to C_j. For example, this rule results in the following chains for the interval order **P** shown in Figure 6.31.

$$C_1 = \{g, h, b\}$$
$$C_2 = \{f, a, e\}$$
$$C_3 = \{c, d\}$$
$$C_4 = \{j\}$$
$$C_5 = \{i\}$$

In this case, it is easy to see that the chain partition is optimal since the width of **P** is 5 and $A = \{a, b, d, i, j\}$ is a 5-element antichain.

However, you should be very careful in applying First Fit to find optimal chain partitions of posets—just as one must be leary of using First Fit to find optimal colorings of graphs.

Example 6.32. The poset on the left side of Figure 6.33 is a height 2 poset on 10 points, and if the poset is partitioned into antichains by applying First Fit and considering the points in the order of their labels, then 5 antichains will be used. Do you see how to extend this poset to force First Fit to use arbitrarily many antichains, while keeping the height of the poset at 2?

On the right side, we show a poset of width 2. Now if this poset is partitioned into chains by applying First Fit and considering the points in the order of their labels, then 4 chains will be used. Do you see how to extend this poset to force First Fit to use arbitrarily many chains while keeping the width of the poset at 2?

Do you get a feeling for why the second problem is a bit harder than the first?

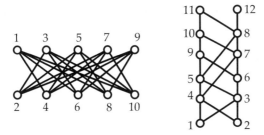

FIGURE 6.33: How First Fit Can Go Wrong

In general, there is always *some* linear order on the ground set of a poset for which First Fit will find an optimal partition into antichains. Also, there is a linear order (in general different from the first) on the ground set for which First Fit will find an optimal partition into chains. However, there is no advantage in searching for such orders, as the algorithms we develop for finding optimal antichain and chain partitions work quite well.

6.9 Discussion

Over coffee, Bob said that he really liked this chapter. "This material was full of cases of very concrete procedures for doing useful things. I like that." Yolanda offered a somewhat different perspective "On the other hand, this last procedure only seems to work with interval orders and we still don't have a clue as to how to find the width of a poset in the general case. This might be very difficult—like the graph coloring problems discussed in the last chapter." Dave weighed in with "Somehow I think there's going to be a fairly efficient process that works for all posets. We may not have all the tools yet, but let's wait a bit."

Not much was said for a while and after a pause, Carlos ventured that there were probably a lot of combinatorial problems for posets that had analogous versions for graphs and in those cases, the poset version would be a bit more complicated, sometime a little bit and sometimes a very big bit. Zori was quiet but she was thinking. These poset structures might even be useful, as she could imagine many settings in which a linear order was impossible or impractical. Maybe there were ways here to earn a few dollars.

6.10 Exercises

1. We say that a relation R on a set X is **symmetric** if $(x, y) \in R$ implies $(y, x) \in R$ for all $x, y \in X$. If $X = \{a, b, c, d, e, f\}$, how many symmetric relations are there on X? How many of these are reflexive?

2. A relation R on a set X is an **equivalence relation** if R is reflexive, symmetric, and transitive. Fix an integer $m \geq 2$. Show that the relation defined on the set $\mathbb{Z}$ of integers by aRb ($a, b \in \mathbb{Z}$) if and only if $a \equiv b \pmod{m}$ is an equivalence relation. (Recall that $a \equiv b \pmod{m}$ means that when dividing a by m and b by m you get the same remainder.)

3. Is the binary relation

$$P = \{(1,1), (2,2), (3,3), (4,4), (1,3), (2,4), (2,5), (4,5), (3,5), (1,5)\}$$

a partial order on the set $X = \{1, 2, 3, 4, 5\}$? If so, discuss what properties you verified and how. If not, list the ordered pairs that must be added to P to make it a partial order or say why it cannot be made a partial order by adding ordered pairs.

4. Draw the diagram of the poset $\mathbf{P} = (X, P)$ where $X = \{1, 2, 3, 5, 6, 10, 15, 30\}$ and $x \leq y$ in P if and only if $x | y$. (Recall that $x | y$ means that x evenly divides y without remainder. Equivalently $x | y$, if and only if $y \equiv 0 \pmod{x}$.)

5. Draw the diagram of the poset $\mathbf{P} = (X, P)$ where

$$X = \{\{1,3,4,5,6\}, \{1,2,4,5,6\}, \{1,2,3,6\}, \{1,2,3\}, \{1,5,6\},$$
$$\{1,3,6\}, \{1,2\}, \{1,6\}, \{3,5\}, \{1\}, \{3\}, \{4\}\}$$

and P is the partial order on X given by the "is a subset of" relationship.

6. A **linear extension** of a poset $\mathbf{P} = (X, P)$ is a total order L on X such that if $x \leq y$ in P, then $x \leq y$ in L. Give linear extension of the three posets shown in Figure 6.8. If you feel very ambitious, try to count the number of linear extensions of the poset on the left side of the figure. Don't list them. Just provide an integer as your answer.

7. Alice and Bob are considering posets $\mathbf{P}$ and $\mathbf{Q}$. They soon realize that $\mathbf{Q}$ is isomorphic to $\mathbf{P}^d$. After 10 minutes of work, they figure out that $\mathbf{P}$ has height 5 and width 3. Bob doesn't want do find the height and width of $\mathbf{Q}$, since he figures it will take (at least) another 10 minutes to answer these questions for $\mathbf{Q}$. Alice says Bob is crazy and that she already knows the height and width of $\mathbf{Q}$. Who's right and why?

8. For this exercise, consider the poset $\mathbf{P}$ in Figure 6.5.

(a) List the maximal elements of **P**.

(b) List the minimal elements of **P**.

(c) Find a maximal chain with two points in **P**.

(d) Find a chain in **P** with three points that is *not* maximal. Say why your chain is not maximal.

(e) Find a maximal antichain with four points in **P**.

9. Find the height h of the poset $\mathbf{P} = (X, P)$ shown below as well as a maximum chain and a partition of X into h antichains using the algorithm from this chapter.

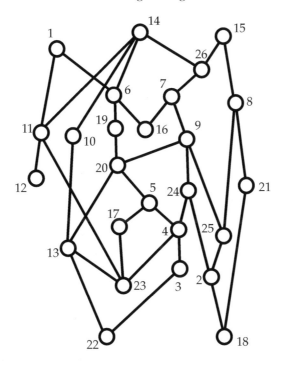

10. For each of the two distinct (up to isomorphism) posets in Figure 6.8, find the width w, an antichain of size w, and a partition of the ground set into w chains.

11. A restaurant chef has designed a new set of dishes for his menu. His set of dishes contains 10 main courses, and he will select a subset of them to place on the menu each night. To ensure variety of main courses for his patrons, he wants to guarantee that a night's menu is neither completely contained in nor completely contains another

Chapter 6 Partially Ordered Sets

night's menu. What is the largest number of menus he can plan using his 10 main courses subject to this requirement?

12. Draw the diagram of the interval order represented in Figure 6.34.

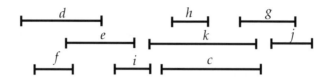

FIGURE 6.34: AN INTERVAL REPRESENTATION

13. Draw the diagram of the interval order represented in Figure 6.35.

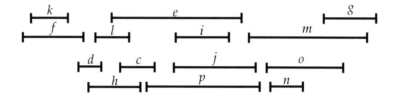

FIGURE 6.35: AN INTERVAL REPRESENTATION

14. Find an interval representation for the poset in Figure 6.36 or give a reason why one does not exist.

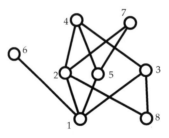

FIGURE 6.36: IS THIS POSET AN INTERVAL ORDER?

15. Find an interval representation for the poset in Figure 6.37 or give a reason why one does not exist.

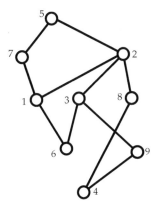

FIGURE 6.37: IS THIS POSET AN INTERVAL ORDER?

16. Find an interval representation for the poset in Figure 6.38 or give a reason why one does not exist.

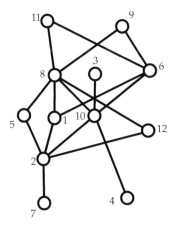

FIGURE 6.38: IS THIS POSET AN INTERVAL ORDER?

17. Find an interval representation for the poset in Figure 6.39 or give a reason why one does not exist.

Chapter 6 Partially Ordered Sets

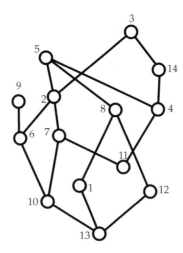

FIGURE 6.39: Is this poset an interval order?

18. Use the First Fit algorithm (ordering by left endpoints) to find the width w of the interval order shown in Figure 6.40 and a partition into w chains. Also give an antichain with w points.

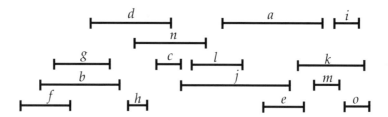

FIGURE 6.40: An interval representation

19. Complete the proof of Theorem 6.30.

Hint. The key idea is to show that if d is the least positive integer for which an interval order **P** has a representation using end points from $\{1, 2, \ldots, n\}$, then every integer i from this set must be both a left end point and a right end point of an interval.

20. Show that every poset is isomorphic to a poset of each of the four types illustrated in Example 6.7.

Hint. For each element x, choose some unique identifying key which is an element/prime/coordinate/observer. Then associate with x a structure that identifies the keys of elements from $D[x]$.

21. The **dimension** of a poset $\mathbf{P} = (X, P)$, denoted $\dim(\mathbf{P})$, is the least t for which P is the intersection of t linear orders on X.

(a) Show that the dimension of a poset $\mathbf{P}$ is the same as the dimension of its dual.

(b) Show that $\mathbf{P}$ is a subposet of $\mathbf{Q}$, then $\dim(\mathbf{P}) \leq \dim(\mathbf{Q})$.

(c) Show that the removal of a point can reduce the dimension by at most 1.

(d) Find the dimension of the posets in Figure 6.8.

(e) Use Dilworth's theorem to show that the dimension of a poset is at most its width.

(f) Use the example on the left side of Figure 6.33 to show that for every $n \geq 2$, there exists a poset $\mathbf{P}_n$ on $2n$ points having width and dimension equal to n.

CHAPTER 7

Inclusion-Exclusion

In this chapter, we study an enumeration technique known as Inclusion-Exclusion. In its simplest case, it is absolutely intuitive. Its power rests in the fact that in many situations, we start with an exponentially large calculation and see it reduce to a manageable size. We focus on three applications that every student of combinatorics should know: (1) counting surjections, (2) derangements, and (3) the Euler ϕ-function.

7.1 Introduction

We start this chapter with an elementary example.

Example 7.1. Let X be the set of 63 students in an applied combinatorics course at a large technological university. Suppose there are 47 computer science majors and 51 male students. Also, we know there are 45 male students majoring in computer science. How many students in the class are female students not majoring in computer science?

Solution. Although the Venn diagrams that you've probably seen drawn many times over the years aren't always the best illustrations (especially if you try to think with some sort of scale), let's use one to get started. In Figure 7.2, we see how the groups in the scenario might overlap. Now we can see that we're after the number of students in the white rectangle but outside the two shaded ovals, which is the female students not majoring in computer science. To compute this, we can start by subtracting the number of male students (the blue region) from the total number of students in the class and then subtracting the number of computer science majors (the yellow region). However, we've now subtracted the overlapping region (the male computer science majors) *twice*, so we must add that number back. Thus, the number of female students in the class who are not majoring in computer science is

$$63 - 51 - 47 + 45 = 10.$$

Chapter 7 Inclusion-Exclusion

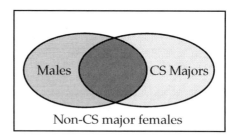

FIGURE 7.2: A VENN DIAGRAM FOR AN APPLIED COMBINATORICS CLASS

Example 7.3. Another type of problem where we can readily see how such a technique is applicable is a generalization of the problem of enumerating integer solutions of equations. In Chapter 2, we discussed how to count the number of solutions to an equation such as

$$x_1 + x_2 + x_3 + x_4 = 100,$$

where $x_1 > 0$, $x_2, x_3 \geq 0$ and $2 \leq x_4 \leq 10$. However, we steered clear of the situation where we add the further restriction that $x_3 \leq 7$. The previous example suggests a way of approaching this modified problem.

First, let's set up the problem so that the lower bound on each variable is of the form $x_i \geq 0$. This leads us to the revised problem of enumerating the integer solutions to

$$x_1' + x_2 + x_3 + x_4' = 97$$

with $x_1', x_2, x_3, x_4' \geq 0$, $x_3 \leq 7$, and $x_4' \leq 8$. (We'll then have $x_1 = x_1' + 1$ and $x_4 = x_4' + 2$ to get our desired solution.) To count the number of integer solutions to this equation with $x_3 \leq 7$ and $x_4' \leq 8$, we must exclude any solution in which $x_3 > 7$ or $x_4' > 8$. There are $C(92, 3)$ solutions with $x_3 > 7$, and the number of solutions in which $x_4' > 8$ is $C(91, 3)$. At this point, it might be tempting to just subtract $C(92, 3)$ and $C(91, 3)$ from $C(100, 3)$, the total number of solutions with all variables nonnegative. However, care is required. If we did that, we would eliminate the solutions with both $x_3 > 7$ *and* $x_4' > 8$ *twice*. To account for this, we notice that there are $C(83, 3)$ solutions with both $x_3 > 7$ and $x_4' > 8$. If we add this number back in after subtracting, we've ensured that the solutions with both $x_3 > 7$ and $x_4' > 8$ are not included in the total count and are not excluded more than once. Thus, the total number of solutions is

$$\binom{100}{3} - \binom{92}{3} - \binom{91}{3} + \binom{83}{3} = 6516.$$

7.1 Introduction

From these examples, you should start to see a pattern emerging that leads to a more general setting. In full generality, we will consider a set X and a family $\mathcal{P} = \{P_1, P_2, \ldots, P_m\}$ of **properties**. We intend that for every $x \in X$ and each $i = 1, 2, \ldots, m$, either x satisfies P_i or it does not. There is no ambiguity. Ultimately, we are interested in determining the number of elements of X which satisfy *none* of the properties in $\mathcal{P}$. In Example 7.1, we could have made property P_1 "is a computer science major" and property P_2 "is male". Then the number of students satisfying *neither P_1 nor P_2* would be the number of female students majoring in something other than computer science, exactly the number we were asked to determine. What would the properties P_1 and P_2 be for Example 7.3?

Let's consider three examples of larger sets of properties. These properties will come back up during the remainder of the chapter as we apply inclusion-exclusion to some more involved situations. Recall that throughout this book, we use the notation $[n]$ for the set $\{1, 2, \ldots, n\}$ when n is a positive integer.

Example 7.4. Let m and n be fixed positive integers and let X consist of all functions from $[n]$ to $[m]$. Then for each $i = 1, 2, \ldots, m$, and each function $f \in X$, we say that f satisfies P_i if there is no j so that $f(j) = i$. In other words, i is not in the image or output of the function f.

As a specific example, suppose that $n = 5$ and $m = 3$. Then the function given by the table below satisfies P_1 but not P_2 or P_3.

i	1	2	3	4	5
$f(i)$	2	3	2	2	3

Example 7.5. Let m be a fixed positive integer and let X consist of all bijections from $[m]$ to $[m]$. Elements of X are called **permutations**. Then for each $i = 1, 2, \ldots, m$, and each permutation $\sigma \in X$, we say that σ satisfies P_i if $\sigma(i) = i$.

For example, the permutation σ of $[5]$ given in by the table below satisfies P_3 and P_5 and no other P_i.

i	1	2	3	4	5
$\sigma(i)$	2	4	3	1	5

Note that in the previous example, we could have said that σ satisfies property P_i if $\sigma(i) \neq i$. But remembering that our goal is to count the number of elements satisfying none of the properties, we would then be counting the number of permutations satisfying $\sigma(i) = i$ for each $i = 1, 2, \ldots, n$, and perhaps we don't need a lot of theory to accomplish this task—the number is one, of course.

Example 7.6. Let m and n be fixed positive integers and let $X = [n]$. Then for each $i = 1, 2, \ldots, m$, and each $j \in X$, we say that j satisfies P_i if i is a divisor of j. Put

Chapter 7 Inclusion-Exclusion

another way, the positive integers that satisfy property P_i are precisely those that are multiples of i.

At first this may appear to be the most complicated of the sets of properties we've discussed thus far. However, being concrete should help clear up any confusion. Suppose that $n = m = 15$. Which properties does 12 satisfy? The divisors of 12 are 1, 2, 3, 4, 6, and 12, so 12 satisfies P_1, P_2, P_3, P_4, P_6, and P_{12}. On the other end of the spectrum, notice that 7 satisfies only properties P_1 and P_7, since those are its only divisors.

7.2 The Inclusion-Exclusion Formula

Now that we have an understanding of what we mean by a property, let's see how we can use this concept to generalize the process we used in the first two examples of the previous section.

Let X be a set and let $\mathcal{P} = \{P_1, P_2, \ldots, P_m\}$ be a family of properties. Then for each subset $S \subseteq [m]$, let $N(S)$ denote the number of elements of X which satisfy property P_i for all $i \in S$. Note that if $S = \emptyset$, then $N(S) = |X|$, as every element of X satisfies every property in S (which contains no actual properties).

Returning for a moment to Example 7.1 with P_1 being "is a computer science major" and P_2 being "is male," we note that $N(\{1\}) = 47$, since there are 47 computer science majors in the class. Also, $N(\{2\}) = 51$ since 51 of the students are male. Finally, $N(\{1, 2\}) = 45$ since there are 45 male computer science majors in the class.

In the examples of the previous section, we subtracted off $N(S)$ for the sets S of size 1 and then added back $N(S)$ for the set of properties of size 2, since we'd subtracted the number of things with both properties (male computer science majors or solutions with both $x_3 > 7$ and $x'_4 > 8$) twice. Symbolically, we determined that the number of objects satisfying none of the properties was

$$N(\emptyset) - N(\{1\}) - N(\{2\}) + N(\{1, 2\}).$$

Suppose that we had three properties P_1, P_2, and P_3. How would we count the number of objects satisfying none of the properties? As before, we start by subtracting for each of P_1, P_2, and P_3. Now we have removed the objects satisfying both P_1 and P_2 twice, so we must add back $N(\{1, 2\})$. similarly, we must do this for the objects satisfying both P_2 and P_3 and both P_1 and P_3. Now let's think about the objects satisfying all three properties. They're counted in $N(\emptyset)$, eliminated *three times* by the $N(\{i\})$ terms, and added back three times by the $N(\{i, j\})$ terms. Thus, they're still being counted! Thus, we must yet subtract $N(\{1, 2, 3\})$ to get the desired number:

$$N(\emptyset) - N(\{1\}) - N(\{2\}) - N(\{3\}) + N(\{1, 2\}) + N(\{2, 3\}) + N(\{1, 3\}) - N(\{1, 2, 3\}).$$

We can generalize this as the following theorem:

Theorem 7.7 (Principle of Inclusion-Exclusion). *The number of elements of X which satisfy none of the properties in $\mathcal{P}$ is given by*

$$\sum_{S \subseteq [m]} (-1)^{|S|} N(S). \tag{7.2.1}$$

Proof. We proceed by induction on the number m of properties. If $m = 1$, then the formula reduces to $N(\emptyset) - N(\{1\})$. This is correct since it says just that the number of elements which do not satisfy property P_1 is the total number of elements minus the number which do satisfy property P_1.

Now assume validity when $m \leq k$ for some $k \geq 1$ and consider the case where $m = k + 1$. Let $X' = \{x \in X : x \text{ satisfies } P_{k+1}\}$ and $X'' = X - X'$ (i.e., X'' is the set of elements that do not satisfy P_{k+1}). Also, let $Q = \{P_1, P_2, \ldots, P_k\}$. Then for each subset $S \subseteq [k]$, let $N'(S)$ count the number of elements of X' satisfying property P_i for all $i \in S$. Also, let $N''(S)$ count the number of elements of X'' satisfying property P_i for each $i \in S$. Note that $N(S) = N'(S) + N''(S)$ for every $S \subseteq [k]$.

Let X'_0 denote the set of elements in X' which satisfy none of the properties in Q (in other words, those that satisfy only P_{k+1} from $\mathcal{P}$), and let X''_0 denote the set of elements of X'' which satisfy none of the properties in Q, and therefore none of the properties in $\mathcal{P}$.

Now by the inductive hypothesis, we know

$$|X'_0| = \sum_{S \subseteq [k]} (-1)^{|S|} N'(S) \quad \text{and} \quad |X''_0| = \sum_{S \subseteq [k]} (-1)^{|S|} N''(S).$$

It follows that

$$|X''_0| = \sum_{S \subseteq [k]} (-1)^{|S|} N''(S) = \sum_{S \subseteq [k]} (-1)^{|S|} (N(S) - N'(S))$$

$$= \sum_{S \subseteq [k]} (-1)^{|S|} N(S) + \sum_{S \subseteq [k]} (-1)^{|S|+1} N(S \cup \{k+1\})$$

$$= \sum_{S \subseteq [k+1]} (-1)^{|S|} N(S). \qquad \square$$

7.3 Enumerating Surjections

As our first example of the power of inclusion-exclusion, consider the following situation: A grandfather has 15 distinct lottery tickets and wants to distribute them to his four grandchildren so that each child gets at least one ticket. In how many ways can he

Chapter 7 Inclusion-Exclusion

make such a distribution? At first, this looks a lot like the problem of enumerating integers solutions of equations, except here the lottery tickets are not identical! A ticket bearing the numbers 1, 3, 10, 23, 47, and 50 will almost surely not pay out the same amount as one with the numbers 2, 7, 10, 30, 31, and 48, so who gets which ticket really makes a difference. Hopefully, you have already recognized that the fact that we're dealing with lottery tickets and grandchildren isn't so important here. Rather, the important fact is that we want to distribute distinguishable objects to distinct entities, which calls for counting functions from one set (lottery tickets) to another (grandchildren). In our example, we don't simply want the total number of functions, but instead we want the number of surjections, so that we can ensure that every grandchild gets a ticket.

For positive integers n and m, let $S(n, m)$ denote the number of surjections from $[n]$ to $[m]$. Note that $S(n, m) = 0$ when $n < m$. In this section, we apply the Inclusion-Exclusion formula to determine a formula for $S(n, m)$. We start by setting X to be the set of all functions from $[n]$ to $[m]$. Then for each $f \in X$ and each $i = 1, 2, \ldots, m$, we say that f satisfies property P_i if i is not in the range of f.

Lemma 7.8. *For each subset $S \subseteq [m]$, $N(S)$ depends only on $|S|$. In fact, if $|S| = k$, then*

$$N(S) = (m - k)^n.$$

Proof. Let $|S| = k$. Then a function f satisfying property P_i for each $i \in S$ is a string of length n from an alphabet consisting of $m - k$ letters. This shows that

$$N(S) = (m - k)^n. \qquad \square$$

Now the following result follows immediately from this lemma by applying the Principle of Inclusion-Exclusion, as there are $C(m, k)$ k-element subsets of $[m]$.

Theorem 7.9. *The number $S(n, m)$ of surjections from $[n]$ to $[m]$ is given by:*

$$S(n, m) = \sum_{k=0}^{m} (-1)^k \binom{m}{k} (m - k)^n.$$

For example,

$$S(5, 3) = \binom{3}{0}(3 - 0)^5 - \binom{3}{1}(3 - 1)^5 + \binom{3}{2}(3 - 2)^5 - \binom{3}{3}(3 - 3)^5$$

$$= 243 - 96 + 3 - 0$$

$$= 150.$$

Returning to our lottery ticket distribution problem at the start of the section, we see that there are $S(15, 4) = 1016542800$ ways for the grandfather to distribute his 15 lottery tickets so that each of the 4 grandchildren receives at least one ticket.

7.4 Derangements

Now let's consider a situation where we can make use of the properties defined in Example 7.5. Fix a positive integer n and let X denote the set of all permutations on $[n]$. A permutation $\sigma \in X$ is called a **derangement** if $\sigma(i) \neq i$ for all $i = 1, 2, \ldots, n$. For example, the permutation σ given below is a derangement, while τ is not.

i	1	2	3	4
$\sigma(i)$	2	4	1	3

i	1	2	3	4
$\tau(i)$	2	4	3	1

If we again let P_i be the property that $\sigma(i) = i$, then the derangements are precisely those permutations which do not satisfy P_i for any $i = 1, 2, \ldots, n$.

Lemma 7.10. *For each subset $S \subseteq [n]$, $N(S)$ depends only on $|S|$. In fact, if $|S| = k$, then*

$$N(S) = (n - k)!$$

Proof. For each $i \in S$, the value $\sigma(i) = i$ is fixed. The other values of σ are a permutation among the remaining $n - k$ positions, and there are $(n - k)!$ of these. $\square$

As before, the principal result of this section follows immediately from the lemma and the Principle of Inclusion-Exclusion.

Theorem 7.11. *For each positive integer n, the number d_n of derangements of $[n]$ satisfies*

$$d_n = \sum_{k=0}^{n} (-1)^k \binom{n}{k} (n-k)!.$$

For example,

$$d_5 = \binom{5}{0}5! - \binom{5}{1}4! + \binom{5}{2}3! - \binom{5}{3}2! + \binom{5}{4}1! - \binom{5}{5}0!$$
$$= 120 - 120 + 60 - 20 + 5 - 1$$
$$= 44.$$

It has been traditional to cast the subject of derangements as a story, called the Hat Check problem. The story belongs to the period of time when men wore top hats. For a fancy ball, 100 men check their top hats with the Hat Check person before entering the ballroom floor. Later in the evening, the mischievous hat check person decides to return hats at random. What is the probability that all 100 men receive a hat other than their own? It turns out that the answer is very close to $1/e$, as the following result shows.

Chapter 7 Inclusion-Exclusion

Theorem 7.12. *For a positive integer n, let d_n denote the number of derangements of $[n]$. Then*

$$\lim_{n \to \infty} \frac{d_n}{n!} = \frac{1}{e}.$$

Equivalently, the fraction of all permutations of $[n]$ that are derangements approaches $1/e$ as n increases.

Proof. It is easy to see that

$$\frac{d_n}{n!} = \frac{\sum_{k=0}^{n}(-1)^k \binom{n}{k}(n-k)!}{n!}$$

$$= \sum_{k=0}^{n}(-1)^k \frac{n!}{k!(n-k)!} \cdot \frac{(n-k)!}{n!}$$

$$= \sum_{k=0}^{n}(-1)^k \frac{1}{k!}.$$

Recall from Calculus that the Taylor series expansion of e^x is given by

$$e^x = \sum_{k=0}^{\infty} \frac{x^k}{k!},$$

and thus the result then follows by substituting $x = -1$. □

Usually we're not as interested in d_n itself as we are in enumerating permutations with certain restrictions, as the following example illustrates.

Example 7.13. Consider the Hat Check problem, but suppose instead of wanting no man to leave with his own hat, we are interested in the number of ways to distribute the 100 hats so that precisely 40 of the men leave with their own hats.

If 40 men leave with their own hats, then there are 60 men who do not receive their own hats. There are $C(100, 60)$ ways to choose the 60 men who will not receive their own hats and d_{60} ways to distribute those hats so that no man receives his own. There's only one way to distribute the 40 hats to the men who must receive their own hats, meaning that there are

$$\binom{100}{60} d_{60} = 4207887349222817212832746283339134521077381515951407221828994446785250023206804862896515376772891317894019692 0$$

such ways to return the hats.

7.5 The Euler ϕ Function

After reading the two previous sections, you're probably wondering why we stated the Principle of Inclusion-Exclusion in such an abstract way, as in those examples $N(S)$ depended only on the size of S and not its contents. In this section, we produce an important example where the value of $N(S)$ *does* depend on S. Nevertheless, we are able to make a reduction to obtain a useful end result. In what follows, let $\mathbb{N}$ denote the set of positive integers.

For a positive integer $n \geq 2$, let

$$\phi(n) = |\{m \in \mathbb{N} : m \leq n, \gcd(m,n) = 1\}|.$$

This function is usually called the **Euler ϕ function** or the **Euler totient function** and has many connections to number theory. We won't focus on the number-theoretic aspects here, only being able to compute $\phi(n)$ efficiently for any n.

For example, $\phi(12) = 4$ since the only numbers from $\{1, 2, \ldots, 12\}$ that are relatively prime to 12 are 1, 5, 7 and 11. As a second example, $\phi(9) = 6$ since 1, 2, 4, 5, 7 and 8 are relatively prime to 9. On the other hand, $\phi(p) = p - 1$ when p is a prime. Suppose you were asked to compute $\phi(321974)$. How would you proceed?

In Chapter 3 we discussed a recursive procedure for determining the greatest common divisor of two integers, and we wrote code for accomplishing this task. Let's assume that we have a function gcd(m,n) that returns the greatest common divisor of the integers m and n. (Conveniently enough, SageMath comes such a function built in.) Then we can calculate $\phi(n)$ with this code snippet:

```
def phi(n):
    answer = 1
    for m in range(2,n):
        if (gcd(m,n) == 1):
            answer += 1
    return(answer)

phi(321974)
```

147744

Running the code above answers almost immediately that $\phi(321974) = 147744$. (As usual, in the web version of the text, you can change the value 321974 to calculate the value of ϕ for other integers. However, if you try to increase the value of n to be too large, you may run into memory issues imposed by the Sage Cell Server used by the text. For instance, attempting to calculate $\phi(319572943)$ results in an error at the time of writing. (You may have better luck running the code directly in the SageMath Cloud or a local installation of SageMath.)

Chapter 7 Inclusion-Exclusion

Given these difficulties, how could we find $\phi(1369122257328767073)$?

Clearly, the program is useless to tackle this beast! It not only iterates $n-2$ times but also invokes a recursion during each iteration. Fortunately, Inclusion-Exclusion comes to the rescue.

Theorem 7.14. *Let $n \geq 2$ be a positive integer and suppose that n has m distinct prime factors: $p_1, p_2, \ldots, p_m$. Then*

$$\phi(n) = n \prod_{i=1}^{m} \frac{p_i - 1}{p_i}. \tag{7.5.1}$$

Our proof of Theorem 7.14 requires the following elementary proposition whose proof we leave as an exercise.

Proposition 7.15. *Let $n \geq 2$, $k \geq 1$, and let $p_1, p_2, \ldots, p_k$ be distinct primes each of which divide n evenly (without remainder). Then the number of integers from $\{1, 2, \ldots, n\}$ which are divisible by each of these k primes is*

$$\frac{n}{p_1 p_2 \cdots p_k}.$$

Proof. We present the argument when $m = 3$. The full result is an easy extension.

In light of Proposition 7.15, the Principle of Inclusion-Exclusion yields:

$$\phi(n) = n - \left(\frac{n}{p_1} + \frac{n}{p_2} + \frac{n}{p_3}\right) + \left(\frac{n}{p_1 p_2} + \frac{n}{p_1 p_3} + \frac{n}{p_2 p_3}\right) - \frac{n}{p_1 p_2 p_3}$$

$$= n \frac{p_1 p_2 p_3 - (p_2 p_3 + p_1 p_3 + p_1 p_2) + (p_3 + p_2 + p_1) - 1}{p_1 p_2 p_3}$$

$$= n \frac{p_1 - 1}{p_1} \frac{p_2 - 1}{p_2} \frac{p_3 - 1}{p_3}. \qquad \square$$

Example 7.16. SageMath reports that

$$1369122257328767073 = (3)^3 (11)(19)^4 (31)^2 (6067)^2$$

is the factorization of 1369122257328767073 into primes. It follows that

$$\phi(1369122257328767073) = 1369122257328767073 \cdot \frac{2}{3} \cdot \frac{10}{11} \cdot \frac{18}{19} \cdot \frac{30}{31} \cdot \frac{6066}{6067}.$$

Thus SageMath quickly reports that

$$\phi(1369122257328767073) = 760615484618973600.$$

Example 7.17. Amanda and Bruce receive the same challenge from their professor, namely to find $\phi(n)$ when

$n =$ 31484972786199768889479107860964368171543984609017931
390019221598516685310407085397223293249028133592410169
3211209710523.

However the Professor also tells Amanda that $n = p_1 p_2$ is the product of two large primes where

$p_1 =$ 470287785858076441566723507866751092927015824834881906763507

and

$p_2 =$ 66948310657809240593656083101755615462290195004890301665 1289.

Is this information of any special value to Amanda? Does it really make her job any easier than Bruce's? Would it level the playing field if the professor told Bruce that n was the product of two primes?

7.6 Discussion

Yolanda said "This seemed like a very short chapter, at least it did to me." Bob agreed "Yes, but the professor indicated that the goal was just provide some key examples. I think he was hinting at more general notions of inversion—although I haven't a clue as to what they might be."

Clearly aggravated, Zori said "I've had all I can stand of this big integer stuff. This won't help me to earn a living." Xing now was uncharacteristically firm in his reply "Zori. You're off base on this issue. Large integers, and specifically integers which are the product of large primes, are central to public key cryptography. If you, or any other citizen, were highly skilled in large integer arithmetic and could quickly factor integers with, say 150 digits, then you would be able to unravel many important secrets. No doubt your life would be in danger."

At first, the group thought that Xing was way out of bounds—but they quickly realized that Xing felt absolutely certain of what he was saying. Zori was quiet for the moment, just reflecting that maybe, just maybe, her skepticism over the relevance of the material in applied combinatorics was unjustified.

7.7 Exercises

1. A school has 147 third graders. The third grade teachers have planned a special treat for the last day of school and brought ice cream for their students. There are

Chapter 7 Inclusion-Exclusion

three flavors: mint chip, chocolate, and strawberry. Suppose that 60 students like (at least) mint chip, 103 like chocolate, 50 like strawberry, 30 like mint chip and strawberry, 40 like mint chip and chocolate, 25 like chocolate and strawberry, and 18 like all three flavors. How many students don't like any of the flavors available?

2. There are 1189 students majoring in computer science at a particular university. They are surveyed about their knowledge of three programming languages: C++, Java, and Python. The survey results reflect that 856 students know C++, 792 know Java, and 692 know Python. Additionally, 639 students know both C++ and Java, 519 know both C++ and Python, and 632 know both Java and Python. There are 488 students who report knowing all three languages. How many students reported that they did not know any of the three programming languages?

3. How many positive integers less than or equal to 100 are divisible by 2? How many positive integers less than or equal to 100 are divisible by 5? Use this information to determine how many positive integers less than or equal to 100 are divisible by *neither* 2 nor 5.

4. How many positive integers less than or equal to 100 are divisible by none of 2, 3, and 5?

5. How many positive integers less than or equal to 1000 are divisible by none of 3, 8, and 25?

6. The State of Georgia is distributing $173 million in funding to Fulton, Gwinnett, DeKalb, Cobb, and Clayton counties (in millions of dollars). In how many ways can this distribution be made, assuming that each county receives at least $1 million, Clayton county receives at most $10 million, and Cobb county receives at most $30 million? (b) What if we add the restriction that Fulton county is to receive at least $5 million (instead of at least $1 million)?

7. How many integer solutions are there to the equation $x_1 + x_2 + x_3 + x_4 = 32$ with $0 \le x_i \le 10$ for $i = 1, 2, 3, 4$?

8. How many integer solutions are there to the inequality

$$y_1 + y_2 + y_3 + y_4 < 184$$

with $y_1 > 0, 0 < y_2 \le 10, 0 \le y_3 \le 17$, and $0 \le y_4 < 19$?

9. A graduate student eats lunch in the campus food court every Tuesday over the course of a 15-week semester. He is joined each week by some subset of a group of six friends from across campus. Over the course of a semester, he ate lunch with each friend 11 times, each pair 9 times, and each triple 6 times. He ate lunch with each

group of four friends 4 times and each group of five friends 4 times. All seven of them ate lunch together only once that semester. Did the graduate student ever eat lunch alone? If so, how many times?

10. A group of 268 students are surveyed about their ability to speak Mandarin, Korean, and Japanese. There are 37 students who do not speak any of the three languages surveyed. Mandarin is spoken by 174 of the students, Japanese is spoken by 139 of the students, and Korean is spoken by 112 of the students. The survey results also reflect that 102 students speak both Mandarin and Japanese, 81 students speak both Mandarin and Korean, and 71 students speak both Japanese and Korean. How many students speak all three languages?

11. As in Example 7.4, let X be the set of functions from $[n]$ to $[m]$ and let a function $f \in X$ satisfy property P_i if there is no j such that $f(j) = i$.

(a) Let the function $f: [8] \to [7]$ be defined by Table 7.18. Does f satisfy property P_2? Why or why not? What about property P_3? List all the properties P_i (with $i \leq 7$) satisfied by f.

(b) Is it possible to define a function $g: [8] \to [7]$ that satisfies no property P_i, $i \leq 7$? If so, give an example. If not, explain why not.

(c) Is it possible to define a function $h: [8] \to [9]$ that satisfies no property P_i, $i \leq 9$? If so, give an example. If not, explain why not.

i	1	2	3	4	5	6	7	8
$f(i)$	4	2	6	1	6	2	4	2

TABLE 7.18: A FUNCTION DEFINED BY A TABLE

12. As in Example 7.5, let X be the set of permutations of $[n]$ and say that $\sigma \in X$ satisfies property P_i if $\sigma(i) = i$.

(a) Let the permutation $\sigma: [8] \to [8]$ be defined by Table 7.19. Does σ satisfy property P_2? Why or why not? What about property P_6? List all the properties P_i (with $i \leq 8$) satisfied by σ.

(b) Give an example of a permutation $\tau: [8] \to [8]$ that satisfies properties P_1, P_4, and P_8 and no other properties P_i with $1 \leq i \leq 8$.

(c) Give an example of a permutation $\pi: [8] \to [8]$ that does not satisfy any property P_i with $1 \leq i \leq 8$.

Chapter 7 Inclusion-Exclusion

i	1	2	3	4	5	6	7	8
$\sigma(i)$	3	1	8	4	7	6	5	2

TABLE 7.19: A PERMUTATION DEFINED BY A TABLE

13. As in Example 7.6, let m and n be positive integers and $X = [n]$. Say that $j \in X$ satisfies property P_i for an i with $1 \leq i \leq m$ if i is a divisor of j.

   (a) Let $m = n = 15$. Does 12 satisfy property P_3? Why or why not? What about property P_5? List the properties P_i with $1 \leq i \leq 15$ that 12 satisfies.

   (b) Give an example of an integer j with $1 \leq j \leq 15$ that satisfies exactly two properties P_i with $1 \leq i \leq 15$.

   (c) Give an example of an integer j with $1 \leq j \leq 15$ that satisfies exactly four properties P_i with $1 \leq i \leq 15$ or explain why such an integer does not exist.

   (d) Give an example of an integer j with $1 \leq j \leq 15$ that satisfies exactly three properties P_i with $1 \leq i \leq 15$ or explain why such an integer does not exist.

14. How many surjections are there from an eight-element set to a six-element set?

15. A teacher has 10 books (all different) that she wants to distribute to John, Paul, Ringo, and George, ensuring that each of them gets at least one book. In how many ways can she do this?

16. A supervisor has nine tasks that must be completed and five employees to whom she may assign them. If she wishes to ensure that each employee is assigned at least one task to perform, how many ways are there to assign the tasks to the employees?

17. A professor is working with six undergraduate research students. He has 12 topics that he would like these students to begin investigating. Since he has been working with Katie for several terms, he wants to ensure that she is given the most challenging topic (and possibly others). Subject to this, in how many ways can he assign the topics to his students if each student must be assigned at least one topic?

18. List all the derangements of $[4]$. (For brevity, you may write a permutation σ as a string $\sigma(1)\sigma(2)\sigma(3)\sigma(4)$.)

19. How many derangements of a nine-element set are there?

20. A soccer team's equipment manager is in a hurry to distribute uniforms to the last six players to show up before a match. Instead of ensuring that each player receives his own uniform, he simply hands a uniform to each of the six players. In how many ways

could he hand out the uniforms so that no player receives his own uniform? (Assume that the six remaining uniforms belong to the last six players to arrive.)

21. A careless payroll clerk is placing employees' paychecks into envelopes that have been pre-labeled. The envelopes are sealed before the clerk realizes he didn't match the names on the paychecks with the names on the envelopes. If there are seven employees, in how many ways could he have placed the paychecks into the envelopes so that exactly three employees receive the correct paycheck?

22. The principle of inclusion-exclusion is not the only approach available for counting derangements. We know that $d_1 = 0$ and $d_2 = 1$. Using this initial information, it is possible to give a recursive form for d_n. In this exercise, we consider two recursions for d_n.

(a) Give a combinatorial argument to prove that the number of derangements satisfies the recursive formula $d_n = (n-1)(d_{n-1} + d_{n-2})$ for $n \geq 2$.

(b) Prove that the number of derangements also satisfies the recursive formula $d_n = nd_{n-1} + (-1)^n$ for $n \geq 2$.

Hint.

(a) For a derangement σ, consider the integer k with $\sigma(k) = 1$. Argue based on the number of choices for k and then whether $\sigma(1) = k$ or not.

(b) You may find it easiest to prove this using the other recursive formula and mathematical induction.

23. Determine $\phi(18)$ by listing the integers it counts as well as by using the formula of Theorem 7.14.

24. Compute $\phi(756)$.

25. Given that $1625190883965792 = (2)^5(3)^4(11)^2(13)(23)^3(181)^2$, compute

$$\phi(1625190883965792).$$

26. Prove Proposition 7.15.

27. At a very small school, there is a class with nine students in it. The students, whom we will denote as A, B, C, D, E, F, G, H, and I, walk from their classroom to the lunchroom in the order $ABCDEFGHI$. (Let's say that A is at the front of the line.) On the way back to their classroom after lunch, they would like to walk in an order so that no student walks immediately behind the same classmate he or she was behind on the way to lunch. (For instance, $ACBDIHGFE$ and $IHGFEDCBA$ would meet their criteria. However, they would not be happy with $CEFGBADHI$ since it contains FG and HI, so G is following F again and I is following H again.)

Chapter 7 Inclusion-Exclusion

(a) One student ponders how many possible ways there would be for them to line up meeting this criterion. Help him out by determining the exact value of this number.

(b) Is this number bigger than, smaller than, or equal to the number of ways they could return so that no student walks in the same position as before (i.e., A is not first, B is not second, ..., and I is not last)?

(c) What fraction (give it as a decimal) of the total number of ways they could line up meet their criterion of no student following immediately behind the same student on the return trip?

CHAPTER 8

Generating Functions

A standard topic of study in first-year calculus is the representation of functions as infinite sums called power series; such a representation has the form $F(x) = \sum_{n=0}^{\infty} a_n x^n$. Perhaps surprisingly these power series can also serve as very powerful enumerative tools. In a combinatorial setting, we consider such power series of this type as another way of encoding the values of a sequence $\{a_n : n \geq 0\}$ indexed by the non-negative integers. The strength of power series as an enumerative technique is that they can be manipulated just like ordinary functions, i.e., they can be added, subtracted and multiplied, and for our purposes, we generally will not care if the power series converges, which anyone who might have found all of the convergence tests studied in calculus daunting will likely find reassuring. However, when we find it convenient to do so, we will use the familiar techniques from calculus and differentiate or integrate them term by term, and for those familiar series that do converge, we will use their representations as functions to facilitate manipulation of the series.

8.1 Basic Notation and Terminology

With a sequence $\sigma = \{a_n : n \geq 0\}$ of real numbers, we associate a "function" $F(x)$ defined by

$$F(x) = \sum_{n=0}^{\infty} a_n x^n.$$

The word "function" is put in quotes as we do not necessarily care about substituting a value of x and obtaining a specific value for $F(x)$. In other words, we consider $F(x)$ as a formal power series and frequently ignore issues of convergence.

It is customary to refer to $F(x)$ as the **generating function** of the sequence σ. As we have already remarked, we are not necessarily interested in calculating $F(x)$ for specific values of x. However, by convention, we take $F(0) = a_0$.

Chapter 8 Generating Functions

Example 8.1. Consider the constant sequence $\sigma = \{a_n : n \geq 0\}$ with $a_n = 1$ for every $n \geq 0$. Then the generating function $F(x)$ of σ is given by

$$F(x) = 1 + x + x^2 + x^3 + x^4 + x^5 + x^6 + \cdots,$$

which is called the **infinite geometric series**.

You may remember that this last expression is the Maclaurin series for the function $F(x) = 1/(1-x)$ and that the series converges when $|x| < 1$. Since we want to think in terms of formal power series, let's see that we can justify the expression

$$\frac{1}{1-x} = 1 + x + x^2 + x^3 + x^4 + x^5 + x^6 + \cdots = \sum_{n=0}^{\infty} x^n$$

without any calculus techniques. Consider the product

$$(1-x)(1 + x + x^2 + x^3 + x^4 + x^5 + x^6 + \cdots)$$

and notice that, since we multiply formal power series just like we multiply polynomials (power series are pretty much polynomials that go on forever), we have that this product is

$$(1 + x + x^2 + x^3 + x^4 + x^5 + x^6 + \cdots) - x(1 + x + x^2 + x^3 + x^4 + x^5 + x^6 + \cdots) = 1.$$

Now we have that

$$(1-x)(1 + x + x^2 + x^3 + x^4 + x^5 + x^6 + \cdots) = 1,$$

or, more usefully, after dividing through by $1-x$,

$$\frac{1}{1-x} = \sum_{n=0}^{\infty} x^n.$$

The method of Example 8.1 can be adapted to address the **finite geometric series** $\sum_{j=0}^{n} x^j$. In that case, we look at

$$(1-x)\sum_{j=0}^{n} x^j = \sum_{j=0}^{n} x^j - \sum_{j=0}^{n} x^{j+1}$$

$$= (1 + x + \cdots + x^n) - (x + x^2 + \cdots x^n + x^{n+1}).$$

$(1-x)(1+x+\cdots+x^n) = (1+x+\cdots+x^n) - (x+x^2+\cdots+x^n+x^{n+1})$

$\Rightarrow (1+x+\cdots+x^n) = \dfrac{1-x^{n+1}}{1-x}$

$\sum_{i=0}^{n} x^i = \dfrac{1-x^{n+1}}{1-x}$

Looking carefully, we see that everying cancels in the final expression except $1 - x^{n+1}$. Dividing both sides by $1 - x$ gives us

$$1 + x + \cdots + x^n = \frac{1 - x^{n+1}}{1 - x} \tag{8.1.1}$$

as the formula for the sum of a finite geometric series.

Example 8.2. Just like you learned in calculus for Maclaurin series, formal power series can be differentiated and integrated term by term. The rigorous mathematical framework that underlies such operations is not our focus here, so take us at our word that this can be done for formal power series without concern about issues of convergence.

To see this in action, consider differentiating the power series of the previous example. This gives

$$\frac{1}{(1-x)^2} = 1 + 2x + 3x^2 + 4x^3 + 5x^4 + 6x^5 + 7x^6 + \cdots = \sum_{n=1}^{\infty} n x^{n-1}.$$

Integration of the series represented by $1/(1 + x) = 1/(1 - (-x))$ yields (after a bit of algebraic manipulation)

$$\log(1 + x) = x - \frac{x^2}{2} + \frac{x^3}{3} - \frac{x^4}{4} + \frac{x^5}{5} - \frac{x^6}{6} + \cdots = \sum_{n=1}^{\infty} (-1)^{n+1} \frac{x^n}{n}.$$

Before you become convinced that we're only going to concern ourselves with generating functions that actually converge, let's see that we can talk about the formal power series

$$F(x) = \sum_{n=0}^{\infty} n! x^n,$$

even though it has radius of convergence 0, i.e., the series $F(x)$ converges only for $x = 0$, so that $F(0) = 1$. Nevertheless, it makes sense to speak of the formal power series $F(x)$ as the generating function for the sequence $\{a_n : n \geq 0\}$, $a_0 = 1$ and a_n is the number of permutations of $\{1, 2, \ldots, n\}$ when $n \geq 1$.

For reference, we state the following elementary result, which emphasizes the form of a product of two power series.

Proposition 8.3. *Let $A(x) = \sum_{n=0}^{\infty} a_n x^n$ and $B(x) = \sum_{n=0}^{\infty} b_n x^n$ be generating functions. Then $A(x)B(x)$ is the generating function of the sequence whose n^{th} term is given by*

$$a_0 b_n + a_1 b_{n-1} + a_2 b_{n-2} + \cdots + a_n b_0 = \sum_{k=0}^{n} a_k b_{n-k}.$$

Chapter 8 Generating Functions

8.2 Another look at distributing apples or folders

A recurring problem so far in this book has been to consider problems that ask about distributing indistinguishable objects (say apples) to distinct entities (say children). We started in Chapter 2 by asking how many ways there were to distribute 40 apples to 5 children so that each child is guaranteed to get at least one apple and saw that the answer was $C(39,4)$. We even saw how to restrict the situation so that one of the children was limited and could receive at most 10 apples. In Chapter 7, we learned how to extend the restrictions so that more than one child had restrictions on the number of apples allowed by taking advantage of the Principle of Inclusion-Exclusion. Before moving on to see how generating functions can allow us to get even more creative with our restrictions, let's take a moment to see how generating functions would allow us to solve the most basic problem at hand.

Example 8.4. We already know that the number of ways to distribute n apples to 5 children so that each child gets at least one apple is $C(n-1,4)$, but it will be instructive to see how we can derive this result using generating functions. Let's start with an even simpler problem: how many ways are there to distribute n apples to *one* child so that each child receives at least one apple? Well, this isn't too hard, there's only one way to do it—give all the apples to the lucky kid! Thus the *sequence* that enumerates the number of ways to do this is $\{a_n : n \geq 1\}$ with $a_n = 1$ for all $n \geq 1$. Then the generating function for this sequence is

$$x + x^2 + x^3 + \cdots = x(1 + x + x^2 + x^3 + \cdots) = \frac{x}{1-x}.$$

How can we get from this fact to the question of five children? Notice what happens when we multiply

$$(x + x^2 + \cdots)(x + x^2 + \cdots)(x + x^2 + \cdots)(x + x^2 + \cdots)(x + x^2 + \cdots).$$

To see what this product represents, first consider how many ways can we get an x^6? We could use the x^2 from the first factor and x from each of the other four, or x^2 from the second factor and x from each of the other four, etc., meaning that the coefficient on x^6 is $5 = C(5,4)$. More generally, what's the coefficient on x^n in the product? In the expansion, we get an x^n for every product of the form $x^{k_1} x^{k_2} x^{k_3} x^{k_4} x^{k_5}$ where $k_1 + k_2 + k_3 + k_4 + k_5 = n$. Returning to the general question here, we're really dealing with distributing n apples to 5 children, and since $k_i > 0$ for $i = 1, 2, \ldots, 5$, we also have the guarantee that each child receives at least one apple, so the product of the generating function for *one* child gives the generating function for *five* children.

Let's pretend for a minute that we didn't know that the coefficients must be $C(n-1,4)$. How could we figure out the coefficients just from the generating function? The

8.2 Another look at distributing apples or folders

generating function we're interested in is $x^5/(1-x)^5$, which you should be able to pretty quickly see satisfies

$$\frac{x^5}{(1-x)^5} = \frac{x^5}{4!}\frac{d^4}{dx^4}\left(\frac{1}{1-x}\right) = \frac{x^5}{4!}\sum_{n=0}^{\infty} n(n-1)(n-2)(n-3)x^{n-4}$$

$$= \sum_{n=0}^{\infty} \frac{n(n-1)(n-2)(n-3)}{4!} x^{n+1} = \sum_{n=0}^{\infty} \binom{n}{4} x^{n+1}.$$

The coefficient on x^n in this series $C(n-1, 4)$, just as we expected.

We could revisit an example from Chapter 7 to see that if we wanted to limit a child to receive at most 4 apples, we would use $(x + x^2 + x^3 + x^4)$ as its generating function instead of $x/(1-x)$, but rather than belabor that here, let's try something a bit more exotic.

Example 8.5. A grocery store is preparing holiday fruit baskets for sale. Each fruit basket will have 20 pieces of fruit in it, chosen from apples, pears, oranges, and grapefruit. How many different ways can such a basket be prepared if there must be at least one apple in a basket, a basket cannot contain more than three pears, and the number of oranges must be a multiple of four?

Solution. In order to get at the number of baskets consisting of 20 pieces of fruit, let's solve the more general problem where each basket has n pieces of fruit. Our method is simple: find the generating function for how to do this with each type of fruit individually and then multiply them. As in the previous example, the product will contain the term x^n for every way of assembling a basket of n pieces of fruit subject to our restrictions. The apple generating function is $x/(1-x)$, since we only want positive powers of x (corresponding to ensuring at least one apple). The generating function for pears is $(1 + x + x^2 + x^3)$, since we can have only zero, one, two, or three pears in basket. For oranges we have $1/(1-x^4) = 1 + x^4 + x^8 + \cdots$, and the unrestricted grapefruit give us a factor of $1/(1-x)$. Multiplying, we have

$$\frac{x}{1-x}(1 + x + x^2 + x^3)\frac{1}{1-x^4}\frac{1}{1-x} = \frac{x}{(1-x)^2(1-x^4)}(1 + x + x^2 + x^3).$$

Now we want to make use of the fact that $(1 + x + x^2 + x^3) = (1 - x^4)/(1-x)$ (by (8.1.1)) to see that our generating function is

$$\frac{x}{(1-x)^3} = \frac{x}{2}\sum_{n=0}^{\infty} n(n-1)x^{n-2} = \sum_{n=0}^{\infty} \frac{n(n-1)}{2} x^{n-1}$$

Chapter 8 Generating Functions

$$= \sum_{n=0}^{\infty} \binom{n}{2} x^{n-1} = \sum_{n=0}^{\infty} \binom{n+1}{2} x^n.$$

Thus, there are $C(n+1, 2)$ possible fruit baskets containing n pieces of fruit, meaning that the answer to the question we originally asked is $C(21, 2) = 210$.

The compact form of the solution to Example 8.5 suggests that perhaps there is a way to come up with this answer without the use of generating functions. Thinking about such an approach would be a good way to solidify your understanding of a variety of the enumerative topics we have already covered.

Example 8.6. Find the number of integer solutions to the equation

$$x_1 + x_2 + x_3 = n$$

($n \geq 0$ an integer) with $x_1 \geq 0$ even, $x_2 \geq 0$, and $0 \leq x_3 \leq 2$.

Solution. Again, we want to look at the generating function we would have if each variable existed individually and take their product. For x_1, we get a factor of $1/(1-x^2)$; for x_2, we have $1/(1-x)$; and for x_3 our factor is $(1 + x + x^2)$. Therefore, the generating function for the number of solutions to the equation above is

$$\frac{1 + x + x^2}{(1-x)(1-x^2)} = \frac{1 + x + x^2}{(1+x)(1-x)^2}.$$

In calculus, when we wanted to integrate a rational function of this form, we would use the method of partial fractions to write it as a sum of "simpler" rational functions whose antiderivatives we recognized. Here, our technique is the same, as we can readily recognize the formal power series for many rational functions. Our goal is to write

$$\frac{1 + x + x^2}{(1+x)(1-x)^2} = \frac{A}{1+x} + \frac{B}{1-x} + \frac{C}{(1-x)^2}.$$

for appropriate constants, A, B, and C. To find the constants, we clear the denominators, giving

$$1 + x + x^2 = A(1-x)^2 + B(1-x^2) + C(1+x).$$

Equating coefficients on terms of equal degree, we have:

$$1 = A + B + C$$
$$1 = -2A + C$$
$$1 = A - B$$

Solving the system, we find $A = 1/4$, $B = -3/4$, and $C = 3/2$. Therefore, our generating function is

$$\frac{1}{4}\frac{1}{1+x} - \frac{3}{4}\frac{1}{1-x} + \frac{3}{2}\frac{1}{(1-x)^2}$$

$$= \frac{1}{4}\sum_{n=0}^{\infty}(-1)^n x^n - \frac{3}{4}\sum_{n=0}^{\infty} x^n + \frac{3}{2}\sum_{n=0}^{\infty} nx^{n-1}.$$

The solution to our question is thus the coefficient on x^n in the above generating function, which is

$$\frac{(-1)^n}{4} - \frac{3}{4} + \frac{3(n+1)}{2},$$

a surprising answer that would not be too easy to come up with via other methods!

The invocation of partial fractions in Example 8.6 is powerful, but solving the necessary system of equations and then hoping that the resulting formal power series have expansions we immediately recognize can be a challenge. If Example 8.6 had not asked about the general case with n on the right-hand side of the equation but instead asked specifically about $n = 30$, you might be wondering if it would just be faster to write some Python code to generate all the solutions or more interesting to huddle up and devise some clever strategy to count them. Fortunately, technology can help us out when working with generating functions. In SageMath, we can use the series() method to get the power series expansion of a given function. The two arguments to series are the variable and the degree of the terms you want to truncate. In the cell below, we ask SageMath to expand the generating function from Example 8.6 by giving us all the terms of degree at most 30 and then collapsing the rest of the series into its form of big-Oh notation, which we discard by storing the output from series() in a polynomial f(x).

```
# Note that SageMath doesn't do well with implied
# multiplication, so use lots of *'s in addition to lots
# of parentheses.
f(x)= ((1+x+x^2)/((1+x)*(1-x)^2)).series(x,31)
f(x)
```

46*x^30 + 44*x^29 + 43*x^28 + 41*x^27 + 40*x^26 + 38*x^25 + 37*x^24 + 35*x^23 + 34*x^22 + 32*x^21 + 31*x^20 + 29*x^19 + 28*x^18 + 26*x^17 + 25*x^16 + 23*x^15 + 22*x^14 + 20*x^13 + 19*x^12 + 17*x^11 + 16*x^10 + 14*x^9 + 13*x^8 + 11*x^7 + 10*x^6 + 8*x^5 + 7*x^4 + 5*x^3 + 4*x^2 + 2*x + 1

Chapter 8 Generating Functions

If all we really want is the coefficient on a specific term, we can use the list() method to turn the polyomial into a list of its coefficients and then index into that list using standard SageMath or Python syntax:

```
coeffs=f(x).list()
coeffs[30]
```

46

Let's see that the answer agrees with what our formula in the solution to Example 8.6 gives us for $n = 30$:

```
n=30
(-1)^n/4 - (3/4) + 3*(n+1)/2
```

46

That's a relief, and so long as we only need a single coefficient, we're now in good shape. But what if we really need a formula for the coefficient on x^n in general? Let's see how we can use SageMath to help us with some of the other steps in Example 8.6. The first thing we'll want is the partial_fraction() method:

```
((1+x+x^2)/((1+x)*(1-x)^2)).partial_fraction()
```

1/4/(x + 1) + 3/4/(x - 1) + 3/2/(x - 1)^2

If you don't like the way that looks, the pretty_print() function can make it easier to read:

```
pretty_print(((1+x+x^2)/((1+x)*(1-x)^2)).partial_fraction())
```

Up to the location of a minus sign, this is what we got by hand, but we get it much faster! From this stage, it's frequently possible to use our knowledge of certain fundamental power series that appear when doing the partial fractions expansion to come up with the general form for the coefficient on an arbitrary term of the power series. To facilitate this, we close this section with an example that illustrates how we can use solutions to counting problems we have already studied in order to figure out the coefficients on generating functions.

Example 8.7. Let n be a positive integer. What is the coefficient on x^k in the generating function

$$\frac{1}{(1-x)^n}?$$

Solution. We have already encountered the case $n = 5$ in the midst of working on Example 8.4, but there we appealed to calculus. Let's take a look at this from the perspective of just counting. The generating function $1/(1-x) = 1 + x + x^2 + \cdots$ encodes the sequence for the number of ways to distribute n apples to *one* child. There's only one way to do that task: give the lucky kid all the apples. Multiplying together a bunch of copies of $1/(1-x)$ then serves to increase the number of children to whom the apples are being distributed, and since each power series being multiplied starts with 1, we are in the situation where the number of apples each child receives must be *nonnegative*. This is therefore a problem from Section 2.5. We have n children and the coefficient on x^k is the number of ways of distributing k apples to them. This requires n artificial apples, so we distribute $k + n$ apples, which determine $k + n - 1$ gaps and we must choose $k - 1$ of them as the locations for dividers. Therefore, we can conclude that

$$\frac{1}{(1-x)^n} = \sum_{k=0}^{\infty} \binom{k+n-1}{k-1} x^k = \sum_{k=0}^{\infty} \binom{k+n-1}{n} x^k.$$

It's possible to arrive at this conclusion using techniques from calculus, but there are a lot of factorials and -1s to monitor, so this combinatorial approach may be less error prone!

8.3 Newton's Binomial Theorem

In Chapter 2, we discussed the binomial theorem and saw that the following formula holds for all integers $p \geq 1$:

$$(1+x)^p = \sum_{n=0}^{p} \binom{p}{n} x^n.$$

You should quickly realize that this formula implies that the generating function for the number of n-element subsets of a p-element set is $(1+x)^p$. The topic of generating functions is what leads us to consider what happens if we encounter $(1+x)^p$ as a generating function with p not a positive integer. It turns out that, by suitably extending the definition of the binomial coefficients to real numbers, we can also extend the binomial theorem in a manner originally discovered by Sir Isaac Newton.

We've seen several expressions that can be used to calculate the binomial coefficients, but in order to extend $C(p, k)$ to real values of p, we will utilize the form

$$\binom{p}{k} = \frac{P(p,k)}{k!},$$

recalling that we've defined $P(p, k)$ recursively as $P(p, 0) = 1$ for all integers $p \geq 0$ and $P(p, k) = pP(p-1, k-1)$ when $p \geq k > 0$ (k an integer). Notice here, however, that the expression for $P(p, k)$ makes sense for any real number p, so long as k is a non-negative integer. We make this definition formal.

Chapter 8 Generating Functions

Definition 8.8. For all real numbers p and nonnegative integers k, the number $P(p,k)$ is defined by

1. $P(p,0) = 1$ for all real numbers p and
2. $P(p,k) = pP(p-1, k-1)$ for all real numbers p and integers $k > 0$.

(Notice that this definition does not require $p \geq k$ as we did with integers.)

We are now prepared to extend the definition of binomial coefficient so that $C(p,k)$ is defined for all real p and nonnegative integer values of k. We do this as follows.

Definition 8.9. For all real numbers p and nonnegative integers k,
$$\binom{p}{k} = \frac{P(p,k)}{k!}.$$

Note that $P(p,k) = C(p,k) = 0$ when p and k are integers with $0 \leq p < k$. On the other hand, we have interesting new concepts such as $P(-5, 4) = (-5)(-6)(-7)(-8)$ and
$$\binom{-7/2}{5} = \frac{(-7/2)(-9/2)(-11/2)(-13/2)(-15/2)}{5!}.$$

With this more general definition of binomial coefficients in hand, we're ready to state Newton's Binomial Theorem for all non-zero real numbers. The proof of this theorem can be found in most advanced calculus books.

Theorem 8.10 (Newton's Binomial Theorem). *For all real p with $p \neq 0$,*
$$(1+x)^p = \sum_{n=0}^{\infty} \binom{p}{n} x^n.$$

Note that the general form reduces to the original version of the binomial theorem when p is a positive integer.

8.4 An Application of the Binomial Theorem

In this section, we see how Newton's Binomial Theorem can be used to derive another useful identity. We begin by establishing a different recursive formula for $P(p,k)$ than was used in our definition of it.

Lemma 8.11. *For each $k \geq 0$, $P(p, k+1) = P(p,k)(p-k)$.*

8.4 An Application of the Binomial Theorem

Proof. When $k = 0$, both sides evaluate to p. Now assume validity when $k = m$ for some non-negative integer m. Then

$$P(p, m+2) = pP(p-1, m+1)$$
$$= p[P(p-1, m)(p-1-m)]$$
$$= [pP(p-1, m)](p-1-m)$$
$$= P(p, m+1)[p-(m+1)].$$

□

Our goal in this section will be to invoke Newton's Binomial Theorem with the exponent $p = -1/2$. To do so in a meaningful manner, we need a simplified expression for $C(-1/2, k)$, which the next lemma provides.

Lemma 8.12. *For each $k \geq 0$,* $\binom{-1/2}{k} = (-1)^k \dfrac{\binom{2k}{k}}{2^{2k}}$.

Proof. We proceed by induction on k. Both sides reduce to 1 when $k = 0$. Now assume validity when $k = m$ for some non-negative integer m. Then

$$\binom{-1/2}{m+1} = \frac{P(-1/2, m+1)}{(m+1)!} = \frac{P(-1/2, m)(-1/2 - m)}{(m+1)m!}$$

$$= \frac{-1/2 - m}{m+1} \binom{-1/2}{m} = (-1)\frac{2m+1}{2(m+1)}(-1)^m \frac{\binom{2m}{m}}{2^{2m}}$$

$$= (-1)^{m+1} \frac{1}{2^{2m}} \frac{(2m+2)(2m+1)}{(2m+2)2(m+1)} \binom{2m}{m} = (-1)^{m+1} \frac{\binom{2m+2}{m+2}}{2^{2m+2}}.$$

□

Theorem 8.13. *The function $f(x) = (1 - 4x)^{-1/2}$ is the generating function of the sequence $\{\binom{2n}{n} : n \geq 0\}$.*

Proof. By Newton's Binomial Theorem and Lemma 8.12, we know that

$$(1-4x)^{-1/2} = \sum_{n=0}^{\infty} \binom{-1/2}{n} (-4x)^n$$

$$= \sum_{n=0}^{\infty} (-1)^n 2^{2n} \binom{-1/2}{n} x^n$$

$$= \sum_{n=0}^{\infty} \binom{2n}{n} x^n.$$

□

We will return to this generating function in Section 9.7, where it will play a role in a seemingly new counting problem that actually is a problem we've already studied in disguise.

Theorem 8.13: $\sum (1-4x)^{-1/2} = \sum \binom{-1/2}{n}(4x)^n = \sum (-1)^n \dfrac{\binom{2n}{n}}{2^{2n}}(4x)^n =$

$\sum (-1)^n \binom{2n}{n} \dfrac{1}{2^{2n}} \cdot 2^{2n} x^n = \sum (-1)^n \binom{2n}{n} x^n$

Chapter 8 Generating Functions

Now recalling Proposition 8.3 about the coefficients in the product of two generating functions, we are able to deduce the following corollary of Theorem 8.13 by squaring the function $f(x) = (1 - 4x)^{-1/2}$.

Corollary 8.14. *For all $n \geq 0$,*

$$2^{2n} = \sum_{k=0}^{n} \binom{2k}{k} \binom{2n-2k}{n-k}.$$

8.5 Partitions of an Integer

A recurring theme in this course has been to count the number of integer solutions to an equation of the form $x_1 + x_2 + \cdots + x_k = n$. What if we wanted to count the number of such solutions but didn't care what k was? How about if we took this new question and required that the x_i be **distinct** (i.e., $x_i \neq x_j$ for $i \neq j$)? What about if we required that each x_i be odd? These certainly don't seem like easy questions to answer at first, but generating functions will allow us to say something very interesting about the answers to the last two questions.

By a **partition** P of an integer, we mean a collection of (not necessarily distinct) positive integers such that $\sum_{i \in P} i = n$. (By convention, we will write the elements of P from largest to smallest.) For example, $2 + 2 + 1$ is a partition of 5. For each $n \geq 0$, let p_n denote the number of partitions of the integer n (with $p_0 = 1$ by convention). Note that $p_8 = 22$ as evidenced by the list in Figure 8.15. Note that there are 6 partitions of 8 into *distinct* parts. Also there are 6 partitions of 8 into *odd* parts. While it might seem that this is a coincidence, it in fact is always the case as Theorem 8.16 states. Before looking at that theorem and its proof, let's think about what a generating function for p_n, the number of partitions of n, would look like. Given a partition of n, we can count how many 1's appear, how many 2's appear, and so on. This suggests a similarity with our fruit basket problems earlier in the chapter, leading to the generating function

$$P(x) = \left(\sum_{m=0}^{\infty} x^m \right) \left(\sum_{m=0}^{\infty} x^{2m} \right) \left(\sum_{m=0}^{\infty} x^{3m} \right) \cdots \left(\sum_{m=0}^{\infty} x^{km} \right) \cdots = \prod_{m=1}^{\infty} \frac{1}{1-x^m}.$$

Here the factor whose sum contains terms x^{km} is accounting for the number of k's in the partition. While $P(x)$ has a quite elegant form, that doesn't mean that it's terribly useful for computing p_n. In fact, providing an asymptotic estimate for p_n was a notoriously difficult problem, finally addressed by Hardy and Ramanujan in 1918. A popular account of this can be found in Robert Kanigel's 1991 book *The Man who Knew Infinity* or the 2016 film with the same title.

8.5 Partitions of an Integer

8 distinct parts	4+1+1+1+1
7+1 distinct parts, odd parts	3+3+2
6+2 distinct parts	3+3+1+1 odd parts
6+1+1	3+2+2+1
5+3 distinct parts, odd parts	3+2+1+1+1
5+2+1 distinct parts	3+1+1+1+1+1 odd parts
5+1+1+1 odd parts	2+2+2+2
4+4	2+2+2+1+1
4+3+1 distinct parts	2+2+1+1+1+1
4+2+2	2+1+1+1+1+1+1
4+2+1+1	1+1+1+1+1+1+1+1 odd parts

FIGURE 8.15: THE PARTITIONS OF 8, NOTING THOSE INTO DISTINCT PARTS AND THOSE INTO ODD PARTS.

Proving the relationship between the number of partitions into distinct parts and the number of partitions into odd parts will involve restricted versions of the generating function $P(x)$ from above.

Theorem 8.16. *For each $n \geq 1$, the number of partitions of n into distinct parts is equal to the number of partitions of n into odd parts.*

Proof. The generating function $D(x)$ for the number of partitions of n into distinct parts is

$$D(x) = \prod_{n=1}^{\infty} (1 + x^n).$$

On the other hand, the generating function $O(x)$ for the number of partitions of n into odd parts is

$$O(x) = \prod_{n=1}^{\infty} \frac{1}{1 - x^{2n-1}}.$$

Chapter 8 Generating Functions

To see that $D(x) = O(x)$, we note that $1-x^{2n} = (1-x^n)(1+x^n)$ for all $n \geq 1$. Therefore,

$$D(x) = \prod_{n=1}^{\infty}(1+x^n) = \prod_{n=1}^{\infty}\frac{1-x^{2n}}{1-x^n} = \frac{\prod_{n=1}^{\infty}(1-x^{2n})}{\prod_{n=1}^{\infty}(1-x^n)}$$

$$= \frac{\prod_{n=1}^{\infty}(1-x^{2n})}{\prod_{n=1}^{\infty}(1-x^{2n-1})\prod_{n=1}^{\infty}(1-x^{2n})} = \prod_{n=1}^{\infty}\frac{1}{1-x^{2n-1}}$$

$$= O(x).$$

$(1-x^{2n-1})(1-x^{2n}) = (1-x^{2n-1} - x^{2n} + x^{2n-1+2n})$

8.6 Exponential generating functions

If we had wanted to be absolutely precise earlier in the chapter, we would have referred to the generating functions we studied as **ordinary generating functions** or even **ordinary power series generating functions**. This is because there are other types of generating functions, based on other types of power series. In this section, we briefly introduce another type of generating function, the **exponential generating function**. While an ordinary generating function has the form $\sum_n a_n x^n$, an exponential generating function is based on the power series for the exponential function e^x. Thus, the exponential generating function for the sequence $\{a_n : n \geq 0\}$ is $\sum_n a_n x^n/n!$. In this section, we will see some ways we can use exponential generating functions to solve problems that we could not tackle with ordinary generating functions. However, we will only scratch the surface of the potential of this type of generating function. We begin with the most fundamental exponential generating function, in analogy with the ordinary generating function $1/(1-x)$ of Example 8.1.

Example 8.17. Consider the constant sequence $1, 1, 1, 1, \ldots$. Then the exponential generating function for this sequence is

$$E(x) = \sum_{n=0}^{\infty}\frac{x^n}{n!}.$$

From calculus, you probably recall that this is the power series for the exponential function e^x, which is why we call this type of generating function an exponential generating function. From this example, we can quickly recognize that the exponential generating function for the number of binary strings of length n is e^{2x} since

$$e^{2x} = \sum_{n=0}^{\infty}\frac{(2x)^n}{n!} = \sum_{n=0}^{\infty}2^n\frac{x^n}{n!}.$$

8.6 Exponential generating functions

In our study of ordinary generating functions earlier in this chapter, we considered examples where quantity (number of apples, etc.) mattered but order did not. One of the areas where exponential generating functions are preferable to ordinary generating functions is in applications where order matters, such as counting strings. For instance, although the bit strings 10001 and 011000 both contain three zeros and two ones, they are not the same strings. On the other hand, two fruit baskets containing two apples and three oranges would be considered equivalent, regardless of how you arranged the fruit. We now consider a couple of examples to illustrate this technique.

Example 8.18. Suppose we wish to find the number of ternary strings in which the number of 0s is even. (There are no restrictions on the number of 1s and 2s.) As with ordinary generating functions, we determine a generating function for each of the digits and multiply them. For 1s and 2s, since we may have any number of each of them, we introduce a factor of e^x for each. For an even number of 0s, we need

$$1 + \frac{x^2}{2!} + \frac{x^4}{4!} + \frac{x^6}{6!} + \cdots = \sum_{n=0}^{\infty} \frac{x^{2n}}{(2n)!}. \quad \text{for 0s}$$

Unlike with ordinary generating functions, we cannot represent this series in a more compact form by simply substituting a function of x into the series for e^y. However, with a small amount of cleverness, we are able to achieve the desired result. To do this, first notice that

$$e^{-x} = 1 - x + \frac{x^2}{2!} - \frac{x^3}{3!} + \cdots = \sum_{n=0}^{\infty} \frac{(-1)^n x^n}{n!}.$$

Thus, when we add the series for e^{-x} to the series for e^x all of the terms with odd powers of x will cancel! We thus find

$$e^x + e^{-x} = 2 + 2\frac{x^2}{2!} + 2\frac{x^4}{4!} + \cdots,$$

which is exactly twice what we need. Therefore, the factor we introduce for 0s is $(e^x + e^{-x})/2$.

Now we have an exponential generating function of

$$\frac{e^x + e^{-x}}{2} e^x e^x = \frac{e^{3x} + e^x}{2} = \frac{1}{2}\left(\sum_{n=0}^{\infty} \frac{3^n x^n}{n!} + \sum_{n=0}^{\infty} \frac{x^n}{n!}\right).$$

To find the number of ternary strings in which the number of 0s is even, we thus need to look at the coefficient on $x^n/n!$ in the series expansion. In doing this, we find that the number of ternary strings with an even number of 0s is $(3^n + 1)/2$.

Chapter 8 Generating Functions

We can also use exponential generating functions when there are bounds on the number of times a symbol appears, such as in the following example.

Example 8.19. How many ternary strings of length n have at least one 0 and at least one 1?

Solution. To ensure that a symbol appears at least once, we need the following exponential generating function

$$x + \frac{x^2}{2!} + \frac{x^3}{3!} + \cdots = \sum_{n=1}^{\infty} \frac{x^n}{n!}.$$

You should notice that this is almost the series for e^x, except it's missing the first term. Thus, $\sum_{n=1}^{\infty} x^n/n! = e^x - 1$. Using this, we now have

$$(e^x - 1)(e^x - 1)e^x = e^{3x} - 2e^{2x} + e^x$$

as the exponential generating function for this problem. Finding the series expansion, we have

$$\sum_{n=0}^{\infty} \frac{3^n x^n}{n!} - 2\sum_{n=0}^{\infty} \frac{2^n x^n}{n!} + \sum_{n=0}^{\infty} \frac{x^n}{n!}.$$

Now we can answer the question by reading off the coefficient on $x^n/n!$, which is $3^n - 2 \cdot 2^n + 1$.

Before proceeding to an additional example, let's take a minute to look at another way to answer the question from the previous example. To count the number of ternary strings of length n with at least one 0 and at least one 1, we can count all ternary strings of length n and use the principle of inclusion-exclusion to eliminate the undesirable strings lacking a 0 and/or a 1. If a ternary string lacks a 0, we're counting all strings made up of 1s and 2s, so there are 2^n strings. Similarly for lacking a 1. However, if we subtract $2 \cdot 2^n$, then we've subtracted the strings that lack both a 0 *and* a 1 twice. A ternary string that has no 0s and no 1s consists only of 2s. There is a single ternary string of length n satisfying this criterion. Thus, we obtain $3^n - 2 \cdot 2^n + 1$ in another way.

Example 8.20. Alice needs to set an eight-digit passcode for her mobile phone. The restrictions on the passcode are a little peculiar. Specifically, it must contain an even number of 0s, at least one 1, and at most three 2s. Bob remarks that although the restrictions are unusual, they don't do much to reduce the number of possible passcodes from the total number of 10^8 eight-digit strings. Carlos isn't convinced that's the case, so he works up an exponential generating function as follows. For the seven digits on which there are no restrictions, a factor of e^{7x} is introduced. To account for an even number of 0s, he uses $(e^x + e^{-x})/2$. For at least one 1, a factor of $e^x - 1$ is required. Finally, $1 + x + x^2/2! + x^3/3!$ accounts for the restriction of at most three 2s. The exponential generating function for the number of n-digit passcodes is thus

$$e^{7x} \frac{e^x + e^{-x}}{2} (e^x - 1)\left(1 + x + \frac{x^2}{2!} + \frac{x^3}{3!}\right).$$

Dave sees this mess written on the whiteboard and groans. He figures they'll be there all day multiplying and making algebra mistakes in trying to find the desired coefficient. Alice points out that they don't really need to find the coefficient on $x^n/n!$ for all n. Instead, she suggests they use SageMath to just find the coefficient on $x^8/8!$.

```
f(x) = (exp(7*x)*((exp(x)+exp(-1*x))/(2))*(exp(x)-1)*\
(1+x+x^2/2+x^3/factorial(3))).series(x,9)
f(x).list()[8]
```

33847837/40320

Since $8! = 40320$, this tells them that there are 33847837 valid passcodes for the mobile phone. A quick calculation shows that Bob was totally off base in claiming that there was no significant reduction in the number of possible strings to use as a passcode. The total number of valid passcodes is only 33.85% of the total number of eight-digit strings!

Exponential generating functions are useful in many other situations beyond enumerating strings. For instance, they can be used to count the number of n-vertex, connected, labeled graphs. However, doing so is beyond the scope of this book. If you are interested in learning much more about generating functions, the book *generatingfunctionology* by Herbert S. Wilf is available online at http://www.math.upenn.edu/~wilf/DownldGF.html.

8.7 Discussion

After studying the proof that the number of partitions of an integer into odd parts is the same as the number of partitions of that integer into distinct parts, Yolanda was beside herself. "Do you guys realize what we just did? We showed that two quantities were equal without saying anything about what those quantities actually were. That's really neat," she said. Nobody said anything for a long time, but after some time Dave said "There might be other instances where you would want to be able to communicate fully, yet hold back on every last detail." Bob said "I don't get it." Alice interjected a comment that was more of question than a statement "Do you mean that parties may want to communicate, while maintaining that the conversation did not occur?" Carlos added "Or maybe they just want to be able to detect whether anyone else was listening." Now Zori was nearly happy. Privacy and security were big ticket items.

8.8 Exercises

Computer algebra systems can be powerful tools for working with generating functions. However, unless an exercise specifically suggests that you use a computer algebra system, we strongly encourage you to solve the problem by hand. This will help you develop a better understanding of how generating functions can be used. You might consider editing the content of the SageMathCells in Section 8.2 to assist with solving problems here where a computer algebra system is suggested, and in some cases, we have included a SageMathCell within the exercise for you to use.

For all exercises in this section, "generating function" should be taken to mean "ordinary generating function." Exponential generating functions are only required in exercises specifically mentioning them.

1. For each *finite* sequence below, give its generating function.

   (a) $1, 4, 6, 4, 1$

   (b) $1, 1, 1, 1, 1, 0, 0, 1$

   (c) $0, 0, 0, 1, 2, 3, 4, 5$

   (d) $1, 1, 1, 1, 1, 1, 1$

   (e) $3, 0, 0, 1, -4, 7$

   (f) $0, 0, 0, 0, 1, 2, -3, 0, 1$

2. For each *infinite* sequence suggested below, give its generating function in closed form, i.e., *not* as an infinite sum. (Use the most obvious choice of form for the general term of each sequence.)

   (a) $0, 1, 1, 1, 1, 1, \ldots$

   (b) $1, 0, 0, 1, 0, 0, 1, 0, 0, 1, 0, 0, 1, \ldots$

   (c) $1, 2, 4, 8, 16, 32, \ldots$

   (d) $0, 0, 0, 0, 1, 1, 1, 1, 1, 1, 1, 1, 1, 1, \ldots$

   (e) $1, -1, 1, -1, 1, -1, 1, -1, 1, -1, \ldots$

   (f) $2^8, 2^7 \binom{8}{1}, 2^6 \binom{8}{2}, \ldots, \binom{8}{8}, 0, 0, 0, \ldots$

   (g) $1, 1, 1, 0, 0, 1, 1, 1, 1, 1, 1, 1, 1, \ldots$

   (h) $0, 0, 0, 1, 2, 3, 4, 5, 6, \ldots$

   (i) $3, 2, 4, 1, 1, 1, 1, 1, 1, \ldots$

   (j) $0, 2, 0, 0, 2, 0, 0, 2, 0, 0, 2, 0, 0, 2, \ldots$

   (k) $6, 0, -6, 0, 6, 0, -6, 0, 6, \ldots$

   (l) $1, 3, 6, 10, 15, \ldots, \binom{n+2}{2}, \ldots$

3. For each generating function below, give a closed form for the nth term of its associated sequence.

(a) $(1+x)^{10}$

(b) $\dfrac{1}{1-x^4}$

(c) $\dfrac{x^3}{1-x^4}$

(d) $\dfrac{1-x^4}{1-x}$

(e) $\dfrac{1+x^2-x^4}{1-x}$

(f) $\dfrac{1}{1-4x}$

(g) $\dfrac{1}{1+4x}$

(h) $\dfrac{x^5}{(1-x)^4}$

(i) $\dfrac{x^2+x+1}{1-x^7}$

(j) $3x^4 + 7x^3 - x^2 + 10 + \dfrac{1}{1-x^3}$

4. Find the coefficient on x^{10} in each of the generating functions below.

(a) $(x^3 + x^5 + x^6)(x^4 + x^5 + x^7)(1 + x^5 + x^{10} + x^{15} + \cdots)$

(b) $(1 + x^3)(x^3 + x^4 + x^5 + \cdots)(x^4 + x^5 + x^6 + x^7 + x^8 + \cdots)$

(c) $(1+x)^{12}$

(d) $\dfrac{x^5}{1-3x^5}$

(e) $\dfrac{1}{(1-x)^3}$

(f) $\dfrac{1}{1-5x^4}$

(g) $\dfrac{x}{1-2x^3}$

(h) $\dfrac{1-x^{14}}{1-x}$

Chapter 8 Generating Functions

5. Find the generating function for the number of ways to create a bunch of n balloons selected from white, gold, and blue balloons so that the bunch contains at least one white balloon, at least one gold balloon, and at most two blue balloons. How many ways are there to create a bunch of 10 balloons subject to these requirements?

6. A volunteer coordinator has 30 identical chocolate chip cookies to distribute to six volunteers. Use a generating function (and computer algebra system) to determine the number of ways she can distribute the cookies so that each volunteer receives at least two cookies and no more than seven cookies.

```
f(x) = x # Generating function on right here
f(x).series(x,n) # Replace n with suitable value
```

7. Consider the inequality
$$x_1 + x_2 + x_3 + x_4 \leq n$$
where $x_1, x_2, x_3, x_4, n \geq 0$ are all integers. Suppose also that $x_2 \geq 2$, x_3 is a multiple of 4, and $0 \leq x_4 \leq 3$. Let c_n be the number of solutions of the inequality subject to these restrictions. Find the generating function for the sequence $\{c_n : n \geq 0\}$ and use it to find a closed formula for c_n.

8. Find the generating function for the number of ways to distribute blank scratch paper to Alice, Bob, Carlos, and Dave so that Alice gets at least two sheets, Bob gets at most three sheets, the number of sheets Carlos receives is a multiple of three, and Dave gets at least one sheet but no more than six sheets of scratch paper. Without finding the power series expansion for this generating function (or using a computer algebra system!), determine the coefficients on x^2 and x^3 in this generating function.

9. What is the generating function for the number of ways to select a group of n students from a class of p students?

10. Using generating functions, find a formula for the number of different types of fruit baskets containing of n pieces of fruit chosen from pomegranates, bananas, apples, oranges, pears, and figs that can be made subject to the following restrictions:
- there are either 0 or 2 pomegranates,
- there is at least 1 banana,
- the number of figs is a multiple of 5,
- there are at most 4 pears, and
- there are no restrictions on the number of apples or oranges.

How many ways are there to form such a fruit basket with $n = 25$ pieces of fruit?

8.8 Exercises

11. Using generating functions, find the number of ways to make change for a 100 dollar bill using only dollar coins and $1, $2, and $5 bills.

```
f(x) = x  # Generating function on right here
pretty_print((f(x)).partial_fraction())
```

Hint. Find the partial fractions expansion for your generating function. Be careful here, as you want a partial fraction expansion in which all coefficients for your denominator polynomials have integer coefficients. The partial_fraction() method in Sage-Math should be useful here, and pretty_print will make it easier to read. Once you have the right partial fractions expansion, you may find the following identity helpful

$$\frac{p(x)}{1 + x + x^2 + \cdots + x^k} = \frac{p(x)(1-x)}{1 - x^{k+1}},$$

where $p(x)$ will be a polynomial in this instance.

12. A businesswoman is traveling in Belgium and wants to buy chocolates for herself, her husband, and their two daughters. A store has dark chocolate truffles (€ 10/box), milk chocolate truffles (€ 8/box), nougat-filled chocolates (€ 5/box), milk chocolate bars (€ 7/bar), and 75% cacao chocolate bars (€ 11/bar). Her purchase is to be subject to the following:

- Only the daughters like dark chocolate truffles, and her purchase must ensure that each daughter gets an equal number of boxes of them (if they get any).
- At least two boxes of milk chocolate truffles must be purchased.
- If she buys any boxes of nougat-filled chocolates, then she buys exactly enough that each family member gets precisely one box of them.
- At most three milk chocolate bars may be purchased.
- There are no restrictions on the number of 75% cacao chocolate bars.

Let s_n be the number of ways the businesswoman can spend exactly € n (*not* buy n items!) at this chocolate shop. Find the generating function for the sequence $\{s_n : n \geq 0\}$. In how many ways can she spend exactly € 100 at the chocolate shop? (A computer algebra system will be helpful for finding coefficients.)

```
f(x) = x  # Generating function on right here
f(x).series(x,n)  # Replace n with suitable value
```

13. Bags of candy are being prepared to distribute to the children at a school. The types of candy available are chocolate bites, peanut butter cups, peppermint candies, and fruit chews. Each bag must contain at least two chocolate bites, an even number of

Chapter 8 Generating Functions

peanut butter cups, and at most six peppermint candies. The fruit chews are available in four different flavors—lemon, orange, strawberry, and cherry. A bag of candy may contain at most two fruit chews, which may be of the same or different flavors. Beyond the number of pieces of each type of candy included, bags of candy are distinguished by using the flavors of the fruit chews included, not just the number. For example, a bag containing two orange fruit chews is different from a bag containing a cherry fruit chew and a strawberry fruit chew, even if the number of pieces of each other type of candy is the same.

(a) Let b_n be the number of different bags of candy with n pieces of candy that can be formed subject to these restrictions. Find the generating function for the sequence $\{b_n : n \geq 0\}$.

(b) Suppose the school has 400 students and the teachers would like to ensure that each student gets a different bag of candy. However, they know there will be fights if the bags do not all contain the same number of pieces of candy. What is the smallest number of pieces of candy they can include in the bags that ensures each student gets a different bag of candy containing the same number of pieces of candy?

14. Make up a combinatorial problem (similar to those found in this chapter) that leads to the generating function

$$\frac{(1 + x^2 + x^4)x^2}{(1 - x)^3(1 - x^3)(1 - x^{10})}.$$

15. Tollbooths in Illinois accept all U.S. coins, including pennies. Carlos has a very large supply of pennies, nickels, dimes, and quarters in his car as he drives on a tollway. He encounters a toll for $ 0.95 and wonders how many different ways he could use his supply of coins to pay the toll without getting change back. (A computer algebra system is probably the best way to get the required coefficient once you have a generating function, since you're not asked for the coefficient on x^n.)

(a) Use a generating function and computer algebra system to determine the number of ways Carlos could pay his $ 0.95 toll by dropping the coins together into the toll bin. (Assume coins of the same denomination cannot be distinguished from each other.)

(b) Suppose that instead of having a bin into which motorists drop the coins to pay their toll, the coins must be inserted one-by-one into a coin slot. In this scenario, Carlos wonders how many ways he could pay the $ 0.95 toll when the order the coins are inserted matters. For instance, in the previous part, the use of three quarters and two dimes would be counted only one time. However, when the coins

8.8 Exercises

must be inserted individually into a slot, there are $10 = C(5,2)$ ways to insert this combination. Use a generating function and computer algebra system to determine the number of ways that Carlos could pay the $ 0.95 toll when considering the order the coins are inserted.

```
f(x) = x  # Generating function on right here
f(x).series(x,n)  # Replace n with suitable value
```

Hint. For part b, you really want an ordinary generating function and not an exponential generating function, despite the fact that order matters. Once you think you have a generating function that works, you might check the coefficients on $x^5, \ldots, x^{10}$ by hand to make sure that you're on the right track.

16. List the partitions of 9. Write a D next to each partition into distinct parts and an O next to each partition into odd parts.

17. Use generating functions to find the number of ways to partition 10 into odd parts.

18. What is the smallest integer that can be partitioned in at least 1000 ways? How many ways can it be partitioned? How many of them are into distinct parts? (A computer algebra system will be helpful for this exercise.)

19. What is the generating function for the number of partitions of an integer into even parts?

20. Find the exponential generating function (in closed form, not as an infinite sum) for each infinite sequence $\{a_n : n \geq 0\}$ whose general term is given below.

(a) $a_n = 5^n$

(b) $a_n = (-1)^n 2^n$

(c) $a_n = 3^{n+2}$

(d) $a_n = n!$

(e) $a_n = n$

(f) $a_n = 1/(n+1)$

21. For each exponential generating function below, give a formula in closed form for the sequence $\{a_n : n \geq 0\}$ it represents.

(a) e^{7x}

(b) $x^2 e^{3x}$

(c) $\dfrac{1}{1+x}$

(d) e^{x^4}

22. Find the coefficient on $x^{10}/10!$ in each of the exponential generating functions below.

179

Chapter 8 Generating Functions

(a) e^{3x}

(b) $\dfrac{e^x - e^{-x}}{2}$

(c) $\dfrac{e^x + e^{-x}}{2}$

(d) $xe^{3x} - x^2$

(e) $\dfrac{1}{1 - 2x}$

(f) e^{x^2}

23. Find the exponential generating function for the number of strings of length n formed from the set $\{a, b, c, d\}$ if there must be at least one a and the number of c's must be even. Find a closed formula for the coefficients of this exponential generating function.

24. Find the exponential generating function for the number of strings of length n formed from the set $\{a, b, c, d\}$ if there must be at least one a and the number of c's must be odd. Find a closed formula for the coefficients of this exponential generating function.

25. Find the exponential generating function for the number of strings of length n formed from the set $\{a, b, c, d\}$ if there must be at least one a, the number of b's must be odd, and the number of d's is either 1 or 2. Find a closed formula for the coefficients of this exponential generating function.

26. Find the exponential generating function for the number of alphanumeric strings of length n formed from the 26 uppercase letters of the English alphabet and 10 decimal digits if

- each vowel must appear at least one time;
- the letter T must appear at least three times;
- the letter Z may appear at most three times;
- each even digit must appear an even number of times; and
- each odd digit must appear an odd number of times.

27. Consider the inequality

$$x_1 + x_2 + x_3 + x_4 \leq n$$

where $x_1, x_2, x_3, x_4, n \geq 0$ are all integers. Suppose also that $x_2 \geq 2$, x_3 is a multiple of 4, and $1 \leq x_4 \leq 3$. Let c_n be the number of solutions of the inequality subject to these restrictions. Find the generating function for the sequence $\{c_n : n \geq 0\}$ and use it to find a closed formula for c_n.

Hint. Yes, this is very close to Exercise 8.8.7. However, the bounds on x_4 are different here. You might try using a computer algebra system to expedite finding the partial fractions expansion, which will have several terms whose power series you can work with quickly. For the term involving $1/(1+x^2)$, work out the series by hand. You may find that your solution to this problem has two parts—one for when n is even and another for when n is odd.

28. Prove Proposition 8.3 about the coefficients in the product of two ordinary generating functions.

CHAPTER 9

Recurrence Equations

We have already seen many examples of recurrence in the definitions of combinatorial functions and expressions. The development of number systems in Appendix B lays the groundwork for recurrence in mathematics. Other examples we have seen include the Collatz sequence of Example 1.8 and the binomial coefficients. In Chapter 3, we also saw how recurrences could arise when enumerating strings with certain restrictions, but we didn't discuss how we might get from a recursive definition of a function to an explicit definition depending only on n, rather than earlier values of the function. In this chapter, we present a more systematic treatment of recurrence with the end goal of finding closed form expressions for functions defined recursively—whenever possible. We will focus on the case of linear recurrence equations. At the end of the chapter, we will also revisit some of what we learned in Chapter 8 to see how generating functions can also be used to solve recurrences.

9.1 Introduction

9.1.1 Fibonacci numbers

One of the most well-known recurrences arises from a simple story. Suppose that a scientist introduces a pair of newborn rabbits to an isolated island. This species of rabbits is unable to reproduce until their third month of life, but after that produces a new pair of rabbits each month. Thus, in the first and second months, there is one pair of rabbits on the island, but in the third month, there are two pairs of rabbits, as the first pair has a pair of offspring. In the fourth month, the original pair of rabbits is still there, as is their first pair of offspring, which are not yet mature enough to reproduce. However, the original pair gives birth to another pair of rabbits, meaning that the island now has three pairs of rabbits. Assuming that there are no rabbit-killing predators on the island and the rabbits have an indefinite lifespan, how many pairs of rabbits are on the island in the tenth month?

Let's see how we can get a recurrence from this story. Let f_n denote the number of pairs rabbits on the island in month n. Thus, $f_1 = 1$, $f_2 = 1$, $f_3 = 2$, and $f_4 = 3$ from our account above. How can we compute f_n? Well, in the nth month we have all the pairs of rabbits that were there during the previous month, which is f_{n-1}; however, some of those pairs of rabbits also reproduce during this month. Only the ones who were born prior to the previous month are able to reproduce during month n, so there are f_{n-2} pairs of rabbits who are able to reproduce, and each produces a new pair of rabbits. Thus, we have that the number of rabbits in month n is $f_n = f_{n-1} + f_{n-2}$ for $n \geq 3$ with $f_1 = f_2 = 1$. The sequence of numbers $\{f_n : n \geq 0\}$ (we take $f_0 = 0$, which satisfies our recurrence) is known as the **Fibonacci sequence** after Leonardo of Pisa, better known as Fibonacci, an Italian mathematician who lived from about 1170 until about 1250. The terms $f_0, f_1, \ldots, f_{20}$ of the Fibonacci sequence are

$$0, 1, 1, 2, 3, 5, 8, 13, 21, 34, 55, 89, 144, 233, 377, 610, 987, 1597, 2584, 4181, 6765.$$

Thus, the answer to our question about the number of pairs of rabbits on the island in the tenth month is 55. That's really easy to compute, but what if we asked for the value of f_{1000} in the Fibonacci sequence? Could you even tell whether the following inequality is true or false—without actually finding f_{1000}?

$$f_{1000} < 2327483838499903832018230933833773932$$

Consider the sequence $\{f_{n+1}/f_n : n \geq 1\}$ of ratios of consecutive terms of the Fibonacci sequence. Figure 9.1 shows these ratios for $n \leq 18$.

$1/1 = 1.0000000000$	$89/55 = 1.6181818182$
$2/1 = 2.0000000000$	$144/89 = 1.6179775281$
$3/2 = 1.5000000000$	$233/144 = 1.6180555556$
$5/3 = 1.6666666667$	$377/233 = 1.6180257511$
$8/5 = 1.6000000000$	$610/377 = 1.6180371353$
$13/8 = 1.6250000000$	$987/610 = 1.6180327869$
$21/13 = 1.6153846154$	$1597/987 = 1.6180344478$
$34/21 = 1.6190476190$	$2584/1597 = 1.6180338134$
$55/34 = 1.6176470588$	$4181/2584 = 1.6180340557$

FIGURE 9.1: THE RATIOS f_{n+1}/f_n FOR $n \leq 18$

The ratios seem to be converging to a number. Can we determine this number? Does this number have anything to do with an explicit formula for f_n (if one even exists)?

Example 9.2. The Fibonacci sequence would not be as well-studied as it is if it were only good for counting pairs of rabbits on a hypothetical island. Here's another instance which again results in the Fibonacci sequence. Let c_n count the number of ways a $2 \times n$ checkerboard can be covered by 2×1 tiles. Then $c_1 = 1$ and $c_2 = 2$ while the recurrence is just $c_{n+2} = c_{n+1} + c_n$, since either the rightmost column of the checkerboard contains a vertical tile (and thus the rest of it can be tiled in c_{n+1} ways) or the rightmost two columns contain two horizontal tiles (and thus the rest of it can be tiled in c_n ways).

9.1.2 Recurrences for strings

In Chapter 3, we saw several times how we could find recurrences that gave us the number of binary or ternary strings of length n when we place a restriction on certain patterns appearing in the string. Let's recall a couple of those types of questions in order to help generate more recurrences to work with.

Example 9.3. Let a_n count the number of binary strings of length n in which no two consecutive characters are 1's. Evidently, $a_1 = 2$ since both binary strings of length 1 are "good." Also, $a_2 = 3$ since only one of the four binary strings of length 2 is "bad,", namely $(1, 1)$. And $a_3 = 5$, since of the 8 binary strings of length 3, the following three strings are "bad":

$$(1,1,0), (0,1,1), (1,1,1).$$

More generally, it is easy to see that the sequence satisfies the recurrence $a_{n+2} = a_{n+1} + a_n$, since we can partition the set of all "good" strings into two sets, those ending in 0 and those ending in 1. If the last bit is 0, then in the first $n + 1$ positions, we can have any "good" string of length $n + 1$. However, if the last bit is 1, then the preceding bit must be 0, and then in the first n positions we can have any "good" string of length n.

As a result, this sequence is just the Fibonacci numbers, albeit offset by 1 position, i.e, $a_n = f_{n+1}$.

Example 9.4. Let t_n count the number of ternary strings in which we never have $(2, 0)$ occurring as a substring in two consecutive positions. Now $t_1 = 3$ and $t_2 = 8$, as of the 9 ternary strings of length 2, exactly one of them is "bad." Now consider the set of all good strings grouped according to the last character. If this character is a 2 or a 1, then the preceding $n + 1$ characters can be any "good" string of length $n + 1$. However, if the last character is a 0, then the first $n + 1$ characters form a good string of length $n + 1$ which does not end in a 2. The number of such strings is $t_{n+1} - t_n$. Accordingly, the recurrence is $t_{n+2} = 3t_{n+1} - t_n$. In particular, $t_3 = 21$.

9.1.3 Lines and regions in the plane

Our next example takes us back to one of the motivating problems discussed in Chapter 1. In Figure 9.5, we show a family of 4 lines in the plane. Each pair of lines intersects and no point in the plane belongs to more than two lines. These lines determine 11 regions.

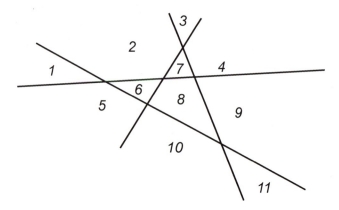

FIGURE 9.5: LINES AND REGIONS

We ask how many regions a family of 1000 lines would determine, given these same restrictions on how the lines intersect. More generally, let r_n denote the number of regions determined by n lines. Evidently, $r_1 = 2$, $r_2 = 4$, $r_3 = 7$ and $r_4 = 11$. Now it is easy to see that we have the recurrence $r_{n+1} = r_n + n + 1$. To see this, choose any one of the $n + 1$ lines and call it l. Line l intersects each of the other lines and since no point in the plane belongs to three or more lines, the points where l intersects the other lines are distinct. Label them consecutively as $x_1, x_2, \ldots, x_n$. Then these points divide line l into $n + 1$ segments, two of which (first and last) are infinite. Each of these segments partitions one of the regions determined by the other n lines into two parts, meaning we have the r_n regions determined by the other n lines and $n + 1$ new regions that l creates.

9.2 Linear Recurrence Equations

What do all of the examples of the previous section have in common? The end result that we were able to achieve is a **linear recurrence**, which tells us how we can compute the nth term of a sequence given some number of previous values (and perhaps also depending nonrecursively on n as well, as in the last example). More precisely a recurrence equation is said to be **linear** when it has the following form

$$c_0 a_{n+k} + c_1 a_{n+k-1} + c_2 a_{n+k-2} + \cdots + c_k a_n = g(n),$$

where $k \geq 1$ is an integer, $c_0, c_1, \ldots, c_k$ are constants with $c_0, c_k \neq 0$, and $g : \mathbb{Z} \to \mathbb{R}$ is a function. (What we have just defined may more properly be called a linear recurrence equation with **constant coefficients**, since we require the c_i to be constants and prohibit them from depending on n. We will avoid this additional descriptor, instead choosing to speak of linear recurrence equations with **nonconstant coefficients** in case we allow the c_i to be functions of n.) A linear equation is **homogeneous** if the function $g(n)$ on the right hand side is the zero function. For example, the Fibonacci sequence satisfies the homogeneous linear recurrence equation

$$a_{n+2} - a_{n+1} - a_n = 0.$$

Note that in this example, $k = 2$, $c_0 = 1$ and $c_k = -1$.

As a second example, the sequence in Example 9.4 satisfies the homogeneous linear recurrence equation

$$t_{n+2} - 3t_{n+1} + t_n = 0.$$

Again, $k = 2$ with $c_0 = c_k = 1$.

On the other hand, the sequence r_n defined in Subsection 9.1.3 satisfies the nonhomogeneous linear recurrence equation

$$r_{n+1} - r_n = n + 1.$$

In this case, $k = 1$, $c_0 = 1$ and $c_k = -1$.

Our immediate goal is to develop techniques for solving linear recurrence equations of both homogeneous and nonhomogeneous types. We will be able to fully resolve the question of solving homogeneous linear recurrence equations and discuss a sort of "guess-and-test" method that can be used to tackle the more tricky nonhomogeneous type.

9.3 Advancement Operators

Much of our motivation for solving recurrence equations comes from an analogous problem in continuous mathematics—differential equations. You don't need to have studied these beasts before in order to understand what we will do in the remainder of this chapter, but if you have, the motivation for how we tackle the problems will be clearer. As their name suggests, differential equations involve derivatives, which we will denote using "operator" notation by Df instead of the Leibniz notation df/dx. In our notation, the second derivative is $D^2 f$, the third is $D^3 f$, and so on. Consider the following example.

Chapter 9 Recurrence Equations

Example 9.6. Solve the equation
$$Df = 3f$$
if $f(0) = 2$.

Solution. Even if you've not studied differential equations, you should recognize that this question is really just asking us to find a function f such that $f(0) = 2$ and its derivative is three times itself. Let's ignore the **initial condition** $f(0) = 2$ for the moment and focus on the meat of the problem. What function, when you take its derivative, changes only by being multiplied by 3? You should quickly think of the function e^{3x}, since $D(e^{3x}) = 3e^{3x}$, which has exactly the property we desire. Of course, for any constant c, the function ce^{3x} also satisfies this property, and this gives us the hook we need in order to satisfy our initial condition. We have $f(x) = ce^{3x}$ and want to find c such that $f(0) = 2$. Now $f(0) = c \cdot 1$, so $c = 2$ does the trick and the solution to this very simple differential equation is $f(x) = 2e^{3x}$.

With differential equations, we apply the differential operator D to differentiable (usually infinitely differentiable) functions. For recurrence equations, we consider the vector space V whose elements are functions from the set $\mathbb{Z}$ of integers to the set $\mathbb{C}$ of complex numbers. We then consider a function $A : V \longrightarrow V$, called the **advancement operator**, and defined by $Af(n) = f(n+1)$. (By various tricks and sleight of hand, we can extend a sequence $\{a_n : n \geq n_0\}$ to be a function whose domain is all of $\mathbb{Z}$, so this technique will apply to our problems.) More generally, $A^p f(n) = f(n+p)$ when p is a positive integer.

Example 9.7. Let $f \in V$ be defined by $f(n) = 7n - 9$. Then we apply the advancement operator polynomial $3A^2 - 5A + 4$ to f with $n = 0$ as follows:
$$(3A^2 - 5A + 4)f(0) = 3f(2) - 5f(1) + 4f(0) = 3(5) - 5(-2) + 4(-9) = -11.$$

As an analogue of Example 9.6, consider the following simple example involving the advancement operator.

Example 9.8. Suppose that the sequence $\{s_n : n \geq 0\}$ satisfies $s_0 = 3$ and $s_{n+1} = 2s_n$ for $n \geq 1$. Find an explicit formula for s_n.

Solution. First, let's write the question in terms of the advancement operator. We can define a function $f(n) = s_n$ for $n \geq 0$, and then the information given becomes that $f(0) = 3$ and
$$Af(n) = 2f(n), \qquad n \geq 0.$$
What function has the property that when we advance it, i.e., evaluate it at $n + 1$, it gives twice the value that it takes at n? The first function that comes into your mind should be 2^n. Of course, just like with our differential equation, for any constant c, $c2^n$

also has this property. This suggests that if we take $f(n) = c2^n$, we're well on our way to solving our problem. Since we know that $f(0) = 3$, we have $f(0) = c2^0 = c$, so $c = 3$. Therefore, $s_n = f(n) = 3 \cdot 2^n$ for $n \geq 0$. This clearly satisfies our initial condition, and now we can check that it also satisfies our advancement operator equation:

$$Af(n) = 3 \cdot 2^{n+1} = 3 \cdot 2 \cdot 2^n = 2 \cdot (3 \cdot 2^n) = 2 \cdot f(n).$$

Before moving on to develop general methods for solving advancement operator equations, let's say a word about why we keep talking in terms of operators and mentioned that we can view any sequence as a function with domain $\mathbb{Z}$. If you've studied any linear algebra, you probably remember learning that the set of all infinitely-differentiable functions on the real line form a vector space and that differentiation is a linear operator on those functions. Our analogy to differential equations holds up just fine here, and functions from $\mathbb{Z}$ to $\mathbb{C}$ form a vector space and A is a linear operator on that space. We won't dwell on the technical aspects of this, and no knowledge of linear algebra is required to understand our development of techniques to solve recurrence equations. However, if you're interested in more placing everything we do on rigorous footing, we discuss this further in Section 9.5.

9.3.1 Constant Coefficient Equations

It is easy to see that a linear recurrence equation can be conveniently rewritten using a polynomial $p(A)$ of the advancement operator:

$$p(A)f = (c_0 A^k + c_1 A^{k-1} + c_2 A^{k-2} + \cdots + c_k)f = g. \tag{9.3.1}$$

In (9.3.1), we intend that $k \geq 1$ is an integer, g is a fixed vector (function) from V, and $c_0, c_1, \ldots, c_k$ are constants with $c_0, c_k \neq 0$. Note that since $c_0 \neq 0$, we can divide both sides by c_0, i.e., we may in fact assume that $c_0 = 1$ whenever convenient to do so.

9.3.2 Roots and Factors

The polynomial $p(A)$ can be analyzed like any other polynomial. It has roots and factors, and although these may be difficult to determine, we know they exist. In fact, if the degree of $p(A)$ is k, we know that over the field of complex numbers, $p(A)$ has k roots, counting multiplicities. Note that since we assume that $c_k \neq 0$, all the roots of the polynomial p are non-zero.

9.3.3 What's Special About Zero?

Why have we limited our attention to recurrence equations of the form $p(A)f = g$ where the constant term in p is non-zero? Let's consider the alternative for a moment.

Chapter 9 Recurrence Equations

Suppose that the constant term of p is zero and that 0 is a root of p of multiplicity m. Then $p(A) = A^m q(A)$ where the constant term of q is non-zero. And the equation $p(A)f = g$ can then be written as $A^m q(A)f = g$. To solve this equation, we consider instead the simpler problem $q(A)f = g$. Then h is a solution of the original problem if and only if the function h' defined by $h'(n) = h(n + m)$ is a solution to the simpler problem. In other words, solutions to the original problem are just translations of solutions to the smaller one, so we will for the most part continue to focus on advancement operator equations where $p(A)$ has nonzero constant term, since being able to solve such problems is all we need in order to solve the larger class of problems.

As a special case, consider the equation $A^m f = g$. This requires $f(n + m) = g(n)$, i.e., f is just a translation of g.

9.4 Solving advancement operator equations

In this section, we will explore some ways of solving advancement operator equations. Some we will make up just for the sake of solving, while others will be drawn from the examples we developed in Section 9.1. Again, readers familiar with differential equations will notice many similarities between the techniques used here and those used to solve linear differential equations with constant coefficients, but we will not give any further examples to make those parallels explicit.

9.4.1 Homogeneous equations

Homogeneous equations, it will turn out, can be solved using very explicit methodology that will work any time we can find the roots of a polynomial. Let's start with another fairly straightforward example.

Example 9.9. Find all solutions to the advancement operator equation

$$(A^2 + A - 6)f = 0. \tag{9.4.1}$$

Solution. Before focusing on finding *all* solutions as we've been asked to do, let's just try to find *some* solution. We start by noticing that here $p(A) = A^2+A-6 = (A+3)(A-2)$. With $p(A)$ factored like this, we realize that we've already solved part of this problem in Example 9.8! In that example, the polynomial of A we encountered was (while not explicitly stated as such there) $A - 2$. The solutions to $(A - 2)f_1 = 0$ are of the form $f_1(n) = c_1 2^n$. What happens if we try such a function here? We have

$$(A + 3)(A - 2)f_1(n) = (A + 3)0 = 0,$$

so that f_1 is a solution to our given advancement operator equation. Of course, it can't be *all* of them. However, it's not hard to see now that $(A+3)f_2 = 0$ has as a solution $f_2(n) = c_2(-3)^n$ by the same reasoning that we used in Example 9.8. Since $(A+3)(A-2) = (A-2)(A+3)$, we see right away that f_2 is also a solution of (9.4.1).

Now we've got two infinite families of solutions to (9.4.1). Do they give us *all* the solutions? It turns out that by combining them, they do in fact give all of the solutions. Consider what happens if we take $f(n) = c_1 2^n + c_2(-3)^n$ and apply $p(A)$ to it. We have

$$\begin{aligned} (A+3)(A-2)f(n) &= (A+3)(c_1 2^{n+1} + c_2(-3)^{n+1} - 2(c_1 2^n + c_2(-3)^n)) \\ &= (A+3)(-5c_2(-3)^n) \\ &= -5c_2(-3)^{n+1} - 15c_2(-3)^n \\ &= 15c_2(-3)^n - 15c_2(-3)^n \\ &= 0. \end{aligned}$$

It's not all that hard to see that since f gives a two-parameter family of solutions to (9.4.1), it gives us all the solutions, as we will show in detail in Section 9.5.

What happened in this example is far from a fluke. If you have an advancement operator equation of the form $p(A)f = 0$ (the constant term of p nonzero) and p has degree k, then the **general solution** of $p(A)f = 0$ will be a k-parameter family (in the previous example, our parameters are the constants c_1 and c_2) whose terms come from solutions to simpler equations arising from the factors of p. We'll return to this thought in a little bit, but first let's look at another example.

Example 9.10. Let's revisit the problem of enumerating ternary strings of length n that do have $(2,0)$ occurring as a substring in two consecutive positions that we encountered in Example 9.4. There we saw that this number satisfies the recurrence equation

$$t_{n+2} = 3t_{n+1} - t_n, \qquad n \geq 1$$

and $t_1 = 3$ and $t_2 = 8$. Before endeavoring to solve this, let's rewrite our recurrence equation as an advancement operator equation. This gives us

$$p(A)t = (A^2 - 3A + 1)t = 0. \tag{9.4.2}$$

The roots of $p(A)$ are $(3 \pm \sqrt{5})/2$. Following the approach of the previous example, our general solution is

$$t(n) = c_1 \left(\frac{3+\sqrt{5}}{2}\right)^n + c_2 \left(\frac{3-\sqrt{5}}{2}\right)^n.$$

Chapter 9 Recurrence Equations

This probably looks suspicious; we're *counting strings* here, so $t(n)$ needs to be a non-negative integer, but the form we've given includes not just fractions but also square roots! However, if you look carefully, you'll see that using the binomial theorem to expand the terms in our expression for $t(n)$ would get rid of all the square roots, so everything is good. (A faster way to convince yourself that this really satisfies (9.4.2) is to mimic the verification we used in the previous example.) Because we have initial values for $t(n)$, we are able to solve for c_1 and c_2 here. Evaluating at $n = 0$ and $n = 1$ we get

$$3 = c_1 + c_2$$
$$8 = c_1 \frac{3 + \sqrt{5}}{2} + c_2 \frac{3 - \sqrt{5}}{2}.$$

A little bit of computation gives

$$c_1 = \frac{7\sqrt{5}}{10} + \frac{3}{2} \quad \text{and} \quad c_2 = -\frac{7\sqrt{5}}{10} + \frac{3}{2}$$

so that

$$t(n) = \left(\frac{7\sqrt{5}}{10} + \frac{3}{2}\right)\left(\frac{3 + \sqrt{5}}{2}\right)^n + \left(-\frac{7\sqrt{5}}{10} + \frac{3}{2}\right)\left(\frac{3 - \sqrt{5}}{2}\right)^n.$$

Example 9.11. Find the general solution to the advancement operator equation

$$(A + 1)(A - 6)(A + 4)f = 0.$$

Solution. By now, you shouldn't be surprised that we immediately make use of the roots of $p(A)$ and have that the solution is

$$f(n) = c_1(-1)^n + c_2 6^n + c_3(-4)^n.$$

By now, you should be able to see most of the pattern for solving homogeneous advancement operator equations. However, the examples we've considered thus far have all had one thing in common: the roots of $p(A)$ were all distinct. Solving advancement operator equations in which this is not the case is not much harder than what we've done so far, but we do need to treat it as a distinct case.

Example 9.12. Find the general solution of the advancement operator equation

$$(A - 2)^2 f = 0.$$

Solution. Here we have the repeated root problem that we mentioned a moment ago. We see immediately that $f_1(n) = c_1 2^n$ is a solution to this equation, but that can't be all,

9.4 Solving advancement operator equations

as we mentioned earlier that we must have a 2-parameter family of solutions to such an equation. You might be tempted to try $f_2(n) = c_2 2^n$ and $f(n) = f_1(n) + f_2(n)$, but then this is just $(c_1 + c_2)2^n$, which is really just a single parameter, $c = c_1 + c_2$.

What can we do to resolve this conundrum? What if we tried $f_2(n) = c_2 n 2^n$? Again, if you're familiar with differential equations, this would be the analogous thing to try, so let's give it a shot. Let's apply $(A - 2)^2$ to this f_2. We have

$$\begin{aligned}(A - 2)^2 f_2(n) &= (A - 2)(c_2(n + 1)2^{n+1} - 2c_2 n 2^n) \\ &= (A - 2)(c_2 2^{n+1}) \\ &= c_2 2^{n+2} - 2c_2 2^{n+1} \\ &= 0.\end{aligned}$$

Since f_2 satisfies our advancement operator equation, we have that the general solution is

$$f(n) = c_1 2^n + c_2 n 2^n.$$

Example 9.13. Consider the recurrence equation

$$f_{n+4} = -2f_{n+3} + 12f_{n+2} + -14f_{n+1} + 5f_n$$

with initial conditions $f_0 = 1$, $f_1 = 2$, $f_2 = 4$, and $f_3 = 4$. Find an explicit formula for f_n.

Solution. We again start by writing the given recurrence equation as an advancement operator equation for a function $f(n)$:

$$(A^4 + 2A^3 - 12A^2 + 14A - 5)f = 0. \tag{9.4.3}$$

Factoring $p(A) = A^4 + 2A^3 - 12A^2 + 14A - 5$ gives $p(A) = (A + 5)(A - 1)^3$. Right away, we see that $f_1(n) = c_1(-5)^n$ is a solution. The previous example should have you convinced that $f_2(n) = c_2 \cdot 1^n = c_2$ and $f_3(n) = c_3 n \cdot 1^n = c_3 n$ are also solutions, and it's not likely to surprise you when we suggest trying $f_4(n) = c_4 n^2$ as another solution. To verify that it works, we see

$$\begin{aligned}(A + 5)(A - 1)^3 f_4(n) &= (A + 5)(A - 1)^2(c_4(n + 1)^2 - c_4 n^2) \\ &= (A + 5)(A - 1)^2(2c_4 n + c_4) \\ &= (A + 5)(A - 1)(2c_4(n + 1) + c_4 - 2c_4 n - c_4) \\ &= (A + 5)(A - 1)(2c_4) \\ &= (A + 5)(2c_4 - 2c_4) \\ &= 0.\end{aligned}$$

Thus, the general solution is

$$f(n) = c_1(-5)^n + c_2 + c_3 n + c_4 n^2.$$

Since we have initial conditions, we see that

$$1 = f(0) = c_1 + c_2$$
$$2 = f(1) = -5c_1 + c_2 + c_3 + c_4$$
$$4 = f(2) = 25c_1 + c_2 + 2c_3 + 4c_4$$
$$4 = f(3) = -125c_1 + c_2 + 3c_3 + 9c_4$$

is a system of equations whose solution gives the values for the c_i. Solving this system gives that the desired solution is

$$f(n) = \frac{1}{72}(-5)^n + \frac{71}{72} + \frac{5}{6}n + \frac{1}{4}n^2.$$

9.4.2 Nonhomogeneous equations

As we mentioned earlier, nonhomogeneous equations are a bit trickier than solving homogeneous equations, and sometimes our first attempt at a solution will not be successful but will suggest a better function to try. Before we're done, we'll revisit the problem of lines in the plane that we've considered a couple of times, but let's start with a more illustrative example.

Example 9.14. Consider the advancement operator equation

$$(A + 2)(A - 6)f = 3^n.$$

Let's try to find the general solution to this, since once we have that, we could find the specific solution corresponding to any given set of initial conditions.

When dealing with nonhomogeneous equations, we proceed in two steps. The reason for this will be made clear in Lemma 9.20, but let's focus on the method for the moment. Our first step is to find the general solution of the homogeneous equation corresponding to the given nonhomogeneous equation. In this case, the homogeneous equation we want to solve is

$$(A + 2)(A - 6)f = 0,$$

for which by now you should be quite comfortable in rattling off a general solution of

$$f_1(n) = c_1(-2)^n + c_2 6^n.$$

9.4 Solving advancement operator equations

Now for the process of actually dealing with the nonhomogeneity of the advancement operator equation. It actually suffices to find *any* solution of the nonhomogeneous equation, which we will call a **particular solution**. Once we have a particular solution f_0 to the equation, the general solution is simply $f = f_0 + f_1$, where f_1 is the general solution to the homogeneous equation.

Finding a particular solution f_0 is a bit trickier than finding the general solution of the homogeneous equation. It's something for which you can develop an intuition by solving lots of problems, but even with a good intuition for what to try, you'll still likely find yourself having to try more than one thing on occasion in order to get a particular solution. What's the best starting point for this intuition? It turns out that the best thing to try is usually (and not terribly surprisingly) something that looks a lot like the right hand side of the equation, but we will want to include one or more new constants to help us actually get a solution. Thus, here we try $f_0(n) = d3^n$. We have

$$(A+2)(A-6)f_0(n) = (A+2)(d3^{n+1} - 6d3^n)$$
$$= (A+2)(-d3^{n+1})$$
$$= -d3^{n+2} - 2d3^{n+1}$$
$$= -5d3^{n+1}.$$

We want f_0 to be a solution to the nonhomogeneous equation, meaning that $(A+2)(A-6)f_0 = 3^n$. This implies that we need to take $d = -1/15$. Now, as we mentioned earlier, the general solution is

$$f(n) = f_0(n) + f_1(n) = -\frac{1}{15}3^n + c_1(-2)^n + c_2 6^n.$$

We leave it to you to verify that this does satisfy the given equation.

You hopefully noticed that in the previous example, we said that the first guess to try for a particular solution looks a lot like right hand side of the equation, rather than exactly like. Our next example will show why we can't always take something that matches exactly.

Example 9.15. Find the solution to the advancement operator equation

$$(A+2)(A-6)f = 6^n$$

if $f(0) = 1$ and $f(1) = 5$.

Solution. The corresponding homogeneous equation here is the same as in the previous example, so its general solution is again $f_1(n) = c_1(-2)^n + c_2 6^n$. Thus, the real work here is finding a particular solution f_0 to the given advancement operator equation. Let's just try what our work on the previous example would suggest here, namely

Chapter 9 Recurrence Equations

$f_0(n) = d6^n$. Applying the advancement operator polynomial $(A+2)(A-6)$ to f_0 then gives, uh, well, zero, since $(A-6)(d6^n) = d6^{n+1} - 6d6^n = 0$. Huh, that didn't work out so well. However, we can take a cue from how we tackled homogeneous advancement operator equations with repeated roots and introduce a factor of n. Let's try $f_0(n) = dn6^n$. Now we have

$$(A+2)(A-6)(dn6^n) = (A+2)(d(n+1)6^{n+1} - 6dn6^n)$$
$$= (A+2)d6^{n+1}$$
$$= d6^{n+2} + 2d6^{n+1}$$
$$= 6^n(36d + 12d) = 48d6^n.$$

We want this to be equal to 6^n, so we have $d = 1/48$. Therefore, the general solution is

$$f(n) = \frac{1}{48}n6^n + c_1(-2)^n + c_26^n.$$

All that remains is to use our initial conditions to find the constants c_1 and c_2. We have that they satisfy the following pair of equations:

$$1 = c_1 + c_2$$
$$5 = \frac{1}{8} - 2c_1 + 6c_2$$

Solving these, we arrive at the desired solution, which is

$$f(n) = \frac{1}{48}n6^n + \frac{9}{64}(-2)^n + \frac{55}{64}6^n.$$

What's the lesson we should take away from this example? When making a guess at a particular solution of a nonhomogeneous advancement operator equation, it does us no good to use any terms that are also solutions of the corresponding homogeneous equation, as they will be annihilated by the advancement operator polynomial. Let's see how this comes into play when finally resolving one of our longstanding examples.

Example 9.16. We're now ready to answer the question of how many regions are determined by n lines in the plane in general position as we discussed in Subsection 9.1.3. We have the recurrence equation

$$r_{n+1} = r_n + n + 1,$$

which yields the nonhomogeneous advancement operator equation $(A-1)r = n+1$. As usual, we need to start with the general solution to the corresponding homogeneous equation. This solution is $f_1(n) = c_1$. Now our temptation is to try $f_0(n) = d_1n + d_2$ as a particular solution. However since the constant term there is a solution to the homogeneous equation, we need a bit more. Let's try increasing the powers of n by 1, giving $f_0(n) = d_1n^2 + d_2n$. Now we have

$$\begin{aligned}(A-1)(d_1n^2 + d_2n) &= d_1(n+1)^2 + d_2(n+1) - d_1n^2 - d_2n \\ &= 2d_1n + d_1 + d_2.\end{aligned}$$

This tells us that we need $d_1 = 1/2$ and $d_2 = 1/2$, giving $f_0(n) = n^2/2 + n/2$. The general solution is then

$$f(n) = c_1 + \frac{n^2 + n}{2}.$$

What is our initial condition here? Well, one line divides the plane into two regions, so $f(1) = 2$. On the other hand, $f(1) = c_1 + 1$, so $c_1 = 1$ and thus

$$f(n) = 1 + \frac{n^2 + n}{2} = \binom{n+1}{2} + 1$$

is the number of regions into which the plane is divided by n lines in general position.

We conclude this section with one more example showing how to deal with a nonhomogeneous advancement operator equation in which the right hand side is of "mixed type".

Example 9.17. Give the general solution of the advancement operator equation

$$(A-2)^2 f = 3^n + 2n.$$

Solution. Finding the solution to the corresponding homogeneous equation is getting pretty easy at this point, so just note that

$$f_1(n) = c_1 2^n + c_2 n 2^n.$$

What should we try as a particular solution? Fortunately, we have no interference from $p(A) = (A-2)^2$ here. Our first instinct is probably to try $f_0(n) = d_1 3^n + d_2 n$. However, this won't actually work. (Try it. You wind up with a leftover constant term that you can't just make zero.) The key here is that if we use a term with a nonzero power of n in it, we need to include the lower order powers as well (so long as they're not superfluous because of $p(A)$). Thus, we try

$$f_0(n) = d_1 3^n + d_2 n + d_3.$$

Chapter 9 Recurrence Equations

This gives

$$(A-2)^2(d_1 3^n + d_2 n + d_3) = (A-2)(d_1 3^{n+1} + d_2(n+1) + d_3 - 2d_1 3^n - 2d_2 n - 2d_3)$$
$$= (A-2)(d_1 3^n - d_2 n + d_2 - d_3)$$
$$= d_1 3^{n+1} - d_2(n+1) + d_2 - d_3 - 2d_1 3^n + 2d_2 n - 2d_2 + 2d_3$$
$$= d_1 3^n + d_2 n - 2d_2 + d_3.$$

We want this to be $3^n + 2n$, so matching coefficients gives $d_1 = 1$, $d_2 = 2$, and $d_3 = 4$. Thus, the general solution is

$$f(n) = 3^n + 2n + 4 + c_1 2^n + c_2 n 2^n.$$

9.5 Formalizing our approach to recurrence equations

So far, our approach to solving recurrence equations has been based on intuition, and we've not given a lot of explanation for why the solutions we've given have been the general solution. In this section, we endeavor to remedy this. Some familiarity with the language of linear algebra will be useful for the remainder of this section, but it is not essential.

Our techniques for solving recurrence equations have their roots in a fundamentally important concept in mathematics, the notion of a vector space. Recall that a vector space[1] consists of a set V of elements called **vectors**; in addition, there is a binary operation called **addition** with the sum of vectors x and y denoted by $x + y$; furthermore, there is an operation called **scalar multiplication** which combines a scalar (real number) α and a vector x to form a product denoted αx. These operations satisfy the following properties:

1. $x + y = y + x$ for every $x, y, \in V$.

2. $x + (y + z) = (x + y) + z$, for every $x, y, z \in V$.

3. There is a vector called **zero** and denoted 0 so that $x + 0 = x$ for every $x \in V$. Note: We are again overloading an operator and using the symbol 0 for something other than a number.

4. For every element $x \in V$, there is an element $y \in V$, called the **additive inverse** of x and denoted $-x$ so that $x + (-x) = 0$. This property enables us to define **subtraction**, i.e., $x - y = x + (-y)$.

---

[1] To be more complete, we should say that we are talking about a vector space over the field of real numbers, but in our course, these are the only kind of vector spaces we will consider. For this reason, we just use the short phrase "vector space".

9.5 Formalizing our approach to recurrence equations

5. $1x = x$ for every $x \in X$.

6. $\alpha(\beta x) = (\alpha\beta)x$, for every $\alpha, \beta \in \mathbb{R}$ and every $x \in V$.

7. $\alpha(x + y) = \alpha x + \alpha y$ for every $\alpha \in \mathbb{R}$ and every $x, y \in V$.

8. $(\alpha + \beta)x = \alpha x + \beta x$, for every $\alpha, \beta \in \mathbb{R}$ and every $x \in V$.

When V is a vector space, a function $\phi: V \to V$ is called an **linear operator**, or just **operator** for short, when $\phi(x + y) = \phi(x) + \phi(y)$ and $\phi(\alpha x) = \alpha \phi(x)$. When $\phi: V \to V$ is an operator, it is customary to write ϕx rather than $\phi(x)$, saving a set of parentheses. The set of all operators over a vector space V is itself a vector space with addition defined by $(\phi + \rho)x = \phi x + \rho x$ and scalar multiplication by $(\alpha \phi)x = \alpha(\phi x)$.

In this chapter, we focus on the real vector space V consisting of all functions of the form $f: \mathbb{Z} \to \mathbb{R}$. Addition is defined by $(f + g)(n) = f(n) + g(n)$ and scalar multiplication is defined by $(\alpha f)(n) = \alpha(f(n))$.

9.5.1 The Principal Theorem

Here is the basic theorem about solving recurrence equations (stated in terms of advancement operator equations)—and while we won't prove the full result, we will provide enough of an outline where it shouldn't be too difficult to fill in the missing details.

Theorem 9.18. *Let k be a positive integer k, and let $c_0, c_1, \ldots, c_k$ be constants with $c_0, c_k \neq 0$. Then the set W of all solutions to the homogeneous linear equation*

$$(c_0 A^k + c_1 A^{k-1} + c_2 A^{k-2} + \cdots + c_k)f = 0 \tag{9.5.1}$$

is a k-dimensional subspace of V.

— Relation to Linear Algebra

The conclusion that the set W of all solutions is a subspace of V is immediate, since

$$p(A)(f + g) = p(A)f + p(A)g \quad \text{and} \quad p(a)(\alpha f) = \alpha p(A)(f).$$

What takes a bit of work is to show that W is a k-dimensional subspace. But once this is done, then to solve the advancement operator equation given in the form of Theorem 9.18, it suffices to find a **basis** for the vector space W. Every solution is just a linear combination of basis vectors. In the next several subsections, we outline how this goal can be achieved.

9.5.2 The Starting Case

The development proceeds by induction (surprise!) with the case $k = 1$ being the base case. In this case, we study a simple equation of the form $(c_0 A + c_1)f = 0$. Dividing by c_0 and rewriting using subtraction rather than addition, it is clear that we are just talking about an equation of the form $(A - r)f = 0$ where $r \neq 0$.

Lemma 9.19. *Let $r \neq 0$, and let f be a solution to the operator equation $(A - r)f = 0$. If $c = f(0)$, then $f(n) = cr^n$ for every $n \in \mathbb{Z}$.*

Proof. We first show that $f(n) = cr^n$ for every $n \geq 0$, by induction on n. The base case is trivial since $c = f(0) = cr^0$. Now suppose that $f(k) = cr^k$ for some non-negative integer k. Then $(A - r)f = 0$ implies that $f(k+1) - rf(k) = 0$, i.e.,

$$f(k+1) = rf(k) = rcr^k = cr^{k+1}.$$

A very similar argument shows that $f(-n) = cr^{-n}$ for every $n \leq 0$. $\square$

Lemma 9.20. *Consider a nonhomogeneous operator equation of the form*

$$p(A)f = (c_0 A^k + c_1 A^{k-1} + c_2 A^{k-2} + \cdots + c_k)f = g, \qquad (9.5.2)$$

with $c_0, c_k \neq 0$, and let W be the subspace of V consisting of all solutions to the corresponding homogeneous equation

$$p(A)f = (c_0 A^k + c_1 A^{k-1} + c_2 A^{k-2} + \cdots + c_k)f = 0. \qquad (9.5.3)$$

If f_0 is a solution to (9.5.2), then every solution f to (9.5.2) has the form $f = f_0 + f_1$ where $f_1 \in W$.

Proof. Let f be a solution of (9.5.2), and let $f_1 = f - f_0$. Then

$$p(A)f_1 = p(A)(f - f_0) = p(A)f - p(A)f_0 = g - g = 0.$$

This implies that $f_1 \in W$ and that $f = f_0 + f_1$ so that all solutions to (9.5.2) do in fact have the desired form. $\square$

Using the preceding two results, we can now provide an outline of the inductive step in the proof of Theorem 9.18, at least in the case where the polynomial in the advancement operator has distinct roots.

Theorem 9.21. *Consider the following advancement operator equation*

$$p(A)f = (A - r_1)(A - r_2)\ldots(A - r_k)f = 0. \qquad (9.5.4)$$

with $r_1, r_2, \ldots, r_k$ distinct non-zero constants. Then every solution to (9.5.4) has the form

$$f(n) = c_1 r_1^n + c_2 r_2^n + c_3 r_3^n + \cdots + c_k r_k^n.$$

9.5 Formalizing our approach to recurrence equations

Proof. The case $k = 1$ is Lemma 9.19. Now suppose we have established the theorem for some positive integer m and consider the case $k = m + 1$. Rewrite (9.5.4) as

$$(A - r_1)(A - r_2)\ldots(A - r_m)[(A - r_{m+1})f] = 0.$$

By the inductive hypothesis, it follows that if f is a solution to (9.5.4), then f is also a solution to the nonhomogeneous equation

$$(A - r_{m+1})f = d_1 r_1^n + d_2 r_2^n + \cdots + d_m r_m^n. \tag{9.5.5}$$

To find a particular solution f_0 to (9.5.5), we look for a solution having the form

$$f_0(n) = c_1 r_1^n + c_2 r_2^n + \cdots + c_m r_m^n. \tag{9.5.6}$$

On the other hand, a simple calculation shows that for each $i = 1, 2, \ldots, m$, we have

$$(A - r_{m+1})c_i r_i^n = c_i r_i^{n+1} - r_{m+1} c_i r_i^n = c_i(r_i - r_{m+1})r_i^n,$$

so it suffices to choose c_i so that $c_i(r_i - r_{m+1}) = d_i$, for each $i = 1, 2, \ldots, m$. This can be done since r_{m+1} is distinct from r_i for $i = 1, 2, \ldots m$.

Now we have a particular solution $f_0(n) = \sum_{i=1}^{m} c_i r_i^n$. Next we consider the corresponding homogeneous equation $(A - r_{m+1})f = 0$. The general solution to this equation has the form $f_1(n) = c_{m+1} r_{m+1}^n$. It follows that every solution to the original equation has the form

$$f(n) = f_0(n) + f_1(n) = c_1 r_1^n + c_2 r_2^n + \cdots + c_m r_m^n + c r_{m+1}^n,$$

which is exactly what we want! □

9.5.3 Repeated Roots

It is straightforward to modify the proof given in the preceding section to obtain the following result. We leave the details as an exercise.

Lemma 9.22. *Let $k \geq 1$ and consider the equation*

$$(A - r)^k f = 0. \tag{9.5.7}$$

Then the general solution to (9.5.7) has the following form

$$f(n) = c_1 r^n + c_2 n r^n + c_3 n^2 r^n + c_4 n^3 r^n + \cdots + c_k n^{k-1} r^n. \tag{9.5.8}$$

9.5.4 The General Case

Combining the results in the preceding sections, we can quickly write the general solution of any homogeneous equation of the form $p(A)f = 0$ *provided* we can factor the polynomial $p(A)$. Note that in general, this solution takes us into the field of *complex numbers*, since the roots of a polynomial with real coefficients are sometimes complex numbers—with non-zero imaginary parts.

We close this section with one more example to illustrate how quickly we can read off the general solution of a homogeneous advancement operator equation $p(A)f = 0$, provided that $p(A)$ is factored.

Example 9.23. Consider the advancement operator equation

$$(A-1)^5(A+1)^3(A-3)^2(A+8)(A-9)^4 f = 0.$$

Then every solution has the following form

$$\begin{aligned} f(n) =& c_1 + c_2 n + c_3 n^2 + c_4 n^3 + c_5 n^4 \\ &+ c_6(-1)^n + c_7 n(-1)^n + c_8 n^2(-1)^n \\ &+ c_9 3^n + c_{10} n 3^n \\ &+ c_{11}(-8)^n \\ &+ c_{12} 9^n + c_{13} n 9^n + c_{14} n^2 9^n + c_{15} n^3 9^n. \end{aligned}$$

9.6 Using generating functions to solve recurrences

The approach we have seen thus far in this chapter is not the only way to solve recurrence equations. Additionally, it really only applies to linear recurrence equations with constant coefficients. In the remainder of the chapter, we will look at some examples of how generating functions can be used as another tool for solving recurrence equations. In this section, our focus will be on linear recurrence equations. In Section 9.7, we will see how generating functions can solve a nonlinear recurrence.

Our first example is the homogeneous recurrence that corresponds to the advancement operator equation in Example 9.9.

Example 9.24. Consider the recurrence equation $r_n + r_{n-1} - 6r_{n-2} = 0$ for the sequence $\{r_n : n \geq 0\}$ with $r_0 = 1$ and $r_1 = 3$. This sequence has generating function

$$f(x) = \sum_{n=0}^{\infty} r_n x^n = r_0 + r_1 x + r_2 x^2 + r_3 x^3 + \cdots.$$

9.6 Using generating functions to solve recurrences

Now consider for a moment what the function $xf(x)$ looks like. It has r_{n-1} as the coefficient on x_n. Similarly, in the function $-6x^2 f(x)$, the coefficient on x^n is $-6r_{n-2}$.

What is our point in all of this? Well, if we add them all up, notice what happens. The coefficient on x_n becomes $r_n + r_{n-1} - 6r_{n-2}$, which is 0 because of the recurrence equation! Now let's see how this all lines up:

$$f(x) = r_0 + r_1 x + r_2 x^2 + r_3 x^3 + \cdots + r_n x^n + \cdots$$
$$xf(x) = 0 + r_0 x + r_1 x^2 + r_2 x^3 + \cdots r_{n-1} x^n + \cdots$$
$$-6x^2 f(x) = 0 + 0 - 6r_0 x^2 - 6r_1 x^3 + \cdots - 6r_{n-2} x^n + \cdots$$

When we add the left-hand side, we get $f(x)(1 + x - 6x^2)$. On the right-hand side, the coefficient on x^n for $n \geq 2$ is 0 because of the recurrence equation. However, we are left with $r_0 + (r_0 + r_1)x = 1 + 4x$, using the initial conditions. Thus, we have the equation

$$f(x)(1 + x - 6x^2) = 1 + 4x,$$

or $f(x) = (1 + 4x)/(1 + x - 6x^2)$. This is a generating function that we can attack using partial fractions in SageMath:

```
f(x) = (1+4*x)/(1+x-6*x^2)
pretty_print(f(x).partial_fraction())
```

This shows us that

$$f(x) = \frac{6}{5} \frac{1}{1 - 2x} - \frac{1}{5} \frac{1}{1 + 3x} = \frac{6}{5} \sum_{n=0}^{\infty} 2^n x^n - \frac{1}{5} \sum_{n=0}^{\infty} (-3)^n x^n.$$

From here, we read off r_n as the coefficient on x^n and have $r_n = (6/5)2^n - (1/5)(-3)^n$.

Although there's a bit more work involved, this method can be used to solve nonhomogeneous recurrence equations as well, as the next example illustrates.

Example 9.25. The recurrence equation $r_n - r_{n-1} - 2r_{n-2} = 2^n$ is nonhomogeneous. Let $r_0 = 2$ and $r_1 = 1$. This time, to solve the recurrence, we start by multiplying both sides by x^n. This gives the equation

$$r_n x^n - r_{n-1} x^n - 2r_{n-2} x^n = 2^n x^n.$$

If we sum this over all values of $n \geq 2$, we have

$$\sum_{n=2}^{\infty} r_n x^n - \sum_{n=2}^{\infty} r_{n-1} x^n - 2 \sum_{n=2}^{\infty} r_{n-2} x^n = \sum_{n=2}^{\infty} 2^n x^n.$$

Chapter 9 Recurrence Equations

The right-hand side you should readily recognize as being almost equal to $1/(1-2x)$. We are missing the 1 and $2x$ terms, however, so must subtract them from the rational function form of the series. On the left-hand side, however, we need to do a bit more work.

The first sum is just missing the first two terms of the series, so we can replace it by $R(x) - (2 + x)$, where $R(x) = \sum_{n=0}^{\infty} r_n x^n$. The second sum is almost $xR(x)$, except it's missing the first term. Thus, it's equal to $xR(x) - 2x$. The sum in the final term is simply $x^2 R(x)$. Thus, the equation can be rewritten as

$$R(x) - (2+x) - (xR(x) - 2x) - 2x^2 R(x) = \frac{1}{1-2x} - 1 - 2x.$$

A little bit of algebra then gets us to the generating function

$$R(x) = \frac{6x^2 - 5x + 2}{(1-2x)(1-x-2x^2)}.$$

This generating function can be expanded using partial fractions in SageMath:

```
f(x) = (6*x^2-5*x+2)/((1-2*x)*(1-x-2*x^2))
pretty_print(f(x).partial_fraction())
```

Therefore, using Example 8.7 for the second rational function, we have

$$R(x) = -\frac{1}{9(1-2x)} + \frac{2}{3(1-2x)^2} + \frac{13}{9(1+x)}$$

$$= -\frac{1}{9} \sum_{n=0}^{\infty} 2^n x^n + \frac{2}{3} \sum_{n=0}^{\infty} \binom{n+2-1}{1} 2^n x^n + \frac{13}{9} \sum_{n=0}^{\infty} (-1)^n x^n.$$

From this generating function, we can read off that

$$r_n = -\frac{1}{9} 2^n + \frac{2(n+1)}{3} 2^n + \frac{13}{9}(-1)^n = \frac{5}{9} 2^n + \frac{2}{3} n 2^n + \frac{13}{9}(-1)^n.$$

The recurrence equations of the two examples in this section can both be solved using the techniques we studied earlier in the chapter. One potential benefit to the generating function approach for nonhomogeneous equations is that it does not require determining an appropriate form for the particular solution. However, the method of generating functions often requires that the resulting generating function be expanded using partial fractions. Both approaches have positives and negatives, so unless instructed to use a specific method, you should choose whichever seems most appropriate for a given situation. In the next section, we will see a recurrence equation that is most easily solved using generating functions because it is nonlinear.

9.7 Solving a nonlinear recurrence

In this section, we will use generating functions to enumerate the a certain type of trees. In doing this, we will see how generating functions can be used in solving a *nonlinear* recurrence equation. We will also make a connection to a counting sequence we encountered back in Chapter 2. To do all of this, we must introduce a bit of terminology. A tree is **rooted** if we have designated a special vertex called its **root**. We will always draw our trees with the root at the top and all other vertices below it. An **unlabeled tree** is one in which we do not make distinctions based upon names given to the vertices. For our purposes, a **binary tree** is one in which each vertex has 0 or 2 children, and an **ordered tree** is one in which the children of a vertex have some ordering (first, second, third, etc.). Since we will be focusing on rooted, unlabeled, binary, ordered trees (RUBOTs for short), we will call the two children of vertices that have children the **left** and **right** children.

In Figure 9.26, we show the rooted, unlabeled, binary, ordered trees with n leaves for $n \leq 4$. Let $C(x) = \sum_{n=0}^{\infty} c_n x^n$ be the generating function for the sequence $\{c_n : n \geq 0\}$ where c_n is the number of RUBOTs with n leaves. (We take $c_0 = 0$ for convenience.) Then we can see from Figure 9.26 that $C(x) = x + x^2 + 2x^3 + 5x^4 + \cdots$. But what are the remaining coefficients? Let's see how we can break a RUBOT with n leaves down into a combination of two smaller RUBOTs to see if we can express c_n in terms of some c_k for $k < n$. When we look at a RUBOT with $n \geq 2$ leaves, we notice that the root vertex must have two children. Those children can be viewed as root nodes of smaller RUBOTs, say the left child roots a RUBOT with k leaves, meaning that the right child roots a RUBOT with $n - k$ leaves. Since there are c_k possible sub-RUBOTs for the left child and c_{n-k} sub-RUBOTs for the right child, there are a total of $c_k c_{n-k}$ RUBOTs in which the root's left child has k leaves on its sub-RUBOT. We can do this for any $k = 1, 2, \ldots, n - 1$, giving us that

$$c_n = \sum_{k=1}^{n-1} c_k c_{n-k}.$$

(This is valid since $n \geq 2$.) Since $c_0 = 0$, we can actually write this as

$$c_n = \sum_{k=0}^{n} c_k c_{n-k}.$$

Let's look at the square of the generating function $C(x)$. By Proposition 8.3, we have

$$\begin{aligned} C^2(x) &= c_0^2 + (c_0 c_1 + c_1 c_0)x + (c_0 c_2 + c_1 c_1 + c_2 c_0)x^2 + \cdots \\ &= 0 + 0 + (c_0 c_2 + c_1 c_1 + c_2 c_0)x^2 + (c_0 c_3 + c_1 c_2 + c_2 c_1 + c_3 c_0)x^3 + \cdots. \end{aligned}$$

Chapter 9 Recurrence Equations

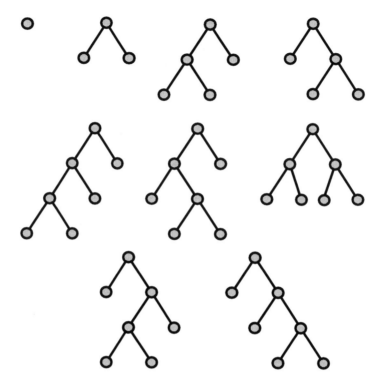

FIGURE 9.26: THE RUBOTS WITH n LEAVES FOR $n \leq 4$

But now we see from our recursion above that the coefficient on x^n in $C^2(x)$ is nothing but c_n for $n \geq 2$. All we're missing is the x term, so adding it in gives us that

$$C(x) = x + C^2(x).$$

Now this is a quadratic equation in $C(x)$, so we can solve for $C(x)$ and have

$$C(x) = \frac{1 \pm \sqrt{1-4x}}{2} = \frac{1 \pm (1-4x)^{1/2}}{2}.$$

Hence, we can use Newton's Binomial Theorem to expand $C(x)$. To do so, we use the following lemma. Its proof is nearly identical to that of Lemma 8.12, and is thus omitted.

Lemma 9.27. *For each $k \geq 1$,*

$$\binom{1/2}{k} = \frac{(-1)^{k-1}}{k} \frac{\binom{2k-2}{k-1}}{2^{2k-1}}.$$

Now we see that

$$C(x) = \frac{1}{2} \pm \frac{1}{2} \sum_{n=0}^{\infty} \binom{1/2}{n}(-4)^n x^n = \frac{1}{2} \pm \frac{1}{2}\left(1 + \sum_{n=1}^{\infty} \frac{(-1)^{n-1}}{n} \frac{\binom{2n-2}{n-1}}{2^{2n-1}}(-4)^n x^n\right)$$

$$= \frac{1}{2} \pm \frac{1}{2} \mp \sum_{n=1}^{\infty} \frac{\binom{2n-2}{n-1}}{n} x^n.$$

Since we need $c_n \geq 0$, we take the "minus" option from the "plus-or-minus" in the quadratic formula and thus have the following theorem.

Theorem 9.28. *The generating function for the number c_n of rooted, unlabeled, binary, ordered trees with n leaves is*

$$C(x) = \frac{1 - \sqrt{1-4x}}{2} = \sum_{n=1}^{\infty} \frac{1}{n}\binom{2n-2}{n-1} x^n.$$

Notice that c_n is a Catalan number, which we first encountered in Chapter 2, where we were counting lattice paths that did not cross the diagonal line $y = x$. (The coefficient c_n is the Catalan number we called $C(n-1)$ in Chapter 2.)

9.8 Discussion

Yolanda took a sip of coffee "I'm glad I paid attention when we were studying vector spaces, bases, and dimension. All this stuff about solutions for recurrence equations made complete sense. And I can really understand why the professor was making a big deal out of factoring. We saw it our first semester when we were learning about partial fractions in calculus. And we saw it again with the differential equations stuff. Isn't it really neat to see how it all fits together?" All this enthusiasm was too much for Alice who was not having a good day. Bob was more sympathetic, saying "Except for the detail about zero as a root of an advancement operator polynomial, I was ok with this chapter." Xing said "Here we learned a precise approach that depended only on factoring. I've been reading on the web and I see that there have been some recent breakthroughs on factoring." Bob jumped back in "But even if you can factor like crazy, if you have a large degree polynomial in the advancement operator equation, then you will have lots of initial conditions. This might be a second major hurdle." Dave

Chapter 9 Recurrence Equations

mumbled "Just do the factoring. The rest is easy." Carlos again was quiet but he knew that Dave was right. Solving big systems of linear equations is relatively easy. The challenge is in the factoring stage.

Despite thinking the material of this chapter was interesting, Bob also wondered if they really needed all of this machinery. "Defining a recursive function is easy in almost all programming languages, so why not just use a computer to calculate the values you need?"[1] Xing started to remark that the techniques of this chapter could provide a good way to understand the growth rate of recursive functions in terms of the big Oh notation of Chapter 4, but Alice interrupted to propose a programming experiment as something that would raise her spirits. (The chance to prove Bob wrong was probably more motivational than the chance to do some coding, but she didn't want to be *too* mean.)

The group decided to take a look at the recurrence in Example 9.25, which they immediately wrote as a recursive function defined on the nonnegative integers by

$$r(n) = \begin{cases} 2^n + r(n-1) + 2r(n-2) & \text{if } n \geq 2; \\ 1 & \text{if } n = 1; \\ 2 & \text{if } n = 0. \end{cases}$$

Alice grabbed her computer and implemented this in SageMath and computed a few test values.

```
def r(n):
    if n == 0:
        return 2
    elif n == 1:
        return 1
    elif n >=2:
        return 2^n + r(n-1)+2*r(n-2)
print(r(1))
print(r(4))
print(r(10))
```

1
53
7397

She then defined a second function s that was the explicit (nonrecursive) solution from Example 9.25 and checked that values matched.

---

[1] The history of how recursion made its way into ALGOL (and therefore most modern programming languages) involved some intrigue. Maarten van Emden recounts this in a blog post entitled "How recursion got into programming: a tale of intrigue, betrayal, and advanced programming-language semantics".

```
s(n)=(5/9)*2^n + (2/3)*n*2^n+(13/9)*(-1)^n
print(s(1))
print(s(4))
print(s(10))
```

1
53
7397

"Is this going somewhere?", Bob asked impatiently. For these values, both r and s seemed to be giving them answers equally quickly. Dave said he'd heard something about a `timeit` command in SageMath that would allow them to compare run times and comandeered Alice's keyboard to type:

```
for n in range(31):
    if n % 5 == 0:
        print("For_n_=_{}:".format(n))
        timeit("r(n)",number=5)
        timeit("s(n)",number=5)
```

```
For n = 0:
5 loops, best of 3: 238 ns per loop
5 loops, best of 3: 44 µs per loop
For n = 5:
5 loops, best of 3: 11.4 µs per loop
5 loops, best of 3: 50.6 µs per loop
For n = 10:
5 loops, best of 3: 127 µs per loop
5 loops, best of 3: 47.6 µs per loop
For n = 15:
5 loops, best of 3: 1.42 ms per loop
5 loops, best of 3: 50 µs per loop
For n = 20:
5 loops, best of 3: 15.7 ms per loop
5 loops, best of 3: 49 µs per loop
For n = 25:
5 loops, best of 3: 133 ms per loop
5 loops, best of 3: 50.4 µs per loop
For n = 30:
5 loops, best of 3: 1.49 s per loop
5 loops, best of 3: 48.8 µs per loop
```

This finally got Bob's attention, since it s seems to be taking a relatively constant time to run even as n increases, while r seems to be taking about 10 times as long to run

Chapter 9 Recurrence Equations

for each increase of 5 in the value of n. As a final test, they execute the SageMath code below, which calculates s(100) almost instantly. On the other hand, it seems like getting a refill on their coffee would be a good way to pass the time waiting on r(40).

```
print(s(100))
print(r(40))
```

85214290348675420878389493250277
29931149867237

9.9 Exercises

1. Write each of the following recurrence equations as advancement operator equations.

   (a) $r_{n+2} = r_{n+1} + 2r_n$

   (b) $r_{n+4} = 3r_{n+3} - r_{n+2} + 2r_n$

   (c) $g_{n+3} = 5g_{n+1} - g_n + 3^n$

   (d) $h_n = h_{n-1} - 2h_{n-2} + h_{n-3}$

   (e) $r_n = 4r_{n-1} + r_{n-3} - 3r_{n-5} + (-1)^n$

   (f) $b_n = b_{n-1} + 3b_{n-2} + 2^{n+1} - n^2$

2. Solve the recurrence equation $r_{n+2} = r_{n+1} + 2r_n$ if $r_0 = 1$ and $r_2 = 3$ (Yes, we specify a value for r_2 but not for r_1).

3. Find the general solution of the recurrence equation $g_{n+2} = 3g_{n+1} - 2g_n$.

4. Solve the recurrence equation $h_{n+3} = 6h_{n+2} - 11h_{n+1} + 6h_n$ if $h_0 = 3$, $h_1 = 2$, and $h_2 = 4$.

5. Find an explicit formula for the nth Fibonacci number f_n. (See Subsection 9.1.1.)

6. For each advancement operator equation below, give its general solution.

   (a) $(A-2)(A+10)f = 0$

   (b) $(A^2 - 36)f = 0$

   (c) $(A^2 - 2A - 5)f = 0$

   (d) $(A^3 - 4A^2 - 20A + 48)f = 0$

   (e) $(A^3 + A^2 - 5A + 3)f = 0$

   (f) $(A^3 + 3A^2 + 3A + 1)f = 0$

7. Solve the advancement operator equation $(A^2 + 3A - 10)f = 0$ if $f(0) = 2$ and $f(1) = 10$.

8. Give the general solution to each advancement operator equation below.

(a) $(A-4)^3(A+1)(A-7)^4(A-1)^2 f = 0$
(b) $(A+2)^4(A-3)^2(A-4)(A+7)(A-5)^3 g = 0$
(c) $(A-5)^2(A+3)^3(A-1)^3(A^2-1)(A-4)^3 h = 0$

9. For each nonhomogeneous advancement operator equation, find its general solution.

(a) $(A-5)(A+2)f = 3^n$
(b) $(A^2+3A-1)g = 2^n + (-1)^n$
(c) $(A-3)^3 f = 3n+1$
(d) $(A^2+3A-1)g = 2n$
(e) $(A-2)(A-4)f = 3n^2 + 9^n$
(f) $(A+2)(A-5)(A-1)f = 5^n$
(g) $(A-3)^2(A+1)g = 2 \cdot 3^n$
(h) $(A-2)(A+3)f = 5n2^n$
(i) $(A-2)^2(A-1)g = 3n^2 2^n + 2^n$
(j) $(A+1)^2(A-3)f = 3^n + 2n^2$

10. Find and solve a recurrence equation for the number g_n of ternary strings of length n that do not contain 102 as a substring.

11. There is a famous puzzle called the Towers of Hanoi that consists of three pegs and n circular discs, all of different sizes. The discs start on the leftmost peg, with the largest disc on the bottom, the second largest on top of it, and so on, up to the smallest disc on top. The goal is to move the discs so that they are stacked in this same order on the rightmost peg. However, you are allowed to move only one disc at a time, and you are never able to place a larger disc on top of a smaller disc. Let t_n denote the fewest moves (a move being taking a disc from one peg and placing it onto another) in which you can accomplish the goal. Determine an explicit formula for t_n.

12. A valid database identifier of length n can be constructed in three ways:
- Starting with A and followed by any valid identifier of length $n-1$.
- Starting with one of the two-character strings 1A, 1B, 1C, 1D, 1E, or 1F and followed by any valid identifier of length $n-2$.
- Starting with 0 and followed by any ternary ($\{0,1,2\}$) string of length $n-1$.

Find a recurrence for the number $g(n)$ of database identifiers of length n and then solve your recurrence to obtain an explicit formula for $g(n)$. (You may consider the empty string of length 0 a valid database identifier, making $g(0) = 1$. This will simplify the arithmetic.)

13. Let t_n be the number of ways to tile a $2 \times n$ rectangle using 1×1 tiles and L-tiles. An L-tile is a 2×2 tile with the upper-right 1×1 square deleted. (An L tile may be rotated so that the "missing" square appears in any of the four positions.) Find a recursive

formula for t_n along with enough initial conditions to get the recursion started. Use this recursive formula to find a closed formula for t_n.

14. Prove Lemma 9.22 about advancement operator equations with repeated roots.

15. Use generating functions to solve the recurrence equation $r_n = r_{n-1} + 6r_{n-2}$ for $n \geq 2$ with $r_0 = 1$ and $r_1 = 3$.

16. Let $a_0 = 0$, $a_1 = 2$, and $a_2 = 5$. Use generating functions to solve the recurrence equation $a_{n+3} = 5a_{n+2} - 7a_{n+1} + 3a_n + 2^n$ for $n \geq 0$.

17. Let $b_0 = 1$, $b_2 = 1$, and $b_3 = 4$. Use generating functions to solve the recurrence equation $b_{n+3} = 4b_{n+2} - b_{n+1} - 6b_n + 3^n$ for $n \geq 0$.

18. Use generating functions to find a closed formula for the Fibonacci numbers f_n.

19. How many rooted, unlabeled, binary, ordered, trees (RUBOTs) with 6 leaves are there? Draw 6 distinct RUBOTs with 6 leaves.

20. In this chapter, we developed a generating function for the Catalan numbers. We first encountered the Catalan numbers in Chapter 2, where we learned they count certain lattice paths. Develop a recurrence for the number l_n of lattice paths similar to the recurrence

$$c_n = \sum_{k=0}^{n} c_k c_{n-k} \quad \text{for } n \geq 2$$

for RUBOTs by thinking of ways to break up a lattice path from $(0,0)$ to (n,n) that does not cross the diagonal $y = x$ into two smaller lattice paths of this type.

CHAPTER 10

Probability

It was a slow day and Dave said he was bored. It was just after lunch, and he complained that there was nothing to do. Nobody really seemed to be listening, although Alice said that Dave might consider studying, even reading ahead in the chapter. Undeterred, Dave said "Hey Alice, how about we play a game. We could take turns tossing a coin, with the other person calling heads or tails. We could keep score with the first one to a hundred being the winner." Alice rolled her eyes at such a lame idea. Sensing Alice's lack of interest, Dave countered "OK, how about a hundred games of Rock, Paper or Scissors?" Zori said "Why play a hundred times? If that's what you're going to do, just play a single game."

Now it was Alice's turn. "If you want to play a game, I've got a good one for you. Just as you wanted, first one to score a hundred wins. You roll a pair of dice. If you roll doubles, I win 2 points. If the two dice have a difference of one, I win 1 point. If the difference is 2, then it's a tie. If the difference is 3, you win one point; if the difference is 4, you win two points; and if the difference is 5, you win three points." Xing interrupted to say "In other words, if the difference is d, then Dave wins $d - 2$ points." Alice continues "Right! And there are three ways Dave can win, with one of them being the biggest prize of all. Also, rolling doubles is rare, so this has to be a good game for Dave."

Zori's ears perked up with Alice's description. She had a gut feeling that this game wasn't really in Dave's favor and that Alice knew what the real situation was. The idea of a payoff with some uncertainty involved seemed very relevant. Carlos was scribbling on a piece of paper, then said politely "Dave, you really should be reading ahead in the chapter".

So what do you think? Is this a fair game? What does it mean for a game to be fair? Should Dave play—independent of the question of whether such silly stuff should occupy one's time? And what does any of this conversation have to do with combinatorics?

Chapter 10 Probability

10.1 An Introduction to Probability

We continue with an informal discussion intended to motivate the more structured development that will follow. Consider the "spinner" shown in Figure 10.1. Suppose we give it a good thwack so that the arrow goes round and round. We then record the number of the region in which the pointer comes to rest. Then observers, none of whom have studied combinatorics, might make the following comments:

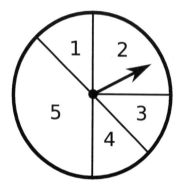

FIGURE 10.1: A Spinner for Games of Chance

1. The odds of landing in region 1 are the same as those for landing in region 3.

2. You are twice as likely to land in region 2 as in region 4.

3. When you land in an odd numbered region, then 60% of the time, it will be in region 5.

We will now develop a more formal framework that will enable us to make such discussions far more precise. We will also see whether Alice is being entirely fair to Bob in her proposed game to one hundred.

We begin by defining a **probability space** as a pair (S, P) where S is a finite set and P is a function that whose domain is the family of all subsets of S and whose range is the set $[0, 1]$ of all real numbers which are non-negative and at most one. Furthermore, the following two key properties must be satisfied:

1. $P(\emptyset) = 0$ and $P(S) = 1$.

2. If A and B are subsets of S, and $A \cap B = \emptyset$, then $P(A \cup B) = P(A) + P(B)$.

10.1 An Introduction to Probability

When (S, P) is a probability space, the function P is called a **probability measure**, the subsets of S are called **events**, and when $E \subseteq S$, the quantity $P(E)$ is referred to as the **probability** of the event E.

Note that we can consider P to be extended to a mapping from S to $[0,1]$ by setting $P(x) = P(\{x\})$ for each element $x \in S$. We call the elements of S **outcomes** (some people prefer to say the elements are **elementary outcomes**) and the quantity $P(x)$ is called the **probability** of x. It is important to realize that if you know $P(x)$ for each $x \in S$, then you can calculate $P(E)$ for any event E, since (by the second property), $P(E) = \sum_{x \in X} P(x)$.

Example 10.2. For the spinner, we can take $S = \{1, 2, 3, 4, 5\}$, with $P(1) = P(3) = P(4) = 1/8$, $P(2) = 2/8 = 1/4$ and $P(5) = 3/8$. So $P(\{2, 3\}) = 1/8 + 2/8 = 3/8$.

Example 10.3. Let S be a finite, nonempty set and let $n = |S|$. For each $E \subseteq S$, set $P(E) = |E|/n$. In particular, $P(x) = 1/n$ for each element $x \in S$. In this trivial example, all outcomes are equally likely.

Example 10.4. If a single six sided die is rolled and the number of dots on the top face is recorded, then the ground set is $S = \{1, 2, 3, 4, 5, 6\}$ and $P(i) = 1/6$ for each $i \in S$. On the other hand, if a pair of dice are rolled and the sum of the dots on the two top faces is recorded, then $S = \{2, 3, 4, \ldots, 11, 12\}$ with $P(2) = P(12) = 1/36$, $P(3) = P(11) = 2/36$, $P(4) = P(10) = 3/36$, $P(5) = P(9) = 4/36$, $P(6) = P(8) = 5/36$ and $P(7) = 6/36$. To see this, consider the two die as distinguished, one die red and the other one blue. Then each of the pairs (i, j) with $1 \leq i, j \leq 6$, the red die showing i spots and the blue die showing j spots is equally likely. So each has probability $1/36$. Then, for example, there are three pairs that yield a total of four, namely $(3, 1)$, $(2, 2)$ and $(1, 3)$. So the probability of rolling a four is $3/36 = 1/12$.

Example 10.5. In Alice's game as described above, the set S can be $\{0, 1, 2, 3, 4, 5\}$, the set of possible differences when a pair of dice are rolled. In this game, we will see that the correct definition of the function P will set $P(0) = 6/36$; $P(1) = 10/36$; $P(2) = 8/36$; $P(3) = 6/36$; $P(4) = 4/36$; and $P(5) = 2/36$. Using Xing's more compact notation, we could say that $P(0) = 1/6$ and $P(d) = 2(6-d)/36$ when $d > 0$.

Example 10.6. A jar contains twenty marbles, of which six are red, nine are blue and the remaining five are green. Three of the twenty marbles are selected at random.[1] Let $X = \{0, 1, 2, 3, 4, 5\}$, and for each $x \in X$, let $P(x)$ denote the probability that the number of blue marbles among the three marbles selected is x. Then $P(i) = C(9, i)C(11, 3-i)/C(20, 3)$ for $i = 0, 1, 2, 3$, while $P(4) = P(5) = 0$. Bob says that it doesn't make sense to have outcomes with probability zero, but Carlos says that it does.

---
[1]This is sometimes called **sampling without replacement**. You should imagine a jar with opaque sides—so you can't see through them. The marbles are stirred/shaken, and you reach into the jar blind folded and draw out three marbles.

Chapter 10 Probability

Example 10.7. In some cards games, each player receives five cards from a standard deck of 52 cards—four suits (spades, hearts, diamonds and clubs) with 13 cards, ace though king in each suit. A player has a **full house** if there are two values x and y for which he has three of the four x's and two of the four y's, e.g. three kings and two eights. If five cards are drawn at random from a standard deck, the probability of a full house is

$$\frac{\binom{13}{1}\binom{12}{1}\binom{4}{3}\binom{4}{2}}{\binom{52}{5}} \approx 0.00144.$$

10.2 Conditional Probability and Independent Events

A jar contains twenty marbles of which six are red, nine are blue and the remaining five are green. While blindfolded, Xing selects two of the twenty marbles random (without replacement) and puts one in his left pocket and one in his right pocket. He then takes off the blindfold.

The probability that the marble in his left pocket is red is 6/20. But Xing first reaches into his right pocket, takes this marble out and discovers that it is blue. Is the probability that the marble in his left pocket is red still 6/20? Intuition says that it's slightly higher than that. Here's a more formal framework for answering such questions.

Let (S, P) be a probability space and let B be an event for which $P(B) > 0$. Then for every event $A \subseteq S$, we define the **probability of A, given B**, denoted $P(A|B)$, by setting $P(A|B) = P(A \cap B)/P(B)$.

Discussion 10.8. Returning to the question raised at the beginning of the section, Bob says that this is just conditional probability. He says let B be the event that the marble in the right pocket is blue and let A be the event that the marble in the left pocket is red. Then $P(B) = 9/20$, $P(A) = 6/20$ and $P(A \cap B) = (9 \cdot 6)/380$, so that $P(A|B) = \frac{54}{380}\frac{20}{9} = 6/19$, which is of course slightly larger than 6/20. Alice is impressed.

Example 10.9. Consider the jar of twenty marbles from the preceding example. A second jar of marbles is introduced. This jar has eighteen marbles: nine red, five blue and four green. A jar is selected at random and from this jar, two marbles are chosen at random. What is the probability that both are green? Bob is on a roll. He says, "Let G be the event that both marbles are green, and let J_1 and J_2 be the event that the marbles come from the first jar and the second jar, respectively. Then $G = (G \cap J_1) \cup (G \cap J_2)$, and $(G \cap J_1) + (G \cap J_2) = \emptyset$. Furthermore, $P(G|J_1) = \binom{5}{2}/\binom{20}{2}$ and $P(G|J_2) = \binom{4}{2}/\binom{18}{2}$, while $P(J_1) = P(J_2) = 1/2$. Also $P(G \cap J_i) = P(J_i)P(G|J_i)$ for each $i = 1, 2$. Therefore,

$$P(G) = \frac{1}{2}\frac{\binom{5}{2}}{\binom{20}{2}} + \frac{1}{2}\frac{\binom{4}{2}}{\binom{18}{2}} = \frac{1}{2}\left(\frac{20}{380} + \frac{12}{306}\right).$$

10.2.1 Independent Events

Let A and B be events in a probability space (S, P). We say A and B are **independent** if $P(A \cap B) = P(A)P(B)$. Note that when $P(B) \neq 0$, A and B are independent if and only if $P(A) = P(A|B)$. Two events that are not independent are said to be **dependent**. Returning to our earlier example, the two events (A: the marble in Xing's left pocket is red and B: the marble in his right pocket is blue) are dependent.

Example 10.10. Consider the two jars of marbles from Example 10.9. One of the two jars is chosen at random and a single marble is drawn from that jar. Let A be the event that the second jar is chosen, and let B be the event that the marble chosen turns out to be green. Then $P(A) = 1/2$ and $P(B) = \frac{1}{2}\frac{5}{20} + \frac{1}{2}\frac{4}{18}$. On the other hand, $P(A \cap B) = \frac{1}{2}\frac{4}{18}$, so $P(A \cap B) \neq P(A)P(B)$, and the two events are not independent. Intuitively, this should be clear, since once you know that the marble is green, it is more likely that you actually chose the first jar.

Example 10.11. A pair of dice are rolled, one red and one blue. Let A be the event that the red die shows either a 3 or a 5, and let B be the event that you get doubles, i.e., the red die and the blue die show the same number. Then $P(A) = 2/6$, $P(B) = 6/36$, and $P(A \cap B) = 2/36$. So A and B are independent.

10.3 Bernoulli Trials

Suppose we have a jar with 7 marbles, four of which are red and three are blue. A marble is drawn at random and we record whether it is red or blue. The probability p of getting a red marble is $4/7$; and the probability of getting a blue is $1 - p = 3/7$.

Now suppose the marble is put back in the jar, the marbles in the jar are stirred, and the experiment is repeated. Then the probability of getting a red marble on the second trial is again $4/7$, and this pattern holds regardless of the number of times the experiment is repeated.

It is customary to call this situation a series of **Bernoulli trials**. More formally, we have an experiment with only two outcomes: **success** and **failure**. The probability of success is p and the probability of failure is $1 - p$. Most importantly, when the experiment is repeated, then the probability of success on any individual test is exactly p.

We fix a positive integer n and consider the case that the experiment is repeated n times. The outcomes are then the binary strings of length n from the two-letter alphabet $\{S, F\}$, for success and failure, respectively. If x is a string with i successes and $n - i$

Chapter 10 Probability

failures, then $P(x) = \binom{n}{i} p^i (1-p)^{n-i}$. Of course, in applications, success and failure may be replaced by: head/tails, up/down, good/bad, forwards/backwards, red/blue, etc.

Example 10.12. When a die is rolled, let's say that we have a success if the result is a two or a five. Then the probability p of success is $2/6 = 1/3$ and the probability of failure is $2/3$. If the die is rolled ten times in succession, then the probability that we get exactly four successes is $C(10, 4)(1/3)^4 (2/3)^6$.

Example 10.13. A fair coin is tossed 100 times and the outcome (heads or tails) is recorded. Then the probability of getting heads 40 times and tails the other 60 times is

$$\binom{100}{40} \left(\frac{1}{2}\right)^{40} \left(\frac{1}{2}\right)^{60} = \frac{\binom{100}{40}}{2^{100}}.$$

Discussion 10.14. Bob says that if a fair coin is tossed 100 times, it is fairly likely that you will get exactly 50 heads and 50 tails. Dave is not so certain this is right. Carlos fires up his computer and in few second, he reports that the probability of getting exactly 50 heads when a fair coin is tossed 100 times is

$$\frac{126114180681955241668515621570}{158456325028528675187087900672}$$

which is .079589, to six decimal places. In other words, not very likely at all. Xing is doing a modestly more complicated calculation, and he reports that you have a 99% chance that the number of heads is at least 20 and at most 80. Carlos adds that when n is very large, then it is increasingly certain that the number of heads in n tosses will be close to $n/2$. Dave asks what do you mean by close, and what do you mean by very large?

10.4 Discrete Random Variables

Let (S, P) be a probability space and let $X : S \longrightarrow \mathbb{R}$ be any function that maps the outcomes in S to real numbers (all values allowed, positive, negative and zero). We call[1] X a **random variable**. The quantity $\sum_{x \in S} X(x) P(x)$, denoted $E(X)$, is called the **expectation** (also called the **mean** or **expected value**) of the random variable X. As the suggestive name reflects, this is what one should expect to be the average behavior of the result of repeated Bernoulli trials.

Note that since we are dealing only with probability spaces (S, P) where S is a finite set, the range of the probability measure P is actually a finite set. Accordingly, we can rewrite the formula for $E(X)$ as $\sum_y y \cdot \text{prob}(X(x) = y)$, where the summation extends over a finite range of values for y.

---

[1] For historical reasons, capital letters, like X and Y are used to denote random variables. They are just functions, so letters like f, g and h might more seem more natural—but maybe not.

10.4 Discrete Random Variables

Example 10.15. For the spinner shown in Figure 10.1, let $X(i) = i^2$ where i is the number of the region. Then

$$E(X) = \sum_{i \in S} i^2 P(i) = 1^2 \frac{1}{8} + 2^2 \frac{2}{8} + 3^2 \frac{1}{8} + 4^2 \frac{1}{8} + 5^2 \frac{3}{8} = \frac{109}{8}.$$

Note that $109/8 = 13.625$. The significance of this quantity is captured in the following statement. If we record the result from the spinner n times in succession as $(i_1, i_2, \ldots, i_n)$ and Xing receives a prize worth i_j^2 for each $j = 1, 2, \ldots, n$), then Xing should "expect" to receive a total prize worth $109n/8 = 13.625n$. Bob asks how this statement can possibly be correct, since $13.625n$ may not even be an integer, and any prize Xing receives will have integral value. Carlos goes on to explain that the concept of expected value provides a formal definition for what is meant by a fair game. If Xing pays 13.625 cents to play the game and is then paid i^2 pennies where i is the number of the region where the spinner stops, then the game is fair. If he pays less, he has an unfair advantage, and if he pays more, the game is biased against him. Bob says "How can Xing pay 13.625 pennies?" Brushing aside Bob's question, Carlos says that one can prove that for every $\epsilon > 0$, there is some n_0 (which depends on ϵ) so that if $n > n_0$, then the probability that Xing's total winnings minus $13.625n$, divided by n is within ϵ of 13.625 is at least $1 - \epsilon$. Carlos turns to Dave and explains politely that this statement gives a precise meaning of what is meant by "close" and "large".

Example 10.16. For Alice's game from the start of the chapter, $S = \{0, 1, 2, 3, 4, 5\}$, we could take X to be the function defined by $X(d) = 2 - d$. Then $X(d)$ records the amount that Bob wins when the difference is d (a negative win for Bob is just a win for Alice in the same amount). We calculate the expectation of X as follows:

$$E(X) = \sum_{d=0}^{5} X(d)p(d) = -2\frac{1}{6} - 1\frac{10}{36} + 0\frac{8}{36} + 1\frac{6}{36} + 2\frac{4}{36} + 3\frac{2}{36} = \frac{-2}{36}.$$

Note that $-2/36 = -.055555\ldots$. So if points were dollars, each time the game is played, Bob should expect to lose slightly more than a nickel. Needless to say, Alice likes to play this game and the more times Bob can be tricked into playing, the more she likes it. On the other hand, by this time in the chapter, Bob should be getting the message and telling Alice to go suck a lemon.

10.4.1 The Linearity of Expectation

The following fundamental property of expectation is an immediate consequence of the definition, but we state it formally because it is so important to discussions to follow.

Proposition 10.17. Let (S, P) be a probability space and let $X_1, X_2, \ldots, X_n$ be random variables. Then
$$E(X_1 + X_2 + \cdots + X_t) = E(X_1) + E(X_2) + \cdots + E(X_n).$$

10.4.2 Implications for Bernoulli Trials

Example 10.18. Consider a series of n Bernoulli trials with p, the probability of success, and let X count the number of successes. Then, we claim that
$$E(X) = \sum_{i=0}^{n} i \binom{n}{i} p^i (1-p)^{n-i} = np$$

To see this, consider the function $f(x) = [px + (1-p)]^n$. Taking the derivative by the chain rule, we find that $f'(x) = np[px + (1-p)]^{n-1}$. Now when $x = 1$, the derivative has value np.

On the other hand, we can use the binomial theorem to expand the function f.
$$f(x) = \sum_{i=0}^{n} \binom{n}{i} x^i p^i (1-p)^{n-i}$$

It follows that
$$f'(x) = \sum_{i=0}^{n} i \binom{n}{i} x^{i-1} p^i (1-p)^{n-i}$$

And now the claim follows by again setting $x = 1$. Who says calculus isn't useful!

Example 10.19. Many states have lotteries to finance college scholarships or other public enterprises judged to have value to the public at large. Although far from a scientific investigation, it seems on the basis of our investigation that many of the games have an expected value of approximately fifty cents when one dollar is invested. So the games are far from fair, and no one should play them unless they have an intrinsic desire to support the various causes for which the lottery profits are targeted.

By contrast, various games of chance played in gambling centers have an expected return of slightly less than ninety cents for every dollar wagered. In this setting, we can only say that one has to place a dollar value on the enjoyment derived from the casino environment. From a mathematical standpoint, you are going to lose. That's how they get the money to build those exotic buildings.

10.5 Central Tendency

Consider the following two situations:

10.5 Central Tendency

- Situation 1. A small town decides to hold a lottery to raise funds for charitable purposes. A total of 10,001 tickets are sold, and the tickets are labeled with numbers from the set {0, 1, 2, ..., 10,000}. At a public ceremony, duplicate tickets are placed in a big box, and the mayor draws the winning ticket from out of the box. Just to heighten the suspense as to who has actually won the prize, the mayor reports that the winning number is at least 7,500. The citizens ooh and aah and they can't wait to see who among them will be the final winner.

- Situation 2. Behind a curtain, a fair coin is tossed 10,000 times, and the number of heads is recorded by an observer, who is reputed to be honest and impartial. Again, the outcome is an integer in the set {0, 1, 2, ..., 10,000}. The observer then emerges from behind the curtain and announces that the number of heads is at least than 7,500. There is a pause and then someone says "What? Are you out of your mind?"

So we have two probability spaces, both with sample space $S = \{0, 1, 2, \ldots, 10,000\}$. For each, we have a random variable X, the winning ticket number in the first situation, and the number of heads in the second. In each case, the expected value, $E(X)$, of the random variable X is 5,000. In the first case, we are not all that surprised at an outcome far from the expected value, while in the second, it seems intuitively clear that this is an extraordinary occurrence. The mathematical concept here is referred to as **central tendency**, and it helps us to understand just how likely a random variable is to stray from its expected value.

For starters, we have the following elementary result.

Theorem 10.20 (Markov's Inequality). *Let X be a random variable in a probability space (S, P). Then for every $k > 0$,*
$$P(|X| \geq k) \leq E(|X|)/k.$$

Proof. Of course, the inequality holds trivially unless $k > E(|X|)$. For k in this range, we establish the equivalent inequality: $kP(|X| \geq k) \leq E(|X|)$.

$$kP(|X| \geq k) = \sum_{r \geq k} kP(|X| = r)$$
$$\leq \sum_{r \geq k} rP(|X| = r)$$
$$\leq \sum_{r > 0} rP(|X| = r)$$
$$= E(|X|). \qquad \square$$

To make Markov's inequality more concrete, we see that on the basis of this trivial result, the probability that either the winning lottery ticket or the number of heads

Chapter 10 Probability

is at least 7,500 is at most 5000/7500 = 2/3. So nothing alarming here in either case. Since we still feel that the two cases are quite different, a more subtle measure will be required.

10.5.1 Variance and Standard Deviation

Again, let (S, P) be a probability space and let X be a random variable. The quantity $E((X - E(X))^2)$ is called the **variance** of X and is denoted var(X). Evidently, the variance of X is a non-negative number. The **standard deviation** of X, denoted σ_X is then defined as the quantity $\sqrt{\text{var}(x)}$, i.e., $\sigma_X^2 = \text{var}(X)$.

Example 10.21. For the spinner shown at the beginning of the chapter, let $X(i) = i^2$ when the pointer stops in region i. Then we have already noted that the expectation $E(X)$ of the random variable X is 109/8. It follows that the variance var(X) is:

$$\text{var}(X) = (1^2 - \frac{109}{8})^2 \frac{1}{8} + (2^2 - \frac{109}{8})^2 \frac{1}{4} + (3^2 - \frac{109}{8})^2 \frac{1}{8} + (4^2 - \frac{109}{8})^2 \frac{1}{8}$$
$$+ (5^2 - \frac{109}{8})^2 \frac{3}{8}$$
$$= (108^2 + 105^2 + 100^2 + 93^2 + 84^2)/512$$
$$= 48394/512$$

It follows that the standard deviation σ_X of X is then $\sqrt{48394/512} \approx 9.722$.

Example 10.22. Suppose that $0 < p < 1$ and consider a series of n Bernoulli trials with the probability of success being p, and let X count the number of successes. We have already noted that $E(X) = np$. Now we claim the variance of X is given by:

$$\text{var}(X) = \sum_{i=0}^{n} (i - np)^2 \binom{n}{i} p^i (1-p)^{n-i} = np(1-p)$$

There are several ways to establish this claim. One way is to proceed directly from the definition, using the same method we used previously to obtain the expectation. But now you need also to calculate the second derivative. Here is a second approach, one that capitalizes on the fact that separate trials in a Bernoulli series are independent.

Let $\mathcal{F} = \{X_1, X_2, \ldots, X_n\}$ be a family of random variables in a probability space (S, P). We say the family $\mathcal{F}$ is **independent** if for each i and j with $1 \leq i < j \leq n$, and for each pair a, b of real numbers with $0 \leq a, b \leq 1$, the following two events are independent: $\{x \in S : X_i(x) \leq a\}$ and $\{x \in S : X_j(x) \leq b\}$. When the family is independent, it is straightforward to verify that

$$\text{var}(X_1 + X_2 + \cdots + X_n) = \text{var}(X_1) + \text{var}(X_2) + \cdots + \text{var}(X_n).$$

10.5 Central Tendency

With the aid of this observation, the calculation of the variance of the random variable X which counts the number of successes becomes a trivial calculation. But in fact, the entire treatment we have outlined here is just a small part of a more complex subject which can be treated more elegantly and ultimately much more compactly—provided you first develop additional background material on families of random variables. For this we will refer you to suitable probability and statistics texts, such as those given in our references.

Proposition 10.23. *Let X be a random variable in a probability space (S, P). Then $\text{var}(X) = E(X^2) - E^2(X)$.*

Proof. Let $E(X) = \mu$. From its definition, we note that

$$\text{var}(X) = \sum_r (r - \mu)^2 \text{prob}(X = r)$$
$$= \sum_r (r^2 - 2r\mu + \mu^2) \text{prob}(X = r)$$
$$= \sum_r r^2 \text{prob}(X = r) - 2\mu \sum_r r \text{prob}(X = r) + \mu^2 \sum_r \text{prob}(X = r)$$
$$= E(X^2) - 2\mu^2 + \mu^2$$
$$= E(X^2) - \mu^2$$
$$= E(X^2) - E^2(X). \qquad \square$$

Variance (and standard deviation) are quite useful tools in discussions of just how likely a random variable is to be near its expected value. This is reflected in the following theorem.

Theorem 10.24 (Chebyshev's Inequality). *Let X be a random variable in a probability space (S, P), and let $k > 0$ be a positive real number. If the expectation $E(X)$ of X is μ and the standard deviation is σ_X, then*

$$\text{prob}(|X - E(X)| \leq k\sigma_X) \geq 1 - \frac{1}{k^2}.$$

Proof. Let $A = \{r \in \mathbb{R} : |r - \mu| > k\sigma_X\}$.
Then we have:

$$\text{var}(X) = E((X - \mu)^2)$$
$$= \sum_{r \in \mathbb{R}} (r - \mu)^2 \text{prob}(X = r)$$
$$\geq \sum_{r \in A} (r - \mu)^2 \text{prob}(X = r)$$

Chapter 10 Probability

$$\geq k^2 \sigma_X^2 \sum_{r \in A} \text{prob}(X = r)$$

$$\geq k^2 \sigma_X^2 \, \text{prob}(|X - \mu| > k\sigma_X).$$

Since $\text{var}(X) = \sigma_X^2$, we may now deduce that $1/k^2 \geq \text{prob}(|X - \mu|) > k\sigma_X)$. Therefore, since $\text{prob}(|X - \mu| \leq k\sigma_X) = 1 - \text{prob}(|X - \mu| > k\sigma_X)$, we conclude that

$$\text{prob}(|X - \mu| \leq k\sigma_X) \geq 1 - \frac{1}{k^2}. \qquad \square$$

Example 10.25. Here's an example of how Chebyshev's Inequality can be applied. Consider n tosses of a fair coin with X counting the number of heads. As noted before, $\mu = E(X) = n/2$ and $\text{var}(X) = n/4$, so $\sigma_X = \sqrt{n}/2$. When $n = 10,000$ and $\mu = 5,000$ and $\sigma_X = 50$. Setting $k = 50$ so that $k\sigma_X = 2500$, we see that the probability that X is within 2500 of the expected value of 5000 is at least 0.9996. So it seems very unlikely indeed that the number of heads is at least 7,500.

Going back to lottery tickets, if we make the rational assumption that all ticket numbers are equally likely, then the probability that the winning number is at least 7,500 is exactly 2501/100001, which is very close to 1/4.

Example 10.26. In the case of Bernoulli trials, we can use basic properties of binomial coefficients to make even more accurate estimates. Clearly, in the case of coin tossing, the probability that the number of heads in 10,000 tosses is at least 7,500 is given by

$$\sum_{i=7,500}^{10,000} \binom{10,000}{i} / 2^{10,000}$$

Now a computer algebra system can make this calculation exactly, and you are encouraged to check it out just to see how truly small this quantity actually is.

10.6 Probability Spaces with Infinitely Many Outcomes

To this point, we have focused entirely on probability spaces (S, P) with S a finite set. More generally, probability spaces are defined where S is an infinite set. When S is countably infinite, we can still define P on the members of S, and now $\sum_{x \in S} P(x)$ is an infinite sum which converges absolutely (since all terms are non-negative) to 1. When S is uncountable, P is not defined on S. Instead, the probability function is defined on a family of subsets of S. Given our emphasis on finite sets and combinatorics, we will discuss the first case briefly and refer students to texts that focus on general concepts from probability and statistics for the second.

Example 10.27. Consider the following game. Nancy rolls a single die. She wins if she rolls a six. If she rolls any other number, she then rolls again and again until the first time that one of the following two situations occurs: (1) she rolls a six, which now this results in a loss or (2) she rolls the same number as she got on her first roll, which results in a win. As an example, here are some sequences of rolls that this game might take:

1. $(4, 2, 3, 5, 1, 1, 1, 4)$. Nancy wins!

2. (6). Nancy wins!

3. $(5, 2, 3, 2, 1, 6)$. Nancy loses. Ouch.

So what is the probability that Nancy will win this game?

Nancy can win with a six on the first roll. That has probability $1/6$. Then she might win on round n where $n \geq 2$. To accomplish this, she has a $5/6$ chance of rolling a number other than six on the first roll; a $4/6$ chance of rolling something that avoids a win/loss decision on each of the rolls, 2 through $n - 1$ and then a $1/6$ chance of rolling the matching number on round n. So the probability of a win is given by:

$$\frac{1}{6} + \sum_{n \geq 2} \frac{5}{6} \left(\frac{4}{6}\right)^{n-2} \frac{1}{6} = \frac{7}{12}.$$

Example 10.28. You might think that something slightly more general is lurking in the background of the preceding example—and it is. Suppose we have two disjoint events A and B in a probability space (S, P) and that $P(A) + P(B) < 1$. Then suppose we make repeated samples from this space with each sample independent of all previous ones. Call it a win if event A holds and a loss if event B holds. Otherwise, it's a tie and we sample again. Now the probability of a win is:

$$P(A) + P(A) \sum_{n \geq 1} (1 - P(A) - P(B))^n = \frac{P(A)}{P(A) + P(B)}.$$

10.7 Discussion

Bob was late for morning coffee and the group was well into dissecting today's applied combinatorics class. As he approached the table, he blurted out "Ok guys, here's a problem that doesn't make any sense to me, except that Nadja, my friend from biology, says that if I have a good feel for probability, then it is transparent." Alice not very softly interjected "Not much goes through six inches of iron." Bob didn't bite "A guy eats lunch at the same diner every day. After lunch, the waiter asks if he wants dessert.

Chapter 10 Probability

He asks for the choices and the waiter replies 'We have three kinds of pie: apple, cherry and pecan.' Then the guy always says 'I'll have pecan pie.' This goes on for six months. Then one day, the waiter says 'I have bad news. Today, we don't have any apple pie, so your only choices are cherry and pecan.' Now the guy says 'In this case, I'll have the cherry pie.' I have to tell you all that this doesn't make any sense to me. Why would the guy ask for cherry pie in preference to pecan pie when he consistently takes pecan pie over both cherry pie and apple pie?"

Zori was the first to say something "Ok guys, I've finally willing to accept the premise that big integer arithmetic, and things that reflect the same flavor, might and I emphasize might, have some relevance in the real world, but this conversation about dessert in some stupid diner is too much." Xing was hesitant but still offered "There's something here. That much I'm sure." Dave said "Yeah, a great dessert. Especially the pecan pie." Alice was not amused. All the while Carlos was thinking. Finally, he said "I think it has something to do with conditional probability. The patron's preference for pecan pie was conditioned on the fact that there were three choices. When there were only two choices, his preferences changed."

Now Yolanda saw more "Doesn't this happen all the time in presidential politics? People prefer candidate A when A, B and C are running, but when candidate C drops out, they shift their preference to candidate B." Alice said "You could say the same thing about close personal relationships." Although she didn't say it, she was thinking that it wouldn't matter how many dropped out if Bob was one of the remaining.

10.8 Exercises

1. Our gang of seven (Alice, Bob, Carlos, Dave, Xing, Yolanda and Zori) are students in a class with a total enrollment of 35. The professor chooses three students at random to go to the board to work challenge problems.

   (a) What is the probability that Yolanda is chosen?

   (b) What is the probability that Yolanda is chosen and Zori is not?

   (c) What is the probability that exactly two members of the club are chosen?

   (d) What is the probability that none of the seven members of club are chosen?

2. Bob says to no one in particular, "Did you know that the probability that you will get at least one '7' in three rolls of a pair of dice is slightly less than 1/2. On the other hand, the probability that you'll get at least one '5' in six rolls of the dice is just over 1/2." Is Bob on target, or out to lunch?

3. Consider the spinner shown in Figure 10.1 at the beginning of the chapter.

(a) What is the probability of getting at least one "5" in three spins?

(b) What is the probability of getting at least one "3" in three spins?

(c) If you keep spinning until you get either a "2" or a "5", what is the probability that you get a "2" first?

(d) If you receive i dollars when the spinner halts in region i, what is the expected value? Since three is right in the middle of the possible outcomes, is it reasonable to pay three dollars to play this game?

4. Alice proposes to Bob the following game. Bob pays one dollar to play. Fifty balls marked $1, 2, \ldots, 50$ are placed in a big jar, stirred around, and then drawn out one by one by Zori, who is wearing a blindfold. The result is a random permutation σ of the integers $1, 2, \ldots, 50$. Bob wins with a payout of two dollars and fifty cents if the permutation σ is a derangement, i.e., $\sigma(i) \neq i$ for all $i = 1, 2, \ldots, n$. Is this a fair game for Bob? If not how should the payoff be adjusted to make it fair?

5. A random graph with vertex set $\{1, 2, \ldots, 10\}$ is constructed using the following method. For each two element subset $\{i, j\}$ from $\{1, 2, \ldots, 10\}$, a fair coin is tossed and the edge $\{i, j\}$ then belongs to the graph when the result is "heads." For each 3-element subset $S \subseteq \{1, 2, \ldots, n\}$, let E_S be the event that S is a complete subgraph in our random graph.

(a) Explain why $P(E_S) = 1/8$ for each 3-element subset S.

(b) Explain why E_S and E_T are independent when $|S \cap T| \leq 1$.

(c) Let $S = \{1, 2, 3\}$, $T = \{2, 3, 4\}$ and $U = \{3, 4, 5\}$. Show that

$$P(E_S|E_T) = P(E_S|E_T E_U).$$

6. Ten marbles labeled $1, 2, \ldots, 10$ are placed in a big jar and then stirred up. Zori, wearing a blindfold, pulls them out of the jar two at a time. Players are allowed to place bets as to whether the sum of the two marbles in a pair is 11. There are $C(10, 2) = 45$ different pairs and exactly 5 of these pairs sums to eleven.

Suppose Zori draws out a pair; the results are observed; then she returns the two balls to the jar and all ten balls are stirred before the next sample is taken. Since the probability that the sum is an "11" is $5/45 = 1/9$, then it would be fair to pay one dollar to play the game if the payoff for an "11" is nine dollars. Similarly, the payoff for a wager of one hundred dollars should be nine hundred dollars.

Now consider an alternative way to play the game. Now Zori draws out a pair; the results are observed; and the marbles are set aside. Next, she draws another pair from the remaining eight marbles, followed by a pair selected from the remaining six, etc.

Chapter 10 Probability

Finally, the fifth pair is just the pair that remains after the fourth pair has been selected. Now players may be free to wager on the outcome of any or all or just some of the five rounds. Explain why either everyone should or no one should wager on the fifth round. Accordingly, the last round is skipped and all marbles are returned to the jar and we start over again.

Also explain why an observant player can make lots of money with a payout ratio of nine to one. Now for a more challenging problem, what is the minimum payout ratio above which a player has a winning strategy?

CHAPTER 11

Applying Probability to Combinatorics

11.1 A First Taste of Ramsey Theory

Bob likes to think of himself as a wild and crazy guy, totally unpredictable. Most guys do. But Alice says that Bob can't change his basic nature, which is excruciatingly boring. Carlos remarks that perhaps we shouldn't be so hard on Bob, because under certain circumstances, we can all be forced to be dull and repetitive.

Recall that when n is a positive integer, we let $[n] = \{1, 2, \ldots, n\}$. In this chapter, when X is a set and k is a non-negative integer with $k \leq |X|$, we borrow from our in-line notation for binomial coefficients and let $C(X, k)$ denote the family of all k-element subsets of X. So $|C([n], k)| = C(n, k)$ whenever $0 \leq k \leq n$.

Recall that the Pigeon Hole Principle asserts that if $n + 1$ pigeons are placed in n holes, then there must be some hole into which two or more pigeons have been placed. More formally, if n and k are positive integers, $t > n(k-1)$ and $f : [t] \longrightarrow [n]$ is any function, then there is a k-element subset $H \subseteq [t]$ and an element $j \in [n]$ so that $f(i) = j$ for every $i \in H$.

We now embark on a study of an elegant extension of this basic result, one that continues to fascinate and challenge.

Returning to the discussion at the start of this section, you might say that an induced subgraph H of a graph G is "boring" if it is either a complete subgraph or an independent set. In either case, exactly every pair of vertices in H behaves in exactly the same boring way. So is boredom inevitable? The answer is yes—at least in a relative sense. As a starter, let's show that any graph on six (or more) vertices has a boring subgraph of size three.

Lemma 11.1. *Let G be any graph with six of more vertices. Then either G contains a complete subgraph of size 3 or an independent set of size 3.*

Proof. Let x be any vertex in G. Then split the remaining vertices into two sets S_1 and

Chapter 11 Applying Probability to Combinatorics

S_2 with S_1 being the neighbors of x and S_2 the non-neighbors. Since G has at least six vertices, we know that either $|S_1| \geq 3$ or $|S_2| \geq 3$. Suppose first that $|S_1| \geq 3$ and let y_1, y_2 and y_3 be distinct vertices from S_1. If $y_i y_j$ is an edge in G for some distinct pair $i, j \in \{1, 23\}$, then $\{x, y_i, y_j\}$ is a complete subgraph of size 3 in G. On the other hand, if there are no edges among the vertices in $\{y_1, y_2, y_3\}$, then we have an independent set of size 3.

The argument when $|S_2| \geq 3$ is dual. □

We note that the bound of six in the preceding lemma is sharp, as a cycle on five vertices does not contain either a complete set of size 3 nor an independent set of size 3.

Next, here is the statement that generalizes this result.

Theorem 11.2 (Ramsey's Theorem for Graphs). *If m and n are positive integers, then there exists a least positive integer $R(m, n)$ so that if G is a graph and G has at least $R(m, n)$ vertices, then either G contains a complete subgraph on m vertices, or G contains an independent set of size n.*

Proof. We show that $R(m, n)$ exists and is at most $\binom{m+n-2}{m-1}$. This claim is trivial when either $m \leq 2$ or $n \leq 2$, so we may assume that $m, n \geq 3$. From this point, we proceed by induction on $t = m + n$ assuming that the result holds when $t \leq 5$.

Now let x be any vertex in G. Then there are at least $\binom{m+n-2}{m-1} - 1$ other vertices, which we partition as $S_1 \cup S_2$, where S_1 are those vertices adjacent to x in G and S_2 are those vertices which are not adjacent to s.

We recall that the binomial coefficients satisfy

$$\binom{m+n-2}{m-1} = \binom{m+n-3}{m-2} + \binom{m+n-3}{m-1} = \binom{m+n-3}{m-2} + \binom{m+n-3}{n-2}$$

So either $|S_1| \geq \binom{m+n-3}{m-2}$ or $|S_1| \geq \binom{m+n-3}{m-1}$. If the first option holds, and S_1 does not have an independent set of size n, then it contains a complete subgraph of size $m - 1$. It follows that we may add x to this set to obtain a complete subgraph of size m in G.

Similarly, if the second option holds, and S_2 does not contain a complete subgraph of size m, then S_2 contains an independent set of size $n - 1$, and we may add x to this set to obtain an independent set of size n in G. □

11.2 Small Ramsey Numbers

Actually determining the **Ramsey numbers** $R(m, n)$ referenced in Theorem 11.2 seems to be a notoriously difficult problem, and only a handful of these values are known precisely. In particular, $R(3, 3) = 6$ and $R(4, 4) = 18$, while $43 \leq R(5, 5) \leq 49$. The distinguished Hungarian mathematician Paul Erdős said on many occasions that it might be possible to determine $R(5,5)$ exactly, if all the world's mathematical talent were to be focused on the problem. But he also said that finding the exact value of $R(6,6)$ might be beyond our collective abilities.

In the following table, we provide information about the Ramsey numbers $R(m,n)$ when m and n are at least 3 and at most 9. When a cell contains a single number, that is the precise answer. When there are two numbers, they represent lower and upper bounds.

m \ n	3	4	5	6	7	8	9
3	6	9	14	18	23	36	39
4		18	25	36, 41	49, 61	58, 84	73, 115
5			43, 49	58, 87	80, 143	101, 216	126, 316
6				102, 165	113, 298	127, 495	169, 780
7					205, 540	217, 1031	241, 1713
8						282, 1870	317, 3583
9							565, 6588

TABLE 11.3: SMALL RAMSEY NUMBERS $R(m,n)$

For additional (or more current) data, see Dynamic Survey #DS1: "Small Ramsey Numbers" by Stanisław Radziszowski in the *Electronic Journal of Combinatorics*. (Table 11.3 was last updated using the 12 January 2014 version of that article.)

11.3 Estimating Ramsey Numbers

We will find it convenient to utilize the following approximation due to Stirling. You can find a proof in almost any advanced calculus book.

$$n! \approx \sqrt{2\pi n} \left(\frac{n}{e}\right)^n \left(1 + \frac{1}{12n} + \frac{1}{288n^2} - \frac{139}{51840n^3} + O\left(\frac{1}{n^4}\right)\right).$$

Of course, we will normally be satisfied with the first term:

$$n! \approx \sqrt{2\pi n} \left(\frac{n}{e}\right)^n$$

Using Stirling's approximation and the binomial coefficients from the proof of Ramsey's Theorem for Graphs, we have the following upper bound:

$$R(n,n) \leq \binom{2n-2}{n-1} \approx \frac{2^{2n}}{4\sqrt{\pi n}}$$

Chapter 11 Applying Probability to Combinatorics

11.4 Applying Probability to Ramsey Theory

The following theorem, due to P. Erdős, is a true classic, and is presented here in a manner that is faithful to how it was first published. As we shall see later, it was subsequently recast—but that's getting the cart ahead of the horse.

Theorem 11.4. *If n is a positive integer. Then*

$$R(n,n) \geq \frac{n}{e\sqrt{2}} 2^{\frac{1}{2}n}.$$

Proof. Let t be an integer with $t > n$ and consider the set $\mathcal{F}$ of all labeled graphs with vertex set $\{1, 2, \ldots, t\}$. Clearly, there are $2^{C(t,2)}$ graphs in this family. Let $\mathcal{F}_1$ denote the subfamily consisting of those graphs which contain a complete subgraph of size n. It is easy to see that

$$|\mathcal{F}_1| \leq \binom{t}{n} 2^{n(t-n)} 2^{C(t-n,2)}.$$

Similarly, let $\mathcal{F}_2$ denote the subfamily consisting of those graphs which contain an independent set of size n. It follows that

$$|\mathcal{F}_2| \leq \binom{t}{n} 2^{n(t-n)} 2^{C(t-n,2)}.$$

We want to take the integer t as large as we can while still guaranteeing that $|\mathcal{F}_1| + |\mathcal{F}_2| \leq |\mathcal{F}|$. This will imply that there is a graph G in $\mathcal{F}$ which does not contain a complete subgraph of size n or an independent set of size n. So consider the following inequality:

$$2\binom{t}{n} 2^{n(t-n)} 2^{C(t-n,2)} < 2^{C(t,2)}. \tag{11.4.1}$$

Now we ask how large can t be without violating inequality (11.4.1)? To answer this, we use the trivial inequality $\binom{t}{n} \leq t^n/n!$ and the use the Stirling approximation for $n!$. After some algebra and taking the nth root of both sides, we see that we need only guarantee that

$$t \leq \frac{n}{e\sqrt{n}} 2^{\frac{1}{2}n} \qquad \square$$

Now let's take a second look at the proof of Theorem 11.4. We consider a probability space (S, P) where the outcomes are graphs with vertex set $\{1, 2, \ldots, t\}$. For each i and

j with $1 \leq i < j \leq t$, edge ij is present in the graph with probability $1/2$. Furthermore, the events for distinct pairs are independent.

Let X_1 denote the random variable which counts the number of n-element subsets of $\{1, 2, \ldots, t\}$ for which all $\binom{n}{2}$ pairs are edges in the graph. Similarly, X_2 is the random variable which counts the number of n-element independent subsets of $\{1, 2, \ldots, t\}$. Then set $X = X_1 + X_2$.

By linearity of expectation, $E(X) = E(X_1) + E(X_2)$ while

$$E(X_1) = E(X_2) = \binom{t}{n} \frac{1}{2^{C(n,2)}}.$$

If $E(X) < 1$, then there must exist a graph with vertex set $\{1, 2, \ldots, t\}$ without a K_n or an I_n. And the question of how large t can be while maintaining $E(X) < 1$ leads to exactly the same calculation we had before.

After more than fifty years and the efforts of many very bright researchers, only marginal improvements have been made on the bounds on $R(n, n)$ from Theorem 11.2 and Theorem 11.4. In particular, no one can settle whether there is some constant $c < 2$ and an integer n_0 so that $R(n, n) < 2^{cn}$ when $n > n_0$. Similarly, no one has been able to answer whether there is some constant $d > 1/2$ and an integer n_1 so that $R(n, n) > 2^{dn}$ when $n > n_1$. We would certainly give you an A for this course if you managed to do either.

Discussion 11.5. Carlos said that he had been trying to prove a good lower bound on $R(n, n)$ using only constructive methods, i.e., no random techniques allowed. But he was having problems. Anything he tried seemed only to show that $R(n, n) \geq n^c$ where c is a constant. That seems so weak compared to the exponential bound which the probabilistic method gives easily. Usually Alice was not very sympathetic to the complaints of others and certainly not from Carlos, who seemed always to be out front. But this time, Alice said to Carlos and in a manner that all could hear "Maybe you shouldn't be so hard on yourself. I read an article on the web that nobody has been able to show that there is a constant $c > 1$ and an integer n_0 so that $R(n, n) > c^n$ when $n > n_0$, provided that only constructive methods are allowed. And maybe, just maybe, saying that you are unable to do something that lots of other famous people seem also unable to do is not so bad." Bob saw a new side of Alice and this too wasn't all bad.

11.5 Ramsey's Theorem

By this time, you are probably not surprised to see that there is a very general form of Ramsey's theorem. We have a bounded number of bins or colors and we are placing the subsets of a fixed size into these categories. The conclusion is that there is a large set which is treated uniformly.

Chapter 11 Applying Probability to Combinatorics

Here's the formal statement.

Theorem 11.6. *Let r and s be positive integers and let $\mathbf{h} = (h_1, h_2, \ldots, h_r)$ be a string of integers with $h_i \geq s$ for each $i = 1, 2, \ldots, s$. Then there exists a least positive integer $R(s : h_1, h_2, \ldots, h_r)$ so that if $n \geq n_0$ and $\phi : C([n], s] \longrightarrow [r]$ is any function, then there exists an integer $\alpha \in [r]$ and a subset $H_\alpha \subseteq [n]$ with $|H_\alpha| = h_\alpha$ so that $\phi(S) = \alpha$ for every $S \in C(H_\alpha, s)$.*

We don't include the proof of this general statement here, but the more ambitious students may attempt it on their own. Note that the case $s = 1$ is just the Pigeon Hole Principle, while the case $s = r = 2$ is just Ramsey's Theorem for Graphs. An argument using double induction is required for the proof in the general case. The first induction is on r and the second is on s.

11.6 The Probabilistic Method

At the outset of this chapter, we presented Erdős' original proof for the lower bound for the Ramsey number $R(n, n)$ using counting. Later, we recast the proof in a probabilistic setting. History has shown that this second perspective is the right one. To illustrate the power of this approach, we present a classic theorem, which is also due to Erdős, showing that there are graphs with large girth and large chromatic number.

The **girth** g of a graph G is the smallest integer for which G contains a cycle on g vertices. The girth of a forest is taken to be infinite, while the girth of a graph is three if and only if it has a triangle. You can check the families of triangle-free, large chromatic number, graphs constructed in Chapter 5 and see that each has girth four.

Theorem 11.7 (Erdős). *For every pair g, t of integers with $g \geq 3$, there exists a graph G with $\chi(G) > t$ and the girth of G greater than g.*

Proof. Before proceeding with the details of the argument, let's pause to get the general idea behind the proof. We choose integers n and s with $n > s$, and it will eventually be clear how large they need to be in terms of g and t. We will then consider a random graph on vertex set $\{1, 2, \ldots, n\}$, and just as before, for each i and j with $1 \leq i < j \leq n$, the probability that the pair ij is an edge is p, but now p will depend on n. Of course, the probability that any given pair is an edge is completely independent of all other pairs.

Our first goal is to choose the values of n, s and p so that with high probability, a random graph does not have an independent set of size s. You might think as a second goal, we would try to get a random graph without small cycles. But this goal is too restrictive. Instead, we just try to get a graph in which there are relatively few small cycles. In fact, we want the number of small cycles to be less than $n/2$. Then we will

234

remove one vertex from each small cycles, resulting in a graph with at least $n/2$ vertices, having no small cycles and no independent set of size s. The chromatic number of this graph is at least $n/2s$, so we will want to have the inequality $n > 2st$.

Now for some details. Let X_1 be the random variable that counts the number of s-element independent sets. Then

$$E(X_1) = \binom{n}{s}(1-p)^{C(s,2)}$$

Now we want $E(X_1) < 1/4$. Since $C(n,s) \leq n^s = e^{s \ln n}$ and $(1-p)^{C(s,2)} \leq e^{-ps^2/2}$, it suffices to set $s = 2\ln n/p$. By Markov's Inequality, the probability that X_1 exceeds $1/2 \geq 2E(X_1)$ is less than $1/2$.

Now let X_2 count the number of cycles in G of size at most g. Then

$$E(X_2) \leq \sum_{i=3}^{g} n(n-1)(n-2)\ldots(n-i+1)p^i \leq g(pn)^g.$$

Now, we want $E(X_2) \leq n/4$, and an easy calculation shows that $g(np)^g \leq n/4$ when $p = n^{1/g-1}/10$. Again by Markov's Inequality, the probability that X_2 exceeds $n/2 \geq 2E(X_2)$ is less than $1/2$.

We conclude that there is a graph G for which $X_1 = 0$ and $X_2 \leq n/2$. Remove a vertex from each of the small cycles in G and let H be the graph that remains. Clearly, H has at least $n/2$ vertices, no cycle of size at most g and no independent set of size s. Finally, the inequality $n > 2st$ requires $n^{1/g}/(40 \ln n) > t$. □

11.6.1 Gaining Intuition with the Probabilistic Method

Experienced researchers are able to simplify the calculations in an argument of this type, as they know what can safely be discarded and what can not. Here's a quick tour of the essential steps. We want $E(X_1)$ to be small, so we set $n^s e^{-ps^2} = 1$ and get $s = \ln n/p$. We want the number of small cycles to be about n so we set $(gp)^g = n$ and get $p = n^{1/g-1}$. Finally, we want $n = st$ which requires $n^{1/g} = t$. The rest is just paying attention to details.

11.7 Discussion

Zori started the conversation with "Who in their right mind would trust their lives to an algorithm that used random methods?" Xing quickly responded "Everyone. At least everyone should. We routinely deal with probabilistic concepts, like getting run over by a bus when crossing the street or having a piano fall on our head. The general public

Chapter 11 Applying Probability to Combinatorics

is much more comfortable with notions of probability, even though they may never know the formal definition of a probability space. I for one am completely comfortable taking an airline flight if I can be assured that the probability of a disaster is less than 10^{-20}."

Dave wasn't biting on this topic. Instead he offered "You have to be struck by the statements that it appears difficult to construct objects which you can prove exist in abundance. I wonder why this is so." Alice said "We all find your brain to be a totally random thing, sometimes making sense but often not." There was laughter or at least some snickering. But after a bit, Carlos said "There's something fundamental here. Maybe one could prove that there are easily stated theorems which only have long proofs." Bob blurted "That doesn't make any sense." Zori saw an opportunity where a client would, at considerable expense, commission her to solve a problem (at least better than the competition) that was readily understood but somehow difficult in the end. She knew about the class $\mathcal{NP}$ but maybe there were even bigger challenges (and bigger paychecks) out there.

11.8 Exercises

1. Consider a random graph with vertex set $\{1, 2, ; n\}$. If the edge probability is $p = 1/2$, then let X denote the number of complete subgraphs of size $t = 2 \log n$ and let Y denote the number of independent sets of size $t = 2 \log n$.

    (a) Show that $E(X + Y) < 1$, when n is sufficiently large.

    (b) Use the result from part a to show that $\omega(G)$ is less than $2 \log n$, while the chromatic number of G is at least $n/(2 \log n)$ (both statements holding with high probability). As a result, the basic inequality $\chi(G) \geq \omega(G)$ is far from being tight for a random graph.

2. We form a random tournament as follows. Start with a complete graph with vertex set $\{1, 2, \ldots, n\}$. For each distinct pair i, j with $1 \leq i < j \leq n$, flip a fair coin. If the result is heads, orient the edge from i to j, which we denote by (x, y). If the toss is tails, then the edge is oriented from j to i, denoted (y, x). Show that when n is large, with high probability, the following statement is true: For every set S of size $\log n/10$, there is a vertex x so that (x, y) in T for every $y \in S$.

3. Let T be a random tournament on n vertices. Show that with high probability, the following statement is true: For every pair x, y of distinct vertices, either (1) (x, y) in T, or (2) there is a vertex z for which both (x, z) and (z, y) are in T.

4. Many statements for random graphs exhibit a **threshold behavior**. Show that a random graph with edge probability $p = 10 \log n/n$ almost certainly has no isolated

vertices, while a random graph with edge probability $p = \log n/(10n)$ almost certainly has at least one isolated vertices.

5. In the sense of the preceding problem, determine the threshold probability for a graph to be connected.

CHAPTER 12

Graph Algorithms

In previous chapters, we have encountered a few algorithms for problems involving discrete structures such as finding euler circuits (Chapter 5) or partitioning a poset into antichains (Chapter 6). This chapter begins a sequence of three chapters that focus on algorithms. In this chapter we explore two minimization problems for graphs in which we assign a weight to each edge of the graph. The first problem studied is determining a spanning tree of minimum weight. The second is of finding shortest paths from a root vertex to each other vertex in a directed graph.

12.1 Minimum Weight Spanning Trees

In this section, we consider pairs $(\mathbf{G}, w)$ where $\mathbf{G} = (V, E)$ is a connected graph and $w \colon E \to \mathbb{N}_0$. For each edge $e \in E$, the quantity $w(e)$ is called the **weight** of e. Given a set S of edges, we define the **weight** of S, denoted $w(S)$, by setting $w(S) = \sum_{e \in S} w(e)$. In particular, the weight of a spanning tree T is just the sum of the weights of the edges in T.

Weighted graphs arise in many contexts. One of the most natural is when the weights on the edges are distances or costs. For example, consider the weighted graph in Figure 12.1. Suppose the vertices represent nodes of a network and the edges represent the ability to establish direct physical connections between those nodes. The weights associated to the edges represent the cost (let's say in thousands of dollars) of building those connections. The company establishing the network among the nodes only cares that there is a way to get data between each pair of nodes. Any additional links would create redundancy in which they are not interested at this time. A spanning tree of the graph ensures that each node can communicate with each of the others and has no redundancy, since removing any edge disconnects it. Thus, to minimize the cost of building the network, we want to find a minimum weight (or cost) spanning tree.

Chapter 12 Graph Algorithms

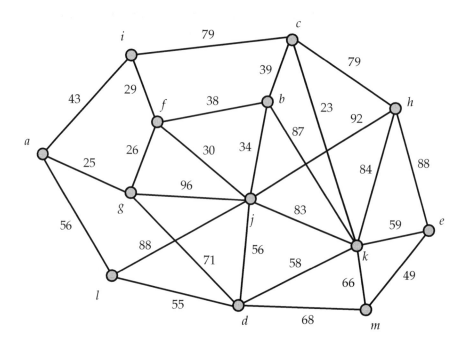

FIGURE 12.1: A WEIGHTED GRAPH

To do this, this section considers the following problem:

Problem 12.2. Find a minimum weight spanning tree **T** of **G**.

To solve this problem, we will develop *two* efficient graph algorithms, each having certain computational advantages and disadvantages. Before developing the algorithms, we need to establish some preliminaries about spanning trees and forests.

12.1.1 Preliminaries

The following proposition about the number of components in a spanning forest of a graph **G** has an easy inductive proof. You are asked to provide it in the exercises.

Proposition 12.3. *Let* $G = (V, E)$ *be a graph on n vertices, and let* $H = (V, S)$ *be a spanning forest. Then* $0 \leq |S| \leq n - 1$. *Furthermore, if* $|S| = n - k$, *then* **H** *has k components. In particular,* **H** *is a spanning tree if and only if it contains* $n - 1$ *edges.*

The following proposition establishes a way to take a spanning tree of a graph, remove an edge from it, and add an edge of the graph that is not in the spanning tree to

240

create a new spanning tree. Effectively, the process exchanges two edges to form the new spanning tree, so we call this the **exchange principle**.

Proposition 12.4 (Exchange Principle). *Let* $\mathbf{T} = (V, S)$ *be spanning tree in a graph* $\mathbf{G}$, *and let* $e = xy$ *be an edge of* $\mathbf{G}$ *which does not belong to* $\mathbf{T}$. *Then*

1. *There is a* unique *path* $P = (x_0, x_1, x_2, \ldots, x_t)$ *with (a)* $x = x_0$; *(b)* $y = x_t$; *and (c)* $x_i x_{i+1} \in S$ *for each* $i = 0, 1, 2, \ldots, t-1$.

2. *For each* $i = 0, 1, 2, \ldots, t-1$, *let* $f_i = x_i x_{i+1}$ *and then set*

$$S_i = \{e\} \cup \{g \in S : g \neq f_i\},$$

i.e., we **exchange** *edge* f_i *for edge* e. *Then* $\mathbf{T}_i = (V, S_i)$ *is a spanning tree of* $\mathbf{G}$.

Proof. For the first fact, it suffices to note that if there were more than one distinct path from x to y in $\mathbf{T}$, we would be able to find a cycle in $\mathbf{T}$. This is impossible since it is a tree. For the second, we refer to Figure 12.5. The black and green edges in the graph shown at the left represent the spanning tree $\mathbf{T}$. Thus, f lies on the unique path from x to y in $\mathbf{T}$ and $e = xy$ is an edge of $\mathbf{G}$ *not* in $\mathbf{T}$. Adding e to $\mathbf{T}$ creates a graph with a unique cycle, since $\mathbf{T}$ had a unique path from x to y. Removing f (which could be any edge f_i of the path, as stated in the proposition) destroys this cycle. Thus $\mathbf{T}_i$ is a connected acyclic subgraph of $\mathbf{G}$ with $n - 1 + 1 - 1 = n - 1$ edges, so it is a spanning tree.

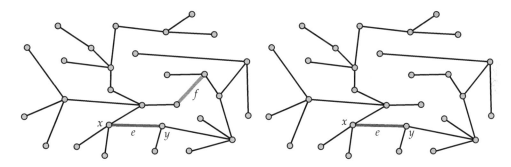

FIGURE 12.5: THE EXCHANGE PRINCIPLE

□

For both of the algorithms we develop, the argument to show that the algorithm is optimal rests on the following technical lemma. To avoid trivialities, we assume $n \geq 3$.

Chapter 12 Graph Algorithms

Lemma 12.6. *Let* **F** *be a spanning forest of* **G** *and let C be a component of* **F**. *Also, let* $e = xy$ *be an edge of minimum weight among all edges with one endpoint in C and the other not in C. Then among all spanning trees of* **G** *that contain the forest* **F**, *there is one of minimum weight that contains the edge e.*

Proof. Let $\mathbf{T} = (V, S)$ be any spanning tree of minimum weight among all spanning trees that contain the forest **F**, and suppose that $e = xy$ is not an edge in **T**. (If it were an edge in **T**, we would be done.) Then let $P = (x_0, x_1, x_2, \ldots, x_t)$ be the unique path in **T** with (a) $x = x_0$; (b) $y = x_t$; and (c) $x_i x_{i+1} \in S$ for each $i = 0, 1, 2, \ldots, t-1$. Without loss of generality, we may assume that $x = x_0$ is a vertex in C while $y = x_t$ does not belong to C. Then there is a least non-negative integer i for which x_i is in C and x_{i+1} is not in C. It follows that x_j is in C for all j with $0 \leq j \leq i$.

Let $f = x_i x_{i+1}$. The edge e has minimum weight among all edges with one endpoint in C and the other not in C, so $w(e) \leq w(f)$. Now let $\mathbf{T}_i$ be the tree obtained by exchanging the edge f for edge e. It follows that $w(\mathbf{T}_i) = w(\mathbf{T}) - w(f) + w(e) \leq w(\mathbf{T})$. Furthermore, $\mathbf{T}_i$ contains the spanning forest **F** as well as the edge e. It is therefore the minimum weight spanning tree we seek. □

Discussion 12.7. Although Bob's combinatorial intuition has improved over the course he doesn't quite understand why we need special algorithms to find minimum weight spanning trees. He figures there can't be that many spanning trees, so he wants to just write them down. Alice groans as she senses that Bob must have been absent when the material from Section 5.6 was discussed. In that section, we learned that a graph on n vertices can have as many as n^{n-2} spanning trees (or horrors, the instructor may have left it off the syllabus). Regardless, this exhaustive approach is already unusable when $n = 20$. Dave mumbles something about being greedy and just adding the lightest edges one-by-one while never adding an edge that would make a cycle. Zori remembers a strategy like this working for finding the height of a poset, but she's worried about the nightmare situation that we learned about with using FirstFit to color graphs. Alice agrees that greedy algorithms have an inconsistent track record but suggests that Lemma 12.6 may be enough to get one to succeed here.

12.1.2 Kruskal's Algorithm

In this section, we develop one of the best known algorithms for finding a minimum weight spanning tree. It is known as **Kruskal's Algorithm**, although some prefer the descriptive label *Avoid Cycles* because of the way it builds the spanning tree.

To start Kruskal's algorithm, we sort the edges according to weight. To be more precise, let m denote the number of edges in $\mathbf{G} = (V, E)$. Then label the edges as $e_1, e_2, e_3, \ldots, e_m$ so that $w(e_1) \leq w(e_2) \leq \cdots \leq w(e_m)$. Any of the many available efficient sorting algorithms can be used to do this step.

12.1 Minimum Weight Spanning Trees

Once the edges are sorted, Kruskal's algorithm proceeds to an initialization step and then inductively builds the spanning tree $\mathbf{T} = (V, S)$:

Algorithm 12.8 (Kruskal's Algorithm).

Initialization. Set $S = \emptyset$ and $i = 0$.

Inductive Step. While $|S| < n - 1$, let j be the least non-negative integer so that $j > i$ and there are no cycles in $S \cup \{e_j\}$. Then (using pseudo-code) set

$$i = j \quad \text{and} \quad S = S \cup \{j\}.$$

The correctness of Kruskal's Algorithm follows from an inductive argument. First, the set S is initialized as the empty set, so there is certainly a minimum weight spanning tree containing all the edges in S. Now suppose that for some i with $0 \le i < n$, $|S| = i$ and there is a minimum weight spanning tree containing all the edges in S. Let $\mathbf{F}$ be the spanning forest determined by the edges in S, and let $C_1, C_2, \ldots, C_s$ be the components of $\mathbf{F}$. For each $k = 1, 2, \ldots, s$, let f_k be a minimum weight edge with one endpoint in C_k and the other not in C_k. Then the edge e added to S by Kruskal's Algorithm is just the edge $\{f_1, f_2, \ldots, f_s\}$ having minimum weight. Applying Lemma 12.6 and the inductive hypothesis, we know that there will still be a minimum weight spanning tree of $\mathbf{G}$ containing all the edges of $S \cup \{e\}$.

Example 12.9 (Kruskal's Algorithm).

Let's see what Kruskal's algorithm does on the weighted graph in Figure 12.1. It first sorts all of the edges by weight. We won't reproduce the list here, since we won't need all of it. The edge of least weight is ck, which has weight 23. It continues adding the edge of least weight, adding ag, fg, fi, fj, and bj. However, after doing this, the edge of lowest weight is fb, which has weight 38. This edge cannot be added, as doing so would make fjb a cycle. Thus, the algorithm bypasses it and adds bc. Edge ai is next inspected, but it, too, would create a cycle and is eliminated from consideration. Then em is added, followed by dl. There are now *two* edges of weight 56 to be considered: al and dj. Our sorting algorithm has somehow decided one of them should appear first, so let's say it's dj. After adding dj, we cannot add al, as $agfjdl$ would form a cycle. Edge dk is next considered, but it would also form a cycle. However, ek can be added. Edges km and dm are then bypassed. Finally, edge ch is added as the twelfth and final edge for this 13-vertex spanning tree. The full list of edges added (in order) is shown to the right. The total weight of this spanning tree is 504.

c	k	23
a	g	25
f	g	26
f	i	29
f	j	30
b	j	34
b	c	39
e	m	49
d	l	55
d	j	56
e	k	59
c	h	79

Chapter 12 Graph Algorithms

12.1.3 Prim's Algorithm

We now develop **Prim's Algorithm** for finding a minimum weight spanning tree. This algorithm is also known by a more descriptive label: *Build Tree*. We begin by choosing a root vertex r. Again, the algorithm proceeds with an initialization step followed by a series of inductive steps.

Algorithm 12.10 (Prim's Algorithm).

Initialization. *Set $W = \{r\}$ and $S = \emptyset$.*

Inductive Step. *While $|W| < n$, let e be an edge of minimum weight among all edges with one endpoint in W and the other not in W. If $e = xy$, $x \in W$ and $y \notin W$, update W and S by setting (using pseudo-code)*

$$W = W \cup \{y\} \quad \text{and} \quad S = S \cup \{e\}.$$

The correctness of Prim's algorithm follows immediately from Lemma 12.6.

Example 12.11 (Prim's Algorithm).

Let's see what Prim's algorithm does on the weighted graph in Figure 12.1. We start with vertex a as the root vertex. The lightest edge connecting a (the only vertex in the tree so far) to the rest of the graph is ag. Next, fg is added. This is followed by fi, fj, bj, and bc. Next, the algorithm identifies ck as the lightest edge connecting $\{a, g, i, f, j, b, c\}$ to the remaining vertices. Notice that this is considerably later than Kruskal's algorithm finds the same edge. The algorithm then determines that al and jd, both of weight 56 are the lightest edges connecting vertices in the tree to the other vertices. It picks arbitrarily, so let's say it takes al. It next finds dl, then ek, and then em. The final edge added is ch. The full list of edges added (in order) is shown to the right. The total weight of this spanning tree is 504. This (not surprisingly) the same weight we obtained using Kruskal's algorithm. However, notice that the spanning tree found is different, as this one contains al instead of dj. This is not an issue, of course, since in both cases an arbitrary choice between two edges of equal weight was made.

a g 25
f g 26
f i 29
f j 30
b j 34
b c 39
c k 23
a l 56
d l 55
e k 59
e m 49
c h 79

12.1.4 Comments on Efficiency

An implementation of Kruskal's algorithm seems to require that the edges be sorted. If the graph has n vertices and m edges, this requires $m \log m$ operations just for the sort. But once the sort is done, the process takes only $n - 1$ steps—provided you keep track of the components as the spanning forest expands. Regardless, it is easy to see that at most $O(n^2 \log n)$ operations are required.

On the other hand, an implementation of Prim's algorithm requires the program to conveniently keep track of the edges incident with each vertex and always be able to identify the edge with least weight among subsets of these edges. In computer science, the data structure that enables this task to be carried out is called a **heap**.

12.2 Digraphs

In this section, we introduce another useful variant of a graph. In a graph, the existence of an edge xy can be used to model a connection between x and y that goes in both ways. However, sometimes such a model is insufficient. For instance, perhaps it is possible to fly from Atlanta directly to Fargo but not possible to fly from Fargo directly to Atlanta. In a graph representing the airline network, an edge between Atlanta and Fargo would lose the information that the flights only operate in one direction. To deal with this problem, we introduce a new discrete structure. A **digraph G** is a pair (V, E) where V is a vertex set and $E \subset V \times V$ with $x \neq y$ for every $(x, y) \in E$. We consider the pair (x, y) as a **directed edge** from x to y. Note that for distinct vertices x and y from V, the ordered pairs (x, y) and (y, x) are distinct, so the digraph may have one, both or neither of the directed edges (x, y) and (y, x). This is in contrast to graphs, where edges are sets, so $\{x, y\}$ and $\{y, x\}$ are the same.

Diagrams of digraphs use arrowheads on the edges to indicate direction. This is illustrated in Figure 12.12. For example, the digraph illustrated there contains the edge (a, f) but not the edge (f, a). It does contain both edges (c, d) and (d, c), however. When **G** is a digraph, a sequence $P = (r = u_0, u_1, \ldots, u_t = x)$ of distinct vertices is called a **directed path** from r to x when $(u_i u_{i+1})$ is a directed edge in **G** for every $i = 0, 1, \ldots, t-1$. A directed path $C = (r = u_0, u_1, \ldots, u_t = x)$ is called a **directed cycle** when (u_t, u_0) is a directed edge of **G**.

12.3 Dijkstra's Algorithm for Shortest Paths

Just as with graphs, it is useful to assign weights to the directed edges of a digraph. Specifically, in this section we consider a pair $(\mathbf{G}, w)$ where $\mathbf{G} = (V, E)$ is a digraph and $w \colon E \to \mathbb{N}_0$ is a function assigning to each directed edge (x, y) a non-negative weight $w(x, y)$. However, in this section, we interpret weight as **distance** so that $w(x, y)$ is now called the **length** of the edge (x, y). If $P = (r = u_0, u_1, \ldots, u_t = x)$ is a directed path from r to x, then the **length** of the path P is just the sum of the lengths of the edges in the path, $\sum_{i=0}^{t-1} w(u_i u_{i+1})$. The **distance** from r to x is then defined to be the

Chapter 12 Graph Algorithms

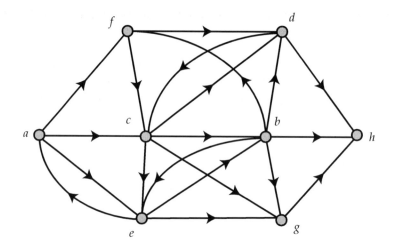

FIGURE 12.12: A DIGRAPH

minimum length of a directed path from r to x. Our goal in this section is to solve the following natural problem, which has many applications:

Problem 12.13. For each vertex x, find the distance from r to x. Also, find a shortest path from r to x.

12.3.1 Description of the Algorithm

To describe **Dijkstra's algorithm** in a compact manner, it is useful to extend the definition of the function w. We do this by setting $w(x, y) = \infty$ when $x \neq y$ and (x, y) is not a directed edge of **G**. In this way, we will treat ∞ as if it were a number (although it is not!).[1]

We are now prepared to describe Dijkstra's Algorithm.

Algorithm 12.14 (Dijkstra's Algorithm). *Let $n = |V|$. At Step i, where $1 \leq i \leq n$, we will have determined:*

1. *A sequence $\sigma = (v_1, v_2, v_3, \ldots, v_i)$ of distinct vertices from **G** with $r = v_1$. These vertices are called **permanent vertices**, while the remaining vertices will be called **temporary vertices**.*

---

[1]This is not an issue for computer implementation of the algorithm, as instead of using ∞, a value given by the product of the number of vertices and the maximum edge weight may be used to simulate infinity.

2. For each vertex $x \in V$, we will have determined a number $\delta(x)$ and a path $P(x)$ from r to x of length $\delta(x)$.

Initialization (Step 1) Set $i = 1$. Set $\delta(r) = 0$ and let $P(r) = (r)$ be the trivial one-point path. Also, set $\sigma = (r)$. For each $x \neq r$, set $\delta(x) = w(r,x)$ and $P(x) = (r,x)$. Let x be a temporary vertex for which $\delta(x)$ is minimum. Set $v_2 = x$, and update σ by appending v_2 to the end of it. Increment i.

Inductive Step (Step i, $i > 1$) If $i < n$, then for each temporary x, let

$$\delta(x) = \min\{\delta(x), \delta(v_i) + w(v_i, x)\}.$$

If this assignment results in a reduction in the value of $\delta(x)$, let $P(x)$ be the path obtained by adding x to the end of $P(v_i)$.

Let x be a temporary vertex for which $\delta(x)$ is minimum. Set $v_{i+1} = x$, and update σ by appending v_{i+1} to it. Increment i.

12.3.2 Example of Dijkstra's Algorithm

Before establishing why Dijkstra's algorithm works, it may be helpful to see an example of how it works. To do this, consider the digraph **G** shown in Figure 12.15. For visual clarity, we have chosen a digraph which is an **oriented graph**, i.e., for each distinct pair x, y of vertices, the graph contains at most one of the two possible directed edges (x, y) and (y, x). Suppose that the root vertex r is the vertex labeled a. The initialization step of Dijkstra's algorithm then results in the following values for δ and P:

Step 1. Initialization

$\sigma = (a)$

$\delta(a) = 0;$ $P(a) = (a)$
$\delta(b) = \infty;$ $P(b) = (a, b)$
$\delta(c) = 47;$ $P(c) = (a, c)$
$\delta(d) = \infty;$ $P(d) = (a, d)$
$\delta(e) = 70;$ $P(e) = (a, e)$
$\delta(f) = 24;$ $P(f) = (a, f)$
$\delta(g) = \infty;$ $P(g) = (a, g)$
$\delta(h) = \infty;$ $P(h) = (a, h)$

Before finishing Step 1, the algorithm identifies vertex f as closest to a and appends it to σ, making a permanent. When entering Step 2, Dijkstra's algorithm attempts to

Chapter 12 Graph Algorithms

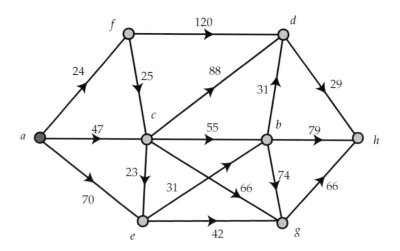

FIGURE 12.15: A DIGRAPH WITH EDGE LENGTHS

find shorter paths from a to each of the temporary vertices by going through f. We call this process "scanning from vertex f." In this scan, the path to vertex d is updated, since $\delta(f) + w(f,d) = 24 + 120 = 144 < \infty = w(a,d)$.

Step 2. Scan from vertex f

$\sigma = (a, f)$

$\delta(a) = 0;$	$P(a) = (a)$
$\delta(b) = \infty;$	$P(b) = (a, b)$
$\delta(c) = 47;$	$P(c) = (a, c)$
$\delta(d) = 144 = 24 + 120 = \delta(f) + w(f, d);$	$P(d) = (a, f, d)$ updated
$\delta(e) = 70;$	$P(e) = (a, e)$
$\delta(f) = 24;$	$P(f) = (a, f)$
$\delta(g) = \infty;$	$P(g) = (a, f)$
$\delta(h) = \infty;$	$P(h) = (a, h)$

Before proceeding to the next step, vertex c is made permanent by making it v_3. In Step 3, therefore, the scan is from vertex c. Vertices b, d, and g have their paths updated. However, although $\delta(c) + w(c, e) = 47 + 23 = 70 = \delta(e)$, we do not change $P(e)$ since $\delta(e)$ is not *decreased* by routing $P(e)$ through c.

12.3 Dijkstra's Algorithm for Shortest Paths

Step 3. Scan from vertex c

$\sigma = (a, f, c)$

$\delta(a) = 0;$ $P(a) = (a)$
$\delta(b) = 102 = 47 + 55 = \delta(c) + w(c,b);$ $P(b) = (a,c,b)$ updated
$\delta(c) = 47;$ $P(c) = (a,c)$
$\delta(d) = 135 = 47 + 88 = \delta(c) + w(c,d);$ $P(d) = (a,c,d)$ updated
$\delta(e) = 70;$ $P(e) = (a,e)$
$\delta(f) = 24;$ $P(f) = (a,f)$
$\delta(g) = 113 = 47 + 66 = \delta(c) + w(c,g);$ $P(g) = (a,c,g)$ updated
$\delta(h) = \infty;$ $P(h) = (a,h)$

Now vertex e is made permanent.

Step 4. Scan from vertex e

$\sigma = (a, f, c, e)$

$\delta(a) = 0;$ $P(a) = (a)$
$\delta(b) = 101 = 70 + 31 = \delta(e) + w(e,b);$ $P(b) = (a,e,b)$ updated
$\delta(c) = 47;$ $P(c) = (a,c)$
$\delta(d) = 135;$ $P(d) = (a,c,d)$
$\delta(e) = 70;$ $P(e) = (a,e)$
$\delta(f) = 24;$ $P(f) = (a,f)$
$\delta(g) = 112 = 70 + 42 = \delta(e) + w(e,g);$ $P(g) = (a,e,g)$ updated
$\delta(h) = \infty;$ $P(h) = (a,h)$

Now vertex b is made permanent.

Step 5. Scan from vertex b

$\sigma = (a, f, c, e, b)$

$\delta(a) = 0;$ $P(a) = (a)$
$\delta(b) = 101;$ $P(b) = (a,e,b)$
$\delta(c) = 47;$ $P(c) = (a,c)$
$\delta(d) = 132 = 101 + 31 = \delta(b) + w(b,d);$ $P(d) = (a,e,b,d)$ updated
$\delta(e) = 70;$ $P(e) = (a,e)$

Chapter 12 Graph Algorithms

$\delta(f) = 24;$ $\qquad$ $P(f) = (a, f)$
$\delta(g) = 112;$ $\qquad$ $P(g) = (a, e, g)$
$\delta(h) = 180 = 101 + 79 = \delta(b) + w(b, h);$ $\qquad$ $P(h) = (a, e, b, h)$ updated

Now vertex g is made permanent.

Step 6. Scan from vertex g

$\sigma = (a, f, c, e, b, g)$
$\delta(a) = 0;$ $\qquad$ $P(a) = (a)$
$\delta(b) = 101;$ $\qquad$ $P(b) = (a, e, b)$
$\delta(c) = 47;$ $\qquad$ $P(c) = (a, c)$
$\delta(d) = 132;$ $\qquad$ $P(d) = (a, e, b, d)$
$\delta(e) = 70;$ $\qquad$ $P(e) = (a, e)$
$\delta(f) = 24;$ $\qquad$ $P(f) = (a, f)$
$\delta(g) = 112;$ $\qquad$ $P(g) = (a, e, g)$
$\delta(h) = 178 = 112 + 66 = \delta(g) + w(g, h);$ $\qquad$ $P(h) = (a, e, g, h)$ updated

Now vertex d is made permanent.

Step 7. Scan from vertex d

$\sigma = (a, f, c, e, b, g, d)$
$\delta(a) = 0;$ $\qquad$ $P(a) = (a)$
$\delta(b) = 101;$ $\qquad$ $P(b) = (a, e, b)$
$\delta(c) = 47;$ $\qquad$ $P(c) = (a, c)$
$\delta(d) = 132;$ $\qquad$ $P(d) = (a, e, b, d)$
$\delta(e) = 70;$ $\qquad$ $P(e) = (a, e)$
$\delta(f) = 24;$ $\qquad$ $P(f) = (a, f)$
$\delta(g) = 112;$ $\qquad$ $P(g) = (a, e, g)$
$\delta(h) = 161 = 132 + 29 = \delta(d) + w(d, h);$ $\qquad$ $P(h) = (a, e, b, d, h)$ updated

Now vertex h is made permanent. Since this is the last vertex, the algorithm halts and returns the following:

Final Results of Dijkstra's Algorithm

$\sigma = (a, f, c, e, b, g, d, h)$

$\delta(a) = 0;$ $P(a) = (a)$

$\delta(b) = 101;$ $P(b) = (a, e, b)$

$\delta(c) = 47;$ $P(c) = (a, c)$

$\delta(d) = 132;$ $P(d) = (a, e, b, d)$

$\delta(e) = 70;$ $P(e) = (a, e)$

$\delta(f) = 24;$ $P(f) = (a, f)$

$\delta(g) = 112;$ $P(g) = (a, e, g)$

$\delta(h) = 161;$ $P(h) = (a, e, b, d, h)$

12.3.3 The Correctness of Dijkstra's Algorithm

Now that we've illustrated Dijkstra's algorithm, it's time to prove that it actually does what we claimed it does: find the distance from the root vertex to each of the other vertices and a path of that length. To do this, we first state two elementary propositions. The first is about shortest paths in general, while the second is specific to the sequence of permanent vertices produced by Dijkstra's algorithm.

Proposition 12.16. *Let x be a vertex and let $P = (r = u_0, u_1, \ldots, u_t = x)$ be a shortest path from r to x. Then for every integer j with $0 < j < t$, $(u_0, u_1, \ldots, u_j)$ is a shortest path from r to u_j and $(u_j, u_{j+1}, \ldots, u_t)$ is a shortest path from u_j to u_t.*

Proposition 12.17. *When the algorithm halts, let $\sigma = (v_1, v_2, v_3, \ldots, v_n)$. Then*

$$\delta(v_1) \leq \delta(v_2) \leq \cdots \leq \delta(v_n).$$

We are now ready to prove the correctness of the algorithm. The proof we give will be inductive, but the induction will have nothing to do with the total number of vertices in the digraph or the step number the algorithm is in.

Theorem 12.18. *Dijkstra's algorithm yields shortest paths for every vertex x in $\mathbf{G}$. That is, when Dijkstra's algorithm terminates, for each $x \in V$, the value $\delta(x)$ is the distance from r to x and $P(x)$ is a shortest path from r to x.*

Proof. The theorem holds trivially when $x = r$. So we consider the case where $x \neq r$. We argue that $\delta(x)$ is the distance from r to x and that $P(x)$ is a shortest path from r to x by induction on the minimum number k of edges in a shortest path from r to x.

Chapter 12 Graph Algorithms

When $k = 1$, the edge (r,x) is a shortest path from r to x. Since $v_1 = r$, we will set $\delta(x) = w(r,x)$ and $P(x) = (r,x)$ at Step 1.

Now fix a positive integer k. Assume that if the minimum number of edges in a shortest path from r to x is at most k, then $\delta(x)$ is the distance from r to x and $P(x)$ is a shortest path from r to x. Let x be a vertex for which the minimum number of edges in a shortest path from r to x is $k+1$. Fix a shortest path $P = (u_0, u_1, u_2, \ldots, u_{k+1})$ from $r = u_0$ to $x = u_{k+1}$. Then $Q = (u_0, u_1, \ldots, u_k)$ is a shortest path from r to u_k. (See Figure 12.19.)

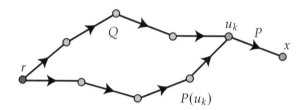

FIGURE 12.19: SHORTEST PATHS

By the inductive hypothesis, $\delta(u_k)$ is the distance from r to u_k, and $P(u_k)$ is a shortest path from r to u_k. Note that $P(u_k)$ need not be the same as path Q, as we suggest in Figure 12.19. However, if distinct, the two paths will have the same length, namely $\delta(u_k)$. Also, the distance from r to x is $\delta(u_k) + w(u_k, x) \geq \delta(u_k)$ since P is a shortest path from r to x and $w(u_k, x) \geq 0$.

Let i and j be the unique integers for which $u_k = v_i$ and $x = v_j$. If $j < i$, then

$$\delta(x) = \delta(v_j) \leq \delta(v_i) = \delta(u_k) \leq \delta(u_k) + w(u_k).$$

Therefore the algorithm has found a path $P(x)$ from r to x having length $\delta(x)$ which is at most the distance from r to x. Clearly, this implies that $\delta(x)$ is the distance from r to x and that $P(x)$ is a shortest path.

On the other hand, if $j > i$, then the inductive step at Step i results in

$$\delta(x) \leq \delta(v_i) + w(v_i, y) = \delta(u_k) + w(u_k, x).$$

As before, this implies that $\delta(x)$ is the distance from r to x and that $P(x)$ is a shortest path. □

12.4 Historical Notes

Kruskal's algorithm was published in 1956 by Joseph B. Kruskal in a three-page paper that appeared in *Proceedings of the American Mathematical Society*. Robert C. Prim published the algorithm that now bears his name the following year in *The Bell System Technical Journal*. Prim's paper focuses on application of the minimum weight (or length or cost) spanning tree problem to telephone networks. He was aware of Kruskal's prior work, as they were colleagues at Bell Laboratories at the time he published his paper. It turns out that Prim had been beaten to the punch by Czech mathematician Vojtěch Jarník in 1929, so some refer to Prim's algorithm as Jarník's algorithm. (It was later rediscovered by Dijkstra, so some attach his name as well, referring to it as the Dijkstra-Jarník-Prim algorithm.) Edsger Dijkstra published his algorithm for finding shortest paths in 1959 in a three-page paper[1] appearing in *Numerische Mathematik*. In fact, Dijkstra's algorithm had been discovered (in an equivalent form) by Edward F. Moore two years earlier. His result appeared in *Proceedings of an International Symposium on the Theory of Switching*.

12.5 Exercises

1. For the graph in Figure 12.20, use Kruskal's algorithm ("avoid cycles") to find a minimum weight spanning tree. Your answer should include a complete list of the edges, indicating which edges you take for your tree and which (if any) you reject in the course of running the algorithm.

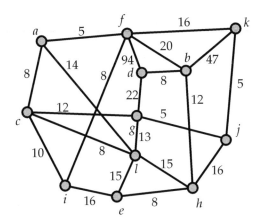

FIGURE 12.20: FIND A MINIMUM WEIGHT SPANNING TREE

---

[1]This is also the paper in which Prim's algorithm was published for the third time. Dijkstra was aware of Kruskal's prior work but argued that his algorithm was preferable because it required that less information about the graph be stored in memory at each step of the algorithm.

Chapter 12 Graph Algorithms

2. For the graph in Figure 12.20, use Prim's algorithm ("build tree") to find a minimum weight spanning tree. Your answer should list the edges selected by the algorithm in the order they were selected.

3. For the graph in Figure 12.21, use Kruskal's algorithm ("avoid cycles") to find a minimum weight spanning tree. Your answer should include a complete list of the edges, indicating which edges you take for your tree and which (if any) you reject in the course of running the algorithm.

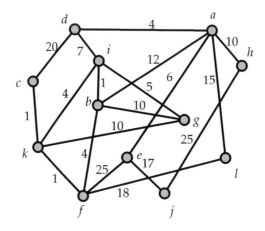

FIGURE 12.21: FIND A MINIMUM WEIGHT SPANNING TREE

4. For the graph in Figure 12.21, use Prim's algorithm ("build tree") to find a minimum weight spanning tree. Your answer should list the edges selected by the algorithm in the order they were selected.

5. For the graph in Figure 12.22, use Kruskal's algorithm ("avoid cycles") to find a minimum weight spanning tree. Your answer should include a complete list of the edges, indicating which edges you take for your tree and which (if any) you reject in the course of running the algorithm.

254

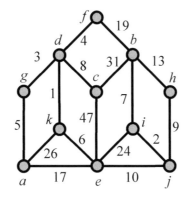

FIGURE 12.22: FIND A MINIMUM WEIGHT SPANNING TREE

6. For the graph in Figure 12.22, use Prim's algorithm ("build tree") to find a minimum weight spanning tree. Your answer should list the edges selected by the algorithm in the order they were selected.

7. A new local bank is being created and will establish a headquarters h, two branches b_1 and b_2, and four ATMs a_1, a_2, a_3, and a_4. They need to build a computer network such that the headquarters, branches, and ATMs can all intercommunicate. Furthermore, they will need to be networked with the Federal Reserve Bank of Atlanta, f. The costs of the feasible network connections (in units of $10,000) are listed below:

hf	80	hb_1	10	hb_2	20	$b_1 b_2$	8
fb_1	12	fa_1	20	$b_1 a_1$	3	$a_1 a_2$	13
ha_2	6	$b_2 a_2$	9	$b_2 a_3$	40	$a_1 a_4$	3
$a_3 a_4$	6						

The bank wishes to minimize the cost of building its network (which must allow for connection, possibly routed through other nodes, from each node to each other node), however due to the need for high-speed communication, they *must* pay to build the connection from h to f as well as the connection from b_2 to a_3. Give a list of the connections the bank should establish in order to minimize their total cost, subject to this constraint. Be sure to explain how you selected the connections and how you know the total cost is minimized.

8. A disconnected weighted graph obviously has no spanning trees. However, it is possible to find a spanning forest of minimum weight in such a graph. Explain how to modify both Kruskal's algorithm and Prim's algorithm to do this.

Chapter 12 Graph Algorithms

9. Prove Proposition 12.3.

10. In the paper where Kruskal's algorithm first appeared, he considered the algorithm a route to a nicer proof that in a connected weighted graph with no two edges having the same weight, there is a *unique* minimum weight spanning tree. Prove this fact using Kruskal's algorithm.

11. Use Dijkstra's algorithm to find the distance from a to each other vertex in the digraph shown in Figure 12.23 and a directed path of that length.

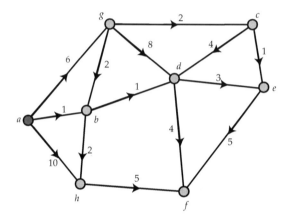

FIGURE 12.23: A DIRECTED GRAPH

12. Table 12.24 contains the length of the directed edge (x, y) in the intersection of *row* x and *column* y in a digraph with vertex set $\{a, b, c, d, e, f\}$. For example, $w(b, d) = 21$. (On the other hand, $w(d, b) = 10$.) Use this data and Dijkstra's algorithm to find the distance from a to each of the other vertices and a directed path of that length from a.

w	a	b	c	d	e	f
a	0	12	8	43	79	35
b	93	0	18	21	60	33
c	17	3	0	37	50	30
d	85	10	91	0	17	7
e	28	47	39	14	0	108
f	31	7	29	73	20	0

TABLE 12.24: A DIGRAPH REPRESENTED AS A TABLE OF DATA

12.5 Exercises

13. Use Dijkstra's algorithm to find the distance from a to each other vertex in the digraph shown in Figure 12.25 and a directed path of that length.

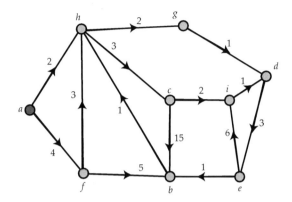

FIGURE 12.25: A DIRECTED GRAPH

14. Table 12.26 contains the length of the directed edge (x, y) in the intersection of *row* x and *column* y in a digraph with vertex set $\{a, b, c, d, e, f\}$. For example, $w(b, d) = 47$. (On the other hand, $w(d, b) = 6$.) Use this data and Dijkstra's algorithm to find the distance from a to each of the other vertices and a directed path of that length from a.

w	a	b	c	d	e	f
a	0	7	17	55	83	42
b	14	0	13	47	27	17
c	37	42	0	16	93	28
d	10	6	8	0	4	32
e	84	19	42	8	0	45
f	36	3	76	5	17	0

TABLE 12.26: A DIGRAPH REPRESENTED AS A TABLE OF DATA

15. Give an example of a digraph having an *undirected* path between each pair of vertices, but having a root vertex r so that Dijkstra's algorithm cannot find a path of finite length from r to some vertex x.

16. Notice that in our discussion of Dijkstra's algorithm, we required that the edge weights be nonnegative. If the edge weights are lengths and meant to model distance, this makes perfect sense. However, in some cases, it might be reasonable to allow negative edge weights. For example, suppose that a positive weight means there is a cost to travel along the directed edge while a negative edge weight means that you make money for traveling along the directed edge. In this case, a directed path with positive total weight results in paying out to travel it, while one with negative total weight results in a profit.

(a) Give an example to show that Dijkstra's algorithm does not always find the path of minimum total weight when negative edge weights are allowed.

(b) Bob and Xing are considering this situation, and Bob suggests that a little modification to the algorithm should solve the problem. He says that if there are negative weights, they just have to find the smallest (i.e., most negative weight) and add the absolute value of that weight to every directed edge. For example, if $w(x, y) \geq -10$ for every directed edge (x, y), Bob is suggesting that they add 10 to every edge weight. Xing is skeptical, and for good reason. Give an example to show why Bob's modification won't work.

CHAPTER 13

Network Flows

This chapter continues our look at the topics of algorithms and optimization. On an intuitive level, networks and network flows are fairly simple. We want to move something (merchandise, water, data) from an initial point to a destination. We have a set of intermediate points (freight terminals, valves, routers) and connections between them (roads, pipes, cables) with each connection able to carry a limited amount. The natural goal is to move as much as possible from the initial point to the destination while respecting each connection's limit. Rather than just guessing at how to perform this maximization, we will develop an algorithm that does it. We'll also see how to easily justify the optimality of our solution though the classic Max Flow-Min Cut Theorem.

13.1 Basic Notation and Terminology

A directed graph in which for each pair of vertices x, y at most one of the directed edges (x, y) and (y, x) between them is present is called an **oriented graph**. The basic setup for a network flow problem begins with an oriented graph **G**, called a **network**, in which we have two special vertices called the **source** and the **sink**. We use the letter S to denote the source, while the letter T is used to denote the sink (terminus). All edges incident with the source are oriented away from the source, while all edges incident with the sink are oriented with the sink. Furthermore, on each edge, we have a non-negative **capacity**, which functions as a constraint on how much can be transmitted via the edge. The capacity of the edge $e = (x, y)$ is denoted $c(e)$ or by $c(x, y)$. In a computer program, the nodes of a network may be identified with integer keys, but in this text, we will typically use letters in labeling the nodes of a network. This helps to distinguish nodes from capacities in diagrams of networks. We illustrate a network in Figure 13.1. The numbers associated with the edges are their capacities, so, for instance, $c(E, B) = 24$ and $c(A, T) = 56$. A **flow** ϕ in a network is a function which assigns to each directed edge $e = (x, y)$ a non-negative value $\phi(e) = \phi(x, y) \leq c(x, y)$ so that the following **conservation laws** hold:

Chapter 13 Network Flows

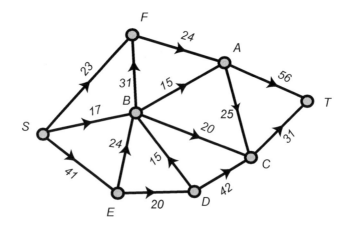

Figure 13.1: A Network

1. $\sum_x \phi(S, x) = \sum_x \phi(x, T)$, i.e., the amount leaving the source is equal to the amount arriving at the sink. This quantity is called the **value** of the flow ϕ.

2. For every vertex y which is neither the source nor the sink the amount leaving y is equal to the amount entering y. That is, $\sum_x \phi(x, y) = \sum_x \phi(y, x)$.

We illustrate a flow in a network in Figure 13.2. In this figure, the numbers associated with each edge are its capacity and the amount of flow that ϕ places on that edge. For example, the edge (E, D) has capacity 20 and currently carries a flow of 8. (Since $\phi(x, y) \le c(x, y)$, it is always easy to determine which number is the capacity and which is the flow.) The value of this flow is $30 = \phi(S, F) + \phi(S, B) + \phi(S, E) = \phi(A, T) + \phi(C, T)$. To see that the second conservation law holds at, for example, vertex B, note that the flow into B is $\phi(S, B) + \phi(E, B) + \phi(D, B) = 20$ and the flow out of B is $\phi(B, F) + \phi(B, A) + \phi(B, C) = 20$.

Given a network, it is very easy to find a flow. We simply assign $\phi(e) = 0$ for every edge e. It is very easy to *underestimate* the importance of this observation, actually. Network flow problems are a special case of a more general class of optimization problems known as **linear programs**, and in general, it may be very difficult to find a feasible solution to a linear programming problem. In fact, conceptually, finding a feasible solution—*any* solution—is just as hard as finding an *optimal* solution.

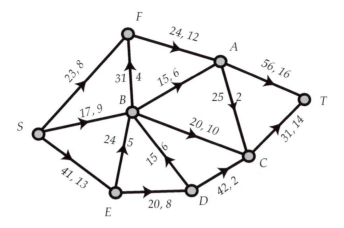

FIGURE 13.2: A NETWORK FLOW

13.2 Flows and Cuts

Considering the applications suggested at the beginning of the chapter, it is natural to ask for the maximum value of a flow in a given network. Put another way, we want to find the largest number v_0 so that there exists a flow ϕ of value v_0 in the network. Of course, we not only want to find the maximum value v_0, but we also want to find a flow ϕ having this value. Although it may seem a bit surprising, we will develop an efficient algorithm which both finds a flow of maximum value *and* finds a certificate verifying the claim of optimality. This certificate makes use of the following important concept.

A partition $V = L \cup U$ of the vertex set V of a network with $S \in L$ and $T \in U$ is called a **cut**.[1] The **capacity** of a cut $V = L \cup U$, denoted $c(L, U)$, is defined by

$$c(L, U) = \sum_{x \in L, y \in U} c(x, y).$$

Put another way, the capacity of the cut $V = L \cup U$ is the total capacity of all edges from L to U. Note that in computing the capacity of the cut $V = L \cup U$, we only add the capacities of the edges from L to U. We do *not* include the edges from U to L in this sum.

---

[1] Our choice of L and U for the names of the two parts of the partition will make more sense later in the chapter.

Chapter 13 Network Flows

Example 13.3. Let's again take a look at the network in Figure 13.2. Let's first consider the cut $V = L_1 \cup U_1$ with

$$L_1 = \{S, F, B, E, D\} \quad \text{and} \quad U_1 = \{A, C, T\}.$$

Here we see that the capacity of the cut is

$$c(L_1, U_1) = c(F, A) + c(B, A) + c(B, C) + c(D, C) = 24 + 15 + 20 + 42 = 101.$$

We must be a bit more careful, however, when we look at the cut $V = L_2 \cup U_2$ with

$$L_2 = \{S, F, B, E\} \quad \text{and} \quad U_2 = \{A, D, C, T\}.$$

Here the capacity of the cut is

$$c(L_2, U_2) = c(F, A) + c(B, A) + c(B, C) + c(E, D) = 24 + 15 + 20 + 20 = 79.$$

Notice that we do not include $c(D, B)$ in the calculation as the directed edge (D, B) is from U_2 to L_2.

The relationship between flows and cuts rests on the following fundamentally important theorem.

Theorem 13.4. *Let $G = (V, E)$ be a network, let ϕ be a flow in G, and let $V = L \cup U$ be a cut. The value of the flow is at most as large as the capacity of the cut.*

Proof. In this proof (and throughout the chapter), we adopt the very reasonable convention that $\phi(x, y) = 0$ if (x, y) is not a directed edge of a network G.

Let ϕ be a flow of value v_0 and let $V = L \cup U$ be a cut. First notice that

$$v_0 = \sum_{y \in V} \phi(S, y) - \sum_{z \in V} \phi(z, S),$$

since the second summation is 0. Also, by the second of our flow conservation laws, we have for any vertex other than the source and the sink,

$$\sum_{y \in V} \phi(x, y) - \sum_{z \in V} \phi(z, x) = 0.$$

Now we have

$$v_0 = \sum_{y \in V} \phi(S, y) - \sum_{z \in V} \phi(z, S)$$

$$= \sum_{y \in V} \phi(S, y) - \sum_{z \in V} \phi(z, S) + \sum_{\substack{x \in L \\ x \neq S}} \left[\sum_{y \in V} \phi(x, y) - \sum_{z \in V} \phi(z, x) \right]$$

$$= \sum_{x \in L} \left[\sum_{y \in V} \phi(x, y) - \sum_{z \in V} \phi(z, x) \right]$$

At this point, we want to pause and look at the last line. Notice that if (a, b) is a directed edge with both endpoints in L, then when the outer sum is conducted for $x = a$, we get an overall contribution of $\phi(a, b)$. On the other hand, when it is conducted for $x = b$, we get a contribution of $-\phi(a, b)$. Thus, the terms cancel out and everything simplifies to

$$\sum_{\substack{x \in L \\ y \in U}} \phi(x, y) - \sum_{\substack{x \in L \\ z \in U}} \phi(z, x) \le \sum_{\substack{x \in L \\ y \in U}} \phi(x, y) \le \sum_{\substack{x \in L \\ y \in U}} c(x, y) = c(L, U).$$

Thus $v_0 \le c(L, U)$. □

Discussion 13.5. Bob's getting a bit of a sense of déjà vu after reading Theorem 13.4. He remembers from Chapter 5 that the maximum size of a clique in a graph is always at most the minimum number of colors required to properly color the graph. However, he also remembers that there are graphs without cliques of size three but with arbitrarily large chromatic number, so he's not too hopeful that this theorem is going to help out much here. Yolanda chimes in with a reminder of Chapter 6, where they learned that the maximum size of an antichain in a poset is equal to the minimum number of chains into which the ground set of the poset can be partitioned. Alice points out that Yolanda's statement is still true if the words "chain" and "antichain" are swapped. This sparks some intense debate about whether the maximum value of a flow in a network must always be equal to the minimum capacity of a cut in that network. After a while, Carlos suggests that continuing to read might be the best idea for resolving their debate.

13.3 Augmenting Paths

In this section, we develop the classic labeling algorithm of Ford and Fulkerson which starts with any flow in a network and proceeds to modify the flow—always increasing the value of the flow—until reaching a step where no further improvements are possible. The algorithm will also help resolve the debate Alice, Bob, Carlos, and Yolanda were having in the previous section.

Our presentation of the labeling algorithm makes use of some natural and quite descriptive terminology. Suppose we have a network $\mathbf{G} = (V, E)$ with a flow ϕ of value

Chapter 13 Network Flows

v. We call ϕ the **current flow** and look for ways to **augment** ϕ by making a relatively small number of changes. An edge (x, y) with $\phi(x, y) > 0$ is said to be **used**, and when $\phi(x, y) = c(x, y) > 0$, we say the edge is **full**. When $\phi(x, y) < c(x, y)$, we say the edge (x, y) has **spare capacity**, and when $0 = \phi(x, y) < c(x, y)$, we say the edge (x, y) is **empty**. Note that we simply ignore edges with zero capacity.

The key tool in modifying a network flow is a special type of path, and these paths are not necessarily directed paths. An **augmenting path** is a sequence $P = (x_0, x_1, \ldots, x_m)$ of distinct vertices in the network such that $x_0 = S$, $x_m = T$, and for each $i = 1, 2, \ldots, m$, either

a (x_{i-1}, x_i) has spare capacity or

b (x_i, x_{i-1}) is used.

When condition (Item a) holds, it is customary to refer to the edge (x_{i-1}, x_i) as a **forward** edge of the augmenting path P. Similarly, if condition (Item b) holds, then the (nondirected) edge (x_{i-1}, x_i) is called a **backward** edge since the path moves from x_{i-1} to x_i, which is opposite the direction of the edge.

Example 13.6. Let's look again at the network and flow in Figure 13.2. The sequence of vertices (S, F, A, T) meets the criteria to be an augmenting path, and each edge in it is a forward edge. Notice that increasing the flow on each of (S, F), (F, A), and (A, T) by any positive amount $\delta \leq 12$ results in increasing the value of the flow and preserves the conservation laws.

If our first example jumped out at you as an augmenting path, it's probably less clear at a quick glance that (S, E, D, C, B, A, T) is also an augmenting path. All of the edges are forward edges except for (C, B), since it's actually (B, C) that is a directed edge in the network. Don't worry if it's not clear how this path can be used to increase the value of the flow in the network, as that's our next topic.

Ignoring, for the moment, the issue of finding augmenting paths, let's see how they can be used to modify the current flow in a way that increases its value by some $\delta > 0$. Here's how for an augmenting path $P = (x_0, x_1, \ldots, x_m)$. First, let δ_1 be the positive number defined by:

$$\delta_1 = \min\{c(x_{i-1}, x_i) - \phi(x_{i-1}, x_i) : (x_{i-1}, x_i) \text{ a forward edge of } P.\}$$

The quantity $c(x_{i-1}, x_i) - \phi(x_{i-1}, x_i)$ is nothing but the spare capacity on the edge (x_{i-1}, x_i), and thus δ_1 is the largest amount by which *all* of the forward edges of P. Note that the edges (x_0, x_1) and (x_{m-1}, x_m) are always forward edges, so the *positive* quantity δ_1 is defined for every augmenting path.

13.3 Augmenting Paths

When the augmenting path P has no backward edges, we set $\delta = \delta_1$. But when P has one or more backward edges, we pause to set

$$\delta_2 = \min\{\phi(x_i, x_{i-1}) : (x_{i-1}, x_i) \text{ a backward edge of } P\}.$$

Since every backward edge is used, $\delta_2 > 0$ whenever we need to define it. We then set $\delta = \min\{\delta_1, \delta_2\}$.

In either case, we now have a positive number δ and we make the following elementary observation, for which you are asked to provide a proof in Exercise 13.7.4.

Proposition 13.7. *Suppose we have an augmenting path $P = (x_0, x_1, \ldots, x_m)$ with $\delta > 0$ calculated as above. Modify the flow ϕ by changing the values along the edges of the path P by an amount which is either $+\delta$ or $-\delta$ according to the following rules:*

1. *Increase the flow along the edges of P which are forwards.*

2. *Decrease the flow along the edges of P which are backwards.*

Then the resulting function $\hat{\phi}$ is a flow and it has value $v + \delta$.

Example 13.8. The network flow shown in Figure 13.2 has many augmenting paths. We already saw two of them in Example 13.6, which we call P_1 and P_3 below. In the list below, be sure you understand why each path is an augmenting path and how the value of δ is determined for each path.

1. $P_1 = (S, F, A, T)$ with $\delta = 12$. All edges are forward.

2. $P_2 = (S, B, A, T)$ with $\delta = 8$. All edges are forward.

3. $P_3 = (S, E, D, C, B, A, T)$ with $\delta = 9$. All edges are forward, except (C, B) which is backward.

4. $P_4 = (S, B, E, D, C, A, T)$ with $\delta = 2$. All edges are forward, except (B, E) and (C, A) which are backward.

In Exercise 13.7.7, you are asked to update the flow in Figure 13.2 for each of these four paths individually.

13.3.1 Caution on Augmenting Paths

Bob's gotten really good at using augmenting paths to increase the value of a network flow. He's not sure how to find them quite yet, but he knows a good thing when he sees it. He's inclined to think that any augmenting path will be a good deal in his quest for a maximum-valued flow. Carlos is pleased about Bob's enthusiasm for network flows

Chapter 13 Network Flows

but is beginning to think that he should warn Bob about the dangers in using just any old augmenting path to update a network flow. They agree that the best situation is when the number of updates that need to be made is small in terms of the number of vertices in the network and that the size of the capacities on the edges and the value of a maximum flow should not have a role in the number of updates.

Bob says he can't see any way that the edge capacities could create a situation where a network with only a few vertices requires many updates, Carlos is thinking that an example is in order. He asks Bob to pick his favorite very large integer and to call it M. He then draws the network on four vertices shown in Figure 13.9. Bob quickly recognizes that the maximum value of a flow in this network is $2M$. He does this using the flow with $\phi(S, A) = M$, $\phi(A, T) = M$, $\phi(S, B) = M$, $\phi(B, T) = M$ and $\phi(A, B) = 0$. Carlos is pleased with Bob's work.

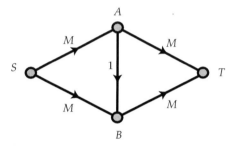

FIGURE 13.9: A SMALL NETWORK

Since this network is really small, it was easy for Bob to find the maximum flow. However, Bob and Carlos agree that "eyeballing" is not an approach that scales well to larger networks, so they need to have an approach to finding that flow using augmenting paths. Bob tells Carlos to give him an augmenting path, and he'll do the updating. Carlos suggests the augmenting path (S, A, B, T), and Bob determines that $\delta = 1$ for this augmenting path. He updates the network (starting from the zero flow, i.e., with $\phi(e) = 0$ for every edge e) and it now has value 1. Bob asks Carlos for another augmenting path, so Carlos gives him (S, B, A, T). Now (B, A) is backward, but that doesn't phase Bob. He performs the update, obtaining a flow of value 2 with (A, B) empty again.

Despite Carlos' hope that Bob could already see where this was heading, Bob eagerly asks for another augmenting path. Carlos promptly gives him (S, A, B, T), which again has $\delta = 1$. Bob's update gives them a flow of value 3. Before Carlos can suggest another augmenting path, Bob realizes what the problem is. He points out that Carlos can just give him (S, B, A, T) again, which will still have $\delta = 1$ and result in the flow value increasing to 4. He says that they could keep alternating between those two augmenting paths, increasing the flow value by 1 each time, until they'd made $2M$ updates to finally have a flow of value $2M$. Since the network only has four vertices and M is very large, he realizes that using any old augmenting path is definitely not a good idea.

Carlos leaves Bob to try to figure out a better approach. He realizes that starting from the zero flow, he'd only need the augmenting paths (S, A, T) and (S, B, T), each with $\delta = M$ to quickly get the maximum flow. However, he's not sure why an algorithm should find those augmenting paths to be preferable. About this time, Dave wanders by and mumbles something about the better augmenting paths using only two edges, while Carlos' two evil augmenting paths each used three. Bob thinks that maybe Dave's onto something, so he decides to go back to reading his textbook.

13.4 The Ford-Fulkerson Labeling Algorithm

In this section, we outline the classic Ford-Fulkerson labeling algorithm for finding a maximum flow in a network. The algorithm begins with a linear order on the vertex set which establishes a notion of **precedence**. Typically, the first vertex in this linear order is the source while the second is the sink. After that, the vertices can be listed in any order. In this book, we will use the following convention: the vertices will be labeled with capital letters of the English alphabet and the linear order will be $(S, T, A, B, C, D, E, F, G, \ldots)$, which we will refer to as the **pseudo-alphabetic order**. Of course, this convention only makes sense for networks with at most 26 vertices, but this limitation will not cramp our style. For real world problems, we take comfort in the fact that computers can deal quite easily with integer keys of just about any size.

Before providing a precise description of the algorithm, let's take a minute to consider a general overview. In carrying out the labeling algorithm, vertices will be classified as either **labeled** or **unlabeled**. At first, we will start with only the source being labeled while all other vertices will be unlabeled. By criteria yet to be spelled out, we will systematically consider unlabeled vertices and determine which should be labeled. If we ever label the sink, then we will have discovered an augmenting path, and the flow will be suitably updated. After updating the flow, we start over again with just the source being labeled.

This process will be repeated until (and we will see that this always occurs) we reach a point where the labeling halts with some vertices labeled (one of these is the source) and some vertices unlabeled (one of these is the sink). We will then note that the partition $V = L \cup U$ into labeled and unlabeled vertices (hence our choice of L and U as names) is a cut whose capacity is exactly equal to the value of the current flow. This resolves the debate from earlier in the chapter and says that the maximum flow/minimum cut question is more like antichains and partitioning into chains than clique

Chapter 13 Network Flows

number and chromatic number. In particular, the labeling algorithm will provide a proof of the following theorem:

Theorem 13.10 (The Max Flow–Min Cut Theorem). *Let $G = (V, E)$ be a network. If v_0 is the maximum value of a flow and c_0 is the minimum capacity c_0 of a cut, then $v_0 = c_0$.*

We're now ready to describe the **Ford-Fulkerson labeling algorithm** in detail.

Algorithm 13.11 (Ford-Fulkerson Labeling Algorithm).

Labeling the Vertices *Vertices will be labeled with ordered triples of symbols. Each time we start the labeling process, we begin by labeling the source with the triple $(*, +, \infty)$. The rules by which we label vertices will be explicit.*

Potential on a Labeled Vertex *Let u be a labeled vertex. The third coordinate of the label given to u will be positive real number—although it may be infinite. We call this quantity the **potential** on u and denote it by $p(u)$. (The potential will serve as the amount that the flow can be updated by.) Note that the potential on the source is infinite.*

First Labeled, First Scanned *The labeling algorithm involves a scan from a labeled vertex u. As the vertices are labeled, they determine another linear order. The source will always be the first vertex in this order. After that, the order in which vertices are labeled will change with time. But the important rule is that we scan vertices in the order that they are labeled—until we label the sink. If for example, the initial scan—always done from the source—results in labels being applied to vertices D, G and M, then we next scan from vertex D. If that scan results in vertices B, F, G and Q being labeled, then we next scan from G, as it was labeled before B, even though B precedes G in the pseudo-alphabetic order. This aspect of the algorithm results in a breadth-first search of the vertices looking for ways to label previously unlabeled vertices.*

Never Relabel a Vertex *Once a vertex is labeled, we do not change its label. We are content to label previously unlabeled vertices—up until the time where we label the sink. Then, after updating the flow and increasing the value, all labels, except of course the special label on the source, are discarded and we start all over again.*

Labeling Vertices Using Forward Edges *Suppose we are scanning from a labeled vertex u with potential $p(u) > 0$. From u, we consider the unlabeled neighbors of u in pseudo-alphabetic order. Now suppose that we are looking at a neighbor v of u with the edge (u, v) belonging to the network. This means that the edge is directed from u to v. If $e = (u, v)$ is not full, then we label the vertex v with the triple $(u, +, p(v))$ where $p(v) = \min\{p(u), c(e) - \phi(e)\}$. We use this definition since the flow cannot be increased by more than the prior potential or the spare capacity on e. Note that the potential $p(v)$ is positive since a is the minimum of two positive numbers.*

Labeling Vertices Using Backward Edges Now suppose that we are looking at a neighbor v of u with the edge (v, u) belonging to the network. This means that the edge is directed from v to u. If $e = (v, u)$ is used, then we label the vertex v with the triple $(u, -, p(v))$ where $p(v) = \min\{p(u), \phi(e)\}$. Here $p(v)$ is defined this way since the flow on e cannot be decreased by more than $\phi(e)$ or $p(u)$. Again, note that the potential $p(v)$ is positive since a is the minimum of two positive numbers.

What Happens When the Sink is Labeled? The labeling algorithm halts if the sink is ever labeled. Note that we are always trying our best to label the sink, since in each scan the sink is the very first vertex to be considered. Now suppose that the sink is labeled with the triple $(u, +, a)$. Note that the second coordinate on the label must be $+$ since all edges incident with the sink are oriented towards the sink.

We claim that we can find an augmenting path P which results in an increased flow with $\delta = a$, the potential on the sink. To see this, we merely back-track. The sink T got its label from $u = u_1$, u_1 got its label from u_2, and so forth. Eventually, we discover a vertex u_m which got its label from the source. The augmenting path is then

$$P = (S, u_m, u_{m-1}, \ldots, u_1, T).$$

The value of δ for this path is the potential $p(T)$ on the sink since we've carefully ensured that $p(u_m) \geq p(u_{m-1}) \geq \cdots \geq p(u_1) \geq p(T)$.

And if the Sink is Not Labeled? On the other hand, suppose we have scanned from every labeled vertex and there are still unlabeled vertices remaining, one of which is the sink. Now we claim victory. To see that we have won, we simply observe that if L is the set of labeled vertices, and U is the set of unlabeled vertices, then every edge $e = (x, y)$ with $x \in L$ and $y \in U$ is full, i.e., $\phi(e) = c(e)$. If this were not the case, then y would qualify for a label with x as the first coordinate. Also, note that $\phi(y, x) = 0$ for every edge e with $x \in L$ and $y \in U$. Regardless, we see that the capacity of the cut $V = L \cup U$ is exactly equal to the value of the current flow, so we have both a maximum flow and minimum cut providing a certificate of optimality.

13.5 A Concrete Example

Let's apply the Labeling Algorithm to the network flow shown in Figure 13.2. Then we start with the source:

$$S: \quad (*, +, \infty)$$

Since the source S is the first vertex labeled, it is also the first one scanned. So we look at the neighbors of S using the pseudo-alphabetic order on the vertices. Thus, the first

Chapter 13 Network Flows

one to be considered is vertex B and since the edge (S, B) is not full, we label B as

$$B : \quad (S, +, 8).$$

We then consider vertex E and label it as

$$E : \quad (S, +, 28).$$

Next is vertex F, which is labeled as

$$F : \quad (S, +, 15).$$

At this point, the scan from S is complete.

The first vertex after S to be labeled was B, so we now scan from B. The (unlabeled) neighbors of B to be considered, in order, are A, C, and D. This results in the following labels:

$$A : \quad (B, +, 8)$$
$$C : \quad (B, +, 8)$$
$$D : \quad (B, -, 6)$$

The next vertex to be scanned is E, but E has no unlabeled neighbors, so we then move on to F, which again has no unlabeled neighbors. Finally, we scan from A, and using the pseudo-alphabetic order, we first consider the sink T (which in this case is the only remaining unlabeled vertex). This results in the following label for T.

$$T : \quad (A, +, 8)$$

Now that the sink is labeled, we know there is an augmenting path. We discover this path by backtracking. The sink T got its label from A, A got its label from B, and B got its label from S. Therefore, the augmenting path is $P = (S, B, A, T)$ with $\delta = 8$. All edges on this path are forward. The flow is then updated by increasing the flow on the edges of P by 8. This results in the flow shown in Figure 13.12. The value of this flow is 38.

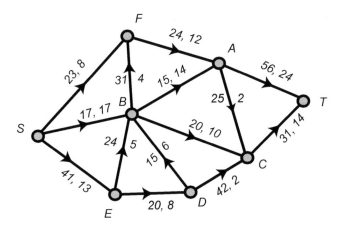

FIGURE 13.12: AN UPDATED NETWORK FLOW

Here is the sequence (reading down the columns) of labels that will be found when the labeling algorithm is applied to this updated flow. (Note that in the scan from S, the vertex B will not be labeled, since now the edge (S, B) is full.)

$S: \quad (*, +, \infty)$ $\qquad$ $D: \quad (E, +, 12)$
$E: \quad (S, +, 28)$ $\qquad$ $A: \quad (F, +, 12)$
$F: \quad (S, +, 15)$ $\qquad$ $C: \quad (B, +, 10)$
$B: \quad (E, +, 19)$ $\qquad$ $T: \quad (A, +, 12)$

This labeling results in the augmenting path $P = (S, F, A, T)$ with $\delta = 12$.

After this update, the value of the flow has been increased and is now $50 = 38 + 12$. We start the labeling process over again and repeat until we reach a stage where some vertices (including the source) are labeled and some vertices (including the sink) are unlabeled.

13.5.1 How the Labeling Algorithm Halts

Consider the network flow in Figure 13.13.

Chapter 13 Network Flows

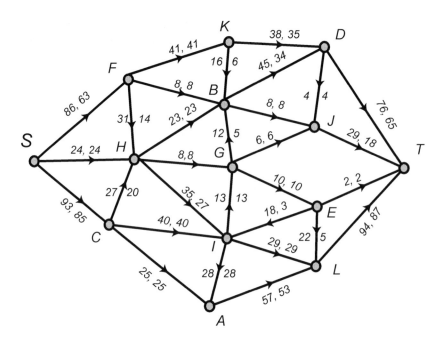

FIGURE 13.13: ANOTHER NETWORK FLOW

The value of the current flow is 172. Applying the labeling algorithm using the pseudo-alphabetic order results in the following labels (reading down the columns):

$$S : (*, +, \infty) \qquad E : (I, -, 3)$$
$$C : (S, +, 8) \qquad G : (E, -, 3)$$
$$F : (S, +, 23) \qquad L : (E, +, 3)$$
$$H : (C, +, 7) \qquad B : (G, +, 3)$$
$$I : (H, +, 7) \qquad T : (L, +, 3)$$

These labels result in the augmenting path $P = (S, C, H, I, E, L, T)$ with $\delta = 3$. After updating the flow and increasing its value to 175, the labeling algorithm halts with the following labels:

$$S : (*, +, \infty) \qquad H : (C, +, 4)$$
$$C : (S, +, 5) \qquad I : (H, +, 4)$$
$$F : (S, +, 23)$$

Now we observe that the labeled and unlabeled vertices are $L = \{S, C, F, H, I\}$ and $U = \{T, A, B, D, E, G, J, K\}$. Furthermore, the capacity of the cut $V = L \cup U$ is

$$41 + 8 + 23 + 8 + 13 + 29 + 28 + 25 = 175.$$

This shows that we have found a cut whose capacity is exactly equal to the value of the current flow. In turn, this shows that the flow is optimal.

13.6 Integer Solutions of Linear Programming Problems

A linear programming problem is an optimization problem that can be stated in the following form: Find the maximum value of a linear function

$$c_1 x_1 + c_2 x_2 + c_3 x_3 + \cdots + c_n x_n$$

subject to m constraints $C_1, C_2, \ldots, C_m$, where each constraint C_i is a linear equation of the form:

$$C_i: \quad a_{i1} x_1 + a_{i2} x_2 + a_{i3} x_3 + \cdots + a_{in} x_n = b_i$$

where all coefficients and constants are real numbers.

While the general subject of linear programming is far too broad for this course, we would be remiss if we didn't point out that:

1. Linear programming problems are a *very* important class of optimization problems and they have many applications in engineering, science, and industrial settings.

2. There are relatively efficient algorithms for finding solutions to linear programming problems.

3. A linear programming problem posed with rational coefficients and constants has an optimal solution with rational values—if it has an optimal solution at all.

4. A linear programming problem posed with integer coefficients and constants need not have an optimal solution with integer values—even when it has an optimal solution with rational values.

5. A very important theme in operations research is to determine when a linear programming problem posed in integers has an optimal solution with integer values. This is a subtle and often very difficult problem.

6. The problem of finding a maximum flow in a network is a special case of a linear programming problem.

Chapter 13 Network Flows

7. A network flow problem in which all capacities are integers has a maximum flow in which the flow on every edge is an integer. The Ford-Fulkerson labeling algorithm guarantees this!

8. In general, linear programming algorithms are not used on networks. Instead, special purpose algorithms, such as Ford-Fulkerson, have proven to be more efficient in practice.

13.7 Exercises

1. Consider the network diagram in Figure 13.14. For each directed edge, the first number is the capacity and the second value is intended to give a flow ϕ in the network. However, the flow suggested is not valid.

(a) Identify the reason(s) ϕ is not valid.

(b) Without changing any of the edge capacities, modify ϕ into a valid flow $\widehat{\phi}$. Try to use as few modifications as possible.

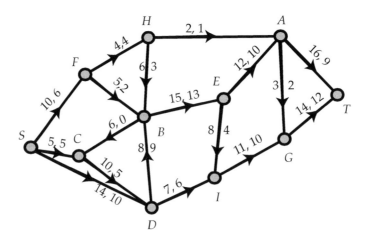

FIGURE 13.14: AN INVALID FLOW IN A NETWORK

2. Alice claims to have found a (valid) network flow of value 20 in the network shown in Figure 13.15. Bob tells her that there's no way she's right, since no flow has value greater than 18. Who's right and why?

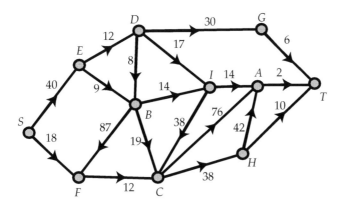

FIGURE 13.15: A NETWORK

3. Find an augmenting path P with *at least one backward edge* for the flow ϕ in the network shown in Figure 13.16. What is the value of δ for P? Carry out an update of ϕ using P to obtain a new flow $\hat{\phi}$. What is the value of $\hat{\phi}$?

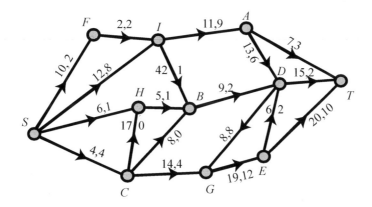

FIGURE 13.16: A NETWORK WITH FLOW

4. Prove Proposition 13.7. You will need to verify that the flow conservation laws hold at each vertex along an augmenting path (other than S and T). There are four cases to consider depending on the forward/backward status of the two edges on the augmenting path that are incident with the vertex.

Chapter 13 Network Flows

5. Find the capacity of the cut (L, U) with

$$L = \{S, F, H, C, B, G, I\} \quad \text{and} \quad U = \{A, D, E, T\}$$

in the network shown in Figure 13.16.

6. Find the capacity of the cut (L, U) with

$$L = \{S, F, D, B, A\} \quad \text{and} \quad U = \{H, C, I, G, E, T\}$$

in the network shown in Figure 13.16.

7. For each of the augmenting paths $P_1, P_2, P_3,$ and P_4 in Example 13.8, update the flow in Figure 13.2. (Note that your solution to this exercise should consist of four network flows. Do not attempt to use the four paths in sequence to create one updated network flow.)

8. Continue running the Ford-Fulkerson labeling algorithm on the network flow in Figure 13.12 until the algorithm halts without labeling the sink. Find the value of the maximum flow as well as a cut of minimum capacity.

9. Use the Ford-Fulkerson labeling algorithm to find a maximum flow and a minimum cut in the network shown in Figure 13.17 by starting from the current flow shown there.

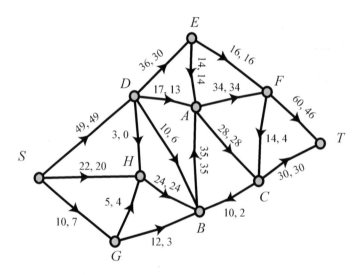

FIGURE 13.17: A NETWORK WITH FLOW

10. Figure 13.18 shows a network. Starting from the zero flow, i.e., the flow with $\phi(e) = 0$ for every directed edge e in the network, use the Ford-Fulkerson labeling algorithm to find a maximum flow and a minimum cut in this network.

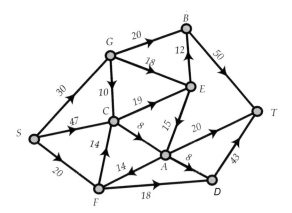

FIGURE 13.18: A NETWORK

11. Consider a network in which the source S has precisely three neighbors: B, E, and F. Suppose also that $c(S, B) = 30$, $c(S, E) = 20$, and $c(S, F) = 25$. You know that there is a flow ϕ on the network but you do not know how much flow is on any edge. You do know, however, that when the Ford-Fulkerson labeling algorithm is run on the network with current flow ϕ, the first two vertices labeled are S with label $(*, +, \infty)$ and F with label $(S, +, 15)$. Use this information to determine the value of the flow ϕ and explain how you do so.

CHAPTER 14

Combinatorial Applications of Network Flows

Clearly finding the maximum flow in a network can have many direct applications to problems in business, engineering, and computer science. However, you may be surprised to learn that finding network flows can also provide reasonably efficient algorithms for solving combinatorial problems. In this chapter, we consider a restricted version of network flows in which each edge has capacity 1. Our goal is to establish algorithms for two combinatorial problems: finding maximum matchings in bipartite graphs and finding the width of a poset as well as a minimal chain partition.

14.1 Introduction

Before delving into the particular combinatorial problems we wish to consider in this chapter, we will state a key theorem. When working with network flow problems, our examples thus far have always had integer capacities and we always found a maximum flow in which every edge carried an integer amount of flow. It is not, however, immediately obvious that this can always be done. Why, for example, could it not be the case that the maximum flow in a particularly pathological network with integer capacities is 23/3? Or how about something even worse, such as $\sqrt{21\pi}$? We can rule out the latter because network flow problems fall into a larger class of problems known as linear programming problems, and a major theorem tells us that if a linear program is posed with all integer constraints (capacities in our case), the solution must be a rational number. However, in the case of network flows, something even stronger is true.

Theorem 14.1. *In a network flow problem in which every edge has integer capacity, there is a maximum flow in which every edge carries an integer amount of flow.*

Notice that the above theorem does not guarantee that every maximum flow has integer flow on every edge, just that we are able to find one. With this theorem in

Chapter 14 Combinatorial Applications of Network Flows

hand, we now see that if we consider network flow problems in which the capacities are all 1 we can find a maximum flow in which every edge carries a flow of either 0 or 1. This can give us a combinatorial interpretation of the flow, in a sense using the full edges as edges that we "take" in some useful sense.

14.2 Matchings in Bipartite Graphs

Recall that a bipartite graph $\mathbf{G} = (V, E)$ is one in which the vertices can be properly colored using only two colors. It is clear that such a coloring then partitions V into two independent sets V_1 and V_2, and so all the edges are between V_1 and V_2. Bipartite graphs have many useful applications, particularly when we have two distinct types of objects and a relationship that makes sense only between objects of distinct types. For example, suppose that you have a set of workers and a set of jobs for the workers to do. We can consider the workers as the set V_1 and the jobs as V_2 and add an edge from worker $w \in V_1$ to job $j \in V_2$ if and only if w is qualified to do j.

For example, the graph in Figure 14.2 is a bipartite graph in which we've drawn V_1 on the bottom and V_2 on the top.

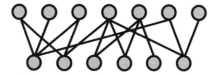

FIGURE 14.2: A BIPARTITE GRAPH

If $\mathbf{G} = (V, E)$ is a graph, a set $M \subseteq E$ is a **matching** in $\mathbf{G}$ if no two edges of M share an endpoint. If v is a vertex that is the endpoint of an edge in M, we say that M saturates v or v is saturated by M. When $\mathbf{G}$ is bipartite with $V = V_1 \cup V_2$, a matching is then a way to pair vertices in V_1 with vertices in V_2 so that no vertex is paired with more than one other vertex. We're usually interested in finding a **maximum matching**, which is a matching that contains the largest number of edges possible, and in bipartite graphs we usually fix the sets V_1 and V_2 and seek a maximum matching from V_1 to V_2. In our workers and jobs example, the matching problem thus becomes trying to find an assignment of workers to jobs such that

  i each worker is assigned to a job for which he is qualified (meaning there's an edge),

  ii each worker is assigned to at most one job, and

  iii each job is assigned at most one worker.

14.2 Matchings in Bipartite Graphs

As an example, in Figure 14.3, the thick edges form a matching from V_1 to V_2. Suppose that you're the manager of these workers (on the bottom) and must assign them to the jobs (on the top). Are you really making the best use of your resources by only putting four of six workers to work? There are no trivial ways to improve the number of busy workers, as the two without responsibilities right now cannot do any of the jobs that are unassigned. Perhaps there's a more efficient assignment that can be made by redoing some of the assignments, however. If there is, how should you go about finding it? If there is not, how would you justify to your boss that there's no better assignment of workers to jobs?

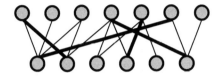

FIGURE 14.3: A MATCHING IN A BIPARTITE GRAPH

At the end of the section, we'll briefly look at a theorem on matchings in bipartite graphs that tells us precisely when an assignment of workers to jobs exists that ensures each worker has a job. First, however, we want to see how network flows can be used to find maximum matchings in bipartite graphs. The algorithm we give, while decent, is not the most efficient algorithm known for this problem. Therefore, it is not likely to be the one used in practice. However, it is a nice example of how network flows can be used to solve a combinatorial problem. The network that we use is formed from a bipartite graph **G** by placing an edge from the source S to each vertex of V_1 and an edge from each vertex of V_2 to the sink T. The edges between V_1 and V_2 are oriented from V_1 to V_2, and *every* edge is given capacity 1. Figure 14.4 contains the network corresponding to our graph from Figure 14.2. Edges in this network are all oriented from bottom to top and all edges have capacity 1. The vertices in V_1 are $x_1, \ldots, x_6$ in order from left to right, while the vertices in V_2 are $y_1, \ldots, y_7$ from left to right.

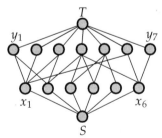

FIGURE 14.4: THE NETWORK CORRESPONDING TO A BIPARTITE GRAPH

Chapter 14 Combinatorial Applications of Network Flows

Now that we have translated a bipartite graph into a network, we need to address the correspondence between matchings and network flows. To turn a matching M into a network flow, we start by placing one unit of flow on the edges of the matching. To have a valid flow, we must also place one unit of flow on the edges from S to the vertices of V_1 saturated by M. Since each of these vertices is incident with a single edge of M, the flow out of each of them is 1, matching the flow in. Similarly, routing one unit of flow to T from each of the vertices of V_2 saturated by M takes care of the conservation laws for the remaining vertices. To go the other direction, simply note that the full edges from V_1 to V_2 in an integer-valued flow is a matching. Thus, we can find a maximum matching from V_1 to V_2 by simply running the labeling algorithm on the associated network in order to find a maximum flow.

In Figure 14.5, we show thick edges to show the edges with flow 1 in the flow corresponding to our guess at a matching from Figure 14.3.

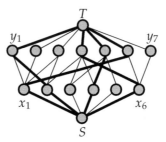

FIGURE 14.5: THE FLOW CORRESPONDING TO A MATCHING

With priority sequence $S, T, x_1, x_2, \ldots, x_6, y_1, y_2, \ldots, y_7$ replacing our usual pseudo-alphabetic order, the labeling algorithm produces the labels shown below.

$$
\begin{aligned}
S &: (*, +, \infty) & y_6 &: (x_6, +, 1) \\
x_3 &: (S, +, 1) & x_1 &: (y_6, -, 1) \\
x_5 &: (S, +, 1) & y_1 &: (x_1, +, 1) \\
y_4 &: (x_3, +, 1) & y_2 &: (x_1, +, 1) \\
y_5 &: (x_3, +, 1) & y_3 &: (x_1, +, 1) \\
x_6 &: (y_4, -, 1) & x_2 &: (y_1, -, 1) \\
x_4 &: (y_5, -, 1) & T &: (y_2, +, 1)
\end{aligned}
$$

This leads us to the augmenting path $S, x_3, y_4, x_6, y_6, x_1, y_2, T$, which gives us the flow shown in Figure 14.6.

14.2 Matchings in Bipartite Graphs

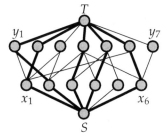

FIGURE 14.6: THE AUGMENTED FLOW

Is this a maximum flow? Another run of the labeling algorithm produces

$$S : (*, +, \infty) \qquad x_4 : (y_5, -, 1)$$
$$x_5 : (S, +, 1) \qquad y_4 : (x_4, +, 1)$$
$$y_5 : (x_5, +, 1) \qquad x_3 : (y_4, -, 1)$$

and then halts. Thus, the flow in Figure 14.6 is a maximum flow.

Now that we know we have a maximum flow, we'd like to be able to argue that the matching we've found is also maximum. After all, the boss isn't going to be happy if he later finds out that this fancy algorithm you claimed gave an optimal assignment of jobs to workers left the fifth worker (x_5) without a job when all six of them could have been put to work. Let's take a look at which vertices were labeled by the Ford-Fulkerson labeling algorithm on the last run. There were three vertices (x_3, x_4, and x_5) from V_1 labeled, while there were only two vertices (y_4 and y_5) from V_2 labeled. Notice that y_4 and y_5 are the only vertices that are neighbors of x_3, x_4, or x_5 in **G**. Thus, no matter how we choose the matching edges from $\{x_3, x_4, x_5\}$, one of these vertices will be left unsaturated. Therefore, one of the workers must go without a job assignment. (In our example, it's the fifth, but it's possible to choose different edges for the matching so another one of them is left without a task.)

The phenomenon we've just observed is not unique to our example. In fact, in *every* bipartite graph **G** $= (V, E)$ with $V = V_1 \cup V_2$ in which we cannot find a matching that saturates all the vertices of V, we will find a similar configuration. This is a famous theorem of Hall, which we state below.

Theorem 14.7 (Hall's Theorem). *Let* **G** $= (V, E)$ *be a bipartite graph with* $V = V_1 \cup V_2$. *There is a matching which saturates all vertices of* V_1 *if and only if for every subset* $A \subseteq V_1$, *the set* $N \subseteq V$ *of neighbors of the vertices in A satisfies* $|N| \geq |A|$.

Chapter 14 Combinatorial Applications of Network Flows

14.3 Chain partitioning

In Chapter 6, we discussed Dilworth's Theorem, which told us that for any poset **P** of width w, there is a partition of **P** into w, but no fewer, chains. However, we were only able to devise an algorithm to find this chain partition (and a maximum antichain) in the special case where **P** was an interval order. Now, through the magic of network flows, we will be able to devise an efficient algorithm that works in general for all posets. However, to do so, we will require a slightly more complicated network than we devised in the previous section.

Suppose that the points of our poset **P** are $\{x_1, x_2, \ldots, x_n\}$. We construct a network from **P** consisting of the source S, sink T, and two points x'_i and x''_i for each point x_i of **P**. All edges in our network will have capacity 1. We add edges from S to x'_i for $1 \leq i \leq n$ and from x''_i to T for $1 \leq i \leq n$. Of course, this network wouldn't be too useful, as it has no edges from the single-prime nodes to the double-prime nodes. To resolve this, we add an edge directed from x'_i to x''_j if and only if $x_i < x_j$ in **P**.

Our running example in this section will be the poset in Figure 14.8(a). We'll discuss the points of the poset as x_i where i is the number printed next to the point in the diagram.

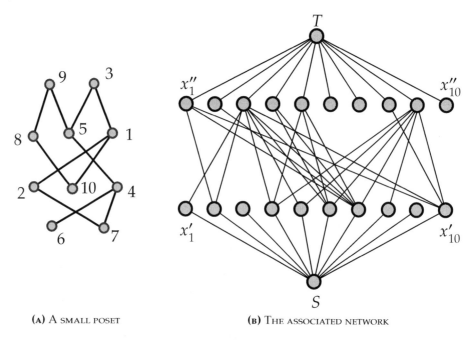

(A) A SMALL POSET   (B) THE ASSOCIATED NETWORK

FIGURE 14.8: A PARTIALLY ORDERED SET (A) AND THE ASSOCIATED NETWORK (B).

14.3 Chain partitioning

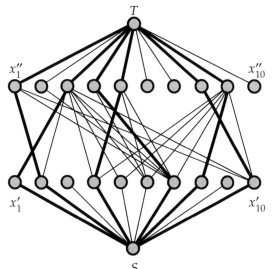

FIGURE 14.9: AN INITIAL FLOW

The first step is to create the network, which we show in Figure 14.8(b). In this network, all capacities are 1, edges are directed from bottom to top, the first row of ten vertices is the x'_i arranged consecutively with x'_1 at the left and x'_{10} at the right, and the second row of ten vertices is the x''_i in increasing order of index. To see how this network is constructed, notice that $x_1 < x_3$ in the poset, so we have the directed edge (x'_1, x''_3). Similarly, x_4 is less than x_3, x_5, and x_9 in the poset, leading to three directed edges leaving x'_4 in the network. As a third example, since x_9 is maximal in the poset, there are no directed edges leaving x'_9.

We have not yet seen how we might turn a maximum flow (or minimum cut) in the network we've just constructed into a minimum chain partition or a maximum antichain. It will be easier to see how this works once we have a confirmed maximum flow. Rather than running the labeling algorithm starting from the zero flow, we eyeball a flow, such as the one shown in Figure 14.9. (Again, we use the convention that thick edges are full, while thin edges are empty.) When we run the labeling algorithm (using priority $S, T, x'_1, \ldots, x'_{10}, x''_1, \ldots, x''_{10}$), we obtain the following list of labels:

$$
\begin{array}{lll}
S: (*, +, \infty) & x''_9: (x'_5, +, 1) & x'_3: (S, +, 1) \\
x'_3: (S, +, 1) & x''_4: (x'_6, +, 1) & x''_1: (x'_7, +, 1) \\
x'_5: (S, +, 1) & x''_5: (x'_6, +, 1) & x''_2: (x'_7, +, 1) \\
x'_6: (S, +, 1) & x'_1: (x''_3, -, 1) & x'_2: (x'_7, +, 1) \\
x'_9: (S, +, 1) & x'_8: (x''_9, -, 1) & T: (x''_2, +, 1)
\end{array}
$$

285

Chapter 14 Combinatorial Applications of Network Flows

$x_3'': (x_5', +, 1)$ $\qquad$ $x_7': (x_4'', -, 1)$

Thus, we find the augmenting path $(S, x_6', x_4'', x_7', x_2'', T)$, and the updated flow can be seen in Figure 14.10.

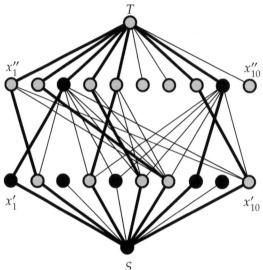

FIGURE 14.10: A BETTER FLOW

If we run the labeling algorithm again, the algorithm assigns the labels below, leaving the sink unlabeled.

$S: (*, +, \infty)$ $\qquad$ $x_5': (S, +, 1)$ $\qquad$ $x_3'': (x_5', +, 1)$ $\qquad$ $x_1': (x_3'', -, 1)$
$x_3': (S, +, 1)$ $\qquad$ $x_9': (S, +, 1)$ $\qquad$ $x_9'': (x_5', +, 1)$ $\qquad$ $x_8': (x_9'', -, 1)$

In Figure 14.10, the black vertices are those the labeled in the final run, while the gold vertices are the unlabeled vertices.

Now that we've gone over the part you already knew how to do, we need to discuss how to translate this network flow and cut into a chain partition and an antichain. If there is a unit of flow on an edge (x_i', x_j''), then a good first instinct is to place x_i and x_j in the same chain of a chain partition. To be able to do this successfully, of course, we need to ensure that this won't result in two incomparable points being placed in a chain. A way to see that everything works as desired is to think of starting with (x_i', x_j'') and then looking for flow leaving x_j'. If there is, it goes to a vertex x_k'', so we may add x_k to the chain since $x_i < x_j < x_k$. Continue in this manner until reaching a vertex in the network that does not have any flow leaving it. Then see if x_i'' has flow coming into it. If it does, it's from a vertex x_m' that can be added since $x_m < x_i < x_j$.

286

Let's see how following this process for the flow in Figure 14.10 leads to a chain partition. If we start with x_1', we see that (x_1', x_3'') is full, so we place x_1 and x_3 in chain C_1. Since x_3' has no flow leaving it, there are no greater elements to add to the chain. However, x_1'' has flow in from x_2', so we add x_2 to C_1. We now see that x_2'' has flow in from x_7', so now $C_1 = \{x_1, x_2, x_3, x_7\}$. Vertex x_7'' has no flow into it, so the building of the first chain stops. The first vertex we haven't placed into a chain is x_4, so we note that (x_4', x_5'') is full, placing x_4 and x_5 in chain C_2. We then look from x_5' and see no flow leaving. However, there is flow into x_4'' from x_6', so x_6 is added to C_2. There is no flow out of x_6'', so $C_2 = \{x_4, x_5, x_6\}$. Now the first point not in a chain is x_8, so we use the flow from x_8' to x_9'' to place x_8 and x_9 in chain C_3. Again, no flow out of x_9', so we look to x_8'', which is receiving flow from x_{10}''. Adding x_{10} to C_3 gives $C_3 = \{x_8, x_9, x_{10}\}$, and since every point is now in a chain, we may stop.

Even once we see that the above process does in fact generate a chain partition, it is not immediately clear that it's a minimum chain partition. For this, we need to find an antichain of as many points as there are chains in our partition. (In the example we've been using, we need to find a three-element antichain.) This is where tracking the labeled vertices comes in handy. Suppose we have determined a chain $C = \{x_1 < x_2 < \cdots < x_k\}$ using the network flow. Since x_1 is the minimal element of this chain, there is no flow into x_1' and hence no flow out of x_1''. Since T is unlabeled, this must mean that x_1'' is unlabeled. Similarly, x_k is the maximal element of C, so there is no flow out of x_k'. Thus, x_k' is labeled. Now considering the sequence of vertices

$$x_k', x_k'', x_{k-1}', x_{k-1}'', \ldots, x_2', x_2'', x_1', x_1'',$$

there must be a place where the vertices switch from being labeled to unlabeled. This must happen with x_i' labeled and x_i'' unlabeled. To see why, suppose that x_i' and x_i'' are both unlabeled while x_{i+1}' and x_{i+1}'' are both labeled. Because x_i and x_{i+1} are consecutive in C, there is flow on (x_i', x_{i+1}''). Therefore, when scanning from x_{i+1}'', the vertex x_i' would be labeled. For each chain of the chain partition, we then take the first element y for which y' is labeled and y'' is unlabeled to form an antichain $A = \{y_1, \ldots, y_w\}$. To see that A is an antichain, notice that if $y_i < y_j$, then (y_i', y_j'') is an edge in the network. Therefore, the scan from y_i' would label y_j''. Using this process, we find that a maximum antichain in our example is $\{x_1, x_5, x_8\}$.

14.4 Exercises

1. Use the techniques of this chapter to find a maximum matching from V_1 to V_2 in the graph shown in Figure 14.11. The vertices on the bottom are the set V_1, while the vertices on the top are the set V_2. If you cannot find a matching that saturates all of the vertices in V_1, explain why.

Chapter 14 Combinatorial Applications of Network Flows

FIGURE 14.11: Is there a matching saturating V_1?

2. Use the techniques of this chapter to find a maximum matching from V_1 to V_2 in the graph shown in Figure 14.12. The vertices on the bottom are the set V_1, while the vertices on the top are the set V_2. If you cannot find a matching that saturates all of the vertices in V_1, explain why.

FIGURE 14.12: Is there a matching saturating V_1?

3. Students are preparing to do final projects for an applied combinatorics course. The five possible topics for their final projects are graph algorithms, posets, induction, graph theory, and generating functions. There are five students in the class, and they have each given their professor the list of topics on which they are willing to do their project. Alice is interested in posets or graphs. Bob would be willing to do his project on graph algorithms, posets, or induction. Carlos will only consider posets or graphs. Dave likes generating functions and induction. Yolanda wants to do her project on either graphs or posets. To prevent unauthorized collaboration, the professor does not want to have two students work on the same topic. Is it possible to assign each student a topic from the lists above so that no two students work on the same project? If so, find such an assignment. If not, find an assignment that maximizes the number of students who have assignments from their lists and explain why you cannot satisfy all the students' requests.

4. Seven colleges and universities are competing to recruit six high school football players to play for their varsity teams. Each school is only allowed to sign one more player, and each player is only allowed to commit to a single school. The table below lists the seven institutions and the students they are trying to recruit, have been admitted, and are also interested in playing for that school. (There's no point in assigning a school a player who cannot meet academic requirements or doesn't want to be part of that team.) The players are identified by the integers 1 through 6. Find a way of assigning the players to the schools that maximizes the number of schools who sign one of the six players.

School	Player numbers
Boston College	1, 3, 4
Clemson University	1, 3, 4, 6
Georgia Institute of Technology	2, 6
University of Georgia	None interested
University of Maryland	2, 3, 5
University of North Carolina	1, 2, 5
Virginia Polytechnic Institute and State University	1, 2, 5, 6

5. The questions in this exercise refer to the network diagram in Figure 14.13. This network corresponds to a poset **P**. As usual, all capacities are assumed to be 1, and all edges are directed upward. Answer the following questions about **P** *without drawing the diagram of the poset*.

(a) Which element(s) are greater than x_1 in **P**?

(b) Which element(s) are less than x_5 in **P**?

(c) Which element(s) are comparable with x_6 in **P**?

(d) List the maximal elements of **P**.

(e) List the minimal elements of **P**.

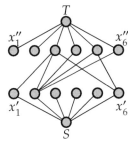

FIGURE 14.13: THE NETWORK CORRESPONDING TO A POSET

6. Draw the diagram of the poset that corresponds to the network in Figure 14.13.

7. Use the methods developed in this chapter to find the width w of the poset corresponding to the network in Figure 14.13. Also find an antichain of size w and a partition into w chains.

Chapter 14 Combinatorial Applications of Network Flows

8. In Figure 14.14 we show a poset **P** and a network used to find a chain partition of **P**. (All edges in the network have a capacity of 1 and are directed from bottom to top. The bold edges currently carry a flow of 1.) Using the network, find the width w of **P**, a partition of **P** into w chains, and an antichain with w elements.

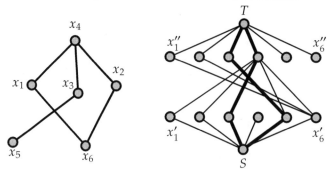

FIGURE 14.14: A POSET AND THE CORRESPONDING NETWORK DIAGRAM

9. Draw the network corresponding to the poset **P** shown in Figure 14.15. Use the network to find the width w of **P**, a partition into w chains, and an antichain of size w.

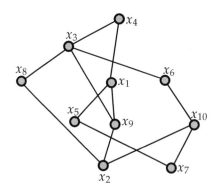

FIGURE 14.15: A POSET

CHAPTER 15

Pólya's Enumeration Theorem

In this chapter, we introduce a powerful enumeration technique generally referred to as Pólya's enumeration theorem[1]. Pólya's approach to counting allows us to use symmetries (such as those of geometric objects like polygons) to form generating functions. These generating functions can then be used to answer combinatorial questions such as

1. How many different necklaces of six beads can be formed using red, blue and green beads? What about 500-bead necklaces?

2. How many musical scales consisting of 6 notes are there?

3. How many isomers of the compound xylenol, $C_6H_3(CH_3)_2(OH)$, are there? What about C_nH_{2n+2}? (In chemistry, **isomers** are chemical compounds with the same number of molecules of each element but with different arrangements of those molecules.)

4. How many nonisomorphic graphs are there on four vertices? How many of them have three edges? What about on 1000 vertices with 257,000 edges? How many r-regular graphs are there on 40 vertices? (A graph is r-**regular** if every vertex has degree r.)

To use Pólya's techniques, we will require the idea of a permutation group. However, our treatment will be self-contained and driven by examples. We begin with a simplified version of the first question above.

---

[1] Like so many results of mathematics, the crux of the result was originally discovered by someone other than the mathematician whose name is associated with it. J.H. Redfield published this result in 1927, 10 years prior to Pólya's work. It would take until 1960 for Redfield's work to be discovered, by which time Pólya's name was firmly attached to the technique.

Chapter 15 Pólya's Enumeration Theorem

15.1 Coloring the Vertices of a Square

Let's begin by coloring the vertices of a square using white and gold. If we fix the position of the square in the plane, there are $2^4 = 16$ different colorings. These colorings are shown in Figure 15.1.

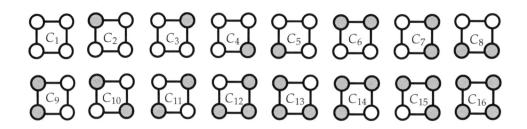

FIGURE 15.1: THE 16 COLORINGS OF THE VERTICES OF A SQUARE.

However, if we think of the square as a metal frame with a white bead or a gold bead at each corner and allow the frame to be rotated and flipped over, we realize that many of these colorings are equivalent. For instance, if we flip coloring C_7 over about the vertical line dividing the square in half, we obtain coloring C_9. If we rotate coloring C_2 clockwise by 90°, we obtain coloring C_3. In many cases, we want to consider such equivalent colorings as a single coloring. (Recall our motivating example of necklaces made of colored beads. It makes little sense to differentiate between two necklaces if one can be rotated and flipped to become the other.)

To systematically determine how many of the colorings shown in Figure 15.1 are not equivalent, we must think about the transformations we can apply to the square and what each does to the colorings. Before examining the transformations' effects on the colorings, let's take a moment to see how they rearrange the vertices. To do this, we consider the upper-left vertex to be 1, the upper-right vertex to be 2, the lower-right vertex to be 3, and the lower-left vertex to be 4. We denote the clockwise rotation by 90° by r_1 and see that r_1 sends the vertex in position 1 to position 2, the vertex in position 2 to position 3, the vertex in position 3 to position 4, and the vertex in position 4 to position 1. For brevity, we will write $r_1(1) = 2$, $r_1(2) = 3$, etc. We can also rotate the square clockwise by 180° and denote that rotation by r_2. In this case, we find that $r_2(1) = 3$, $r_2(2) = 4$, $r_2(3) = 1$, and $r_2(4) = 2$. Notice that we can achieve the transformation r_2 by doing r_1 twice in succession. Furthermore, the clockwise rotation by 270°, r_3, can be achieved by doing r_1 three times in succession. (Counterclockwise rotations can be avoided by noting that they have the same effect as a clockwise rotation, although

15.1 Coloring the Vertices of a Square

by a different angle.)

When it comes to flipping the square, there are four axes about which we can flip it: vertical, horizontal, positive-slope diagonal, and negative-slope diagonal. We denote these flips by v, h, p, and n, respectively. Now notice that $v(1) = 2$, $v(2) = 1$, $v(3) = 4$, and $v(4) = 3$. For the flip about the horizontal axis, we have $h(1) = 4$, $h(2) = 3$, $h(3) = 2$, and $h(4) = 1$. For p, we have $p(1) = 3$, $p(2) = 2$, $p(3) = 1$, and $p(4) = 4$. Finally, for n we find $n(1) = 1$, $n(2) = 4$, $n(3) = 3$, and $n(4) = 2$. There is one more transformation that we must mention; the transformation that does nothing to the square is called the **identity transformation**, denoted ι. It has $\iota(1) = 1$, $\iota(2) = 2$, $\iota(3) = 3$, and $\iota(4) = 4$.

Now that we've identified the eight transformations of the square, let's make a table showing which colorings from Figure 15.1 are left unchanged by the application of each transformation. Not surprisingly, the identity transformation leaves all of the colorings unchanged. Because r_1 moves the vertices cyclically, we see that only C_1 and C_{16} remain unchanged when it is applied. Any coloring with more than one color would have a vertex of one color moved to one of the other color. Let's consider which colorings are fixed by v, the flip about the vertical axis. For this to happen, the color at position 1 must be the same as the color at position 2, and the color at position 3 must be the same as the color at position 4. Thus, we would expect to find $2 \cdot 2 = 4$ colorings unchanged by v. Examining Figure 15.1, we see that these colorings are C_1, C_6, C_8, and C_{16}. Performing a similar analysis for the remaining five transformations leads to Table 15.2.

Transformation	Fixed colorings
ι	All 16
r_1	C_1, C_{16}
r_2	$C_1, C_{10}, C_{11}, C_{16}$
r_3	C_1, C_{16}
v	C_1, C_6, C_8, C_{16}
h	C_1, C_7, C_9, C_{16}
p	$C_1, C_3, C_5, C_{10}, C_{11}, C_{13}, C_{15}, C_{16}$
n	$C_1, C_2, C_4, C_{10}, C_{11}, C_{12}, C_{14}, C_{16}$

TABLE 15.2: COLORINGS FIXED BY TRANSFORMATIONS OF THE SQUARE

At this point, it's natural to ask where this is going. After all, we're trying to count the number of *nonequivalent* colorings, and Table 15.2 makes no effort to group colorings based on how a transformation changes one coloring to another. It turns out that there is a useful connection between counting the nonequivalent colorings and determining the number of colorings fixed by each transformation. To develop this

Chapter 15 Pólya's Enumeration Theorem

connection, we first need to discuss the equivalence relation created by the action of the transformations of the square on the set C of all 2-colorings of the square. (Refer to Section B.13 for a refresher on the definition of equivalence relation.) To do this, notice that applying a transformation to a square with colored vertices results in another square with colored vertices. For instance, applying the transformation r_1 to a square colored as in C_{12} results in a square colored as in C_{13}. We say that the transformations of the square **act** on the set C of colorings. We denote this action by adding a star to the transformation name. For instance, $r_1^*(C_{12}) = C_{13}$ and $v^*(C_{10}) = C_{11}$.

If τ is a transformation of the square with $\tau^*(C_i) = C_j$, then we say colorings C_i and C_j are **equivalent** and write $C_i \sim C_j$. Since $\iota^*(C) = C$ for all $C \in \mathcal{C}$, $\sim$ is reflexive. If $\tau_1^*(C_i) = C_j$ and $\tau_2^*(C_j) = C_k$, then $\tau_2^*(\tau_1^*(C_i)) = C_k$, so $\sim$ is transitive. To complete our verification that $\sim$ is an equivalence relation, we must establish that it is symmetric. For this, we require the notion of the **inverse** of a transformation τ, which is simply the transformation τ^{-1} that undoes whatever τ did. For instance, the inverse of r_1 is the *counter*clockwise rotation by 90°, which has the same effect on the location of the vertices as r_3. If $\tau^*(C_i) = C_j$, then $\tau^{-1*}(C_j) = C_i$, so $\sim$ is symmetric.

Before proceeding to establish the connection between the number of nonequivalent colorings (equivalence classes under $\sim$) and the number of colorings fixed by a transformation in full generality, let's see how it looks for our example. In looking at Figure 15.1, you should notice that $\sim$ partitions C into six equivalence classes. Two contain one coloring each (the all white and all gold colorings). One contains two colorings (C_{10} and C_{11}). Finally, three contain four colorings each (one gold vertex, one white vertex, and the remaining four with two vertices of each color). Now look again at Table 15.2 and add up the number of colorings fixed by each transformation. In doing this, we obtain 48, and when 48 is divided by the number of transformations (8), we get 6 (the number of equivalence classes)! It turns out that this is far from a fluke, as we will soon see. First, however, we introduce the concept of a permutation group to generalize our set of transformations of the square.

15.2 Permutation Groups

Entire books have been written on the theory of the mathematical structures known as **groups**. However, our study of Pólya's enumeration theorem requires only a few facts about a particular class of groups that we introduce in this section. First, recall that a bijection from a set X to itself is called a **permutation**. A **permutation group** is a set P of permutations of a set X so that

1. the identity permutation ι is in P;

2. if $\pi_1, \pi_2 \in P$, then $\pi_2 \circ \pi_1 \in P$; and

3. if $\pi_1 \in P$, then $\pi_1^{-1} \in P$.

For our purposes, X will always be finite and we will usually take $X = [n]$ for some positive integer n. The **symmetric group on n elements**, denoted S_n, is the set of all permutations of $[n]$. Every finite permutation group (and more generally every finite group) is a subgroup of S_n for some positive integer n.

As our first example of a permutation group, consider the set of permutations we discussed in Section 15.1, called the **dihedral group of the square**. We will denote this group by D_8. We denote by D_{2n} the similar group of transformations for a regular n-gon, using $2n$ as the subscript because there are $2n$ permutations in this group.[1] The first criterion to be a permutation group is clearly satisfied by D_8. Verifying the other two is quite tedious, so we only present a couple of examples. First, notice that $r_2 \circ r_1 = r_3$. This can be determined by carrying out the composition of these functions as permutations or by noting that rotating 90° clockwise and then 180° clockwise is the same as rotating 270° clockwise. For $v \circ r$, we find $v \circ r(1) = 1$, $v \circ r(3) = 3$, $v \circ r(2) = 4$, and $v \circ r(4) = 2$, so $v \circ r = n$. For inverses, we have already discussed that $r_1^{-1} = r_3$. Also, $v^{-1} = v$, and more generally, the inverse of *any* flip is that same flip.

15.2.1 Representing permutations

The way a permutation rearranges the elements of X is central to Pólya's enumeration theorem. A proper choice of representation for a permutation is very important here, so let's discuss how permutations can be represented. One way to represent a permutation π of $[n]$ is as a $2 \times n$ matrix in which the first row represents the domain and the second row represents π by putting $\pi(i)$ in position i. For example,

$$\pi = \begin{pmatrix} 1 & 2 & 3 & 4 & 5 \\ 2 & 4 & 3 & 5 & 1 \end{pmatrix}$$

is the permutation of $[5]$ with $\pi(1) = 2$, $\pi(2) = 4$, $\pi(3) = 3$, $\pi(4) = 5$, and $\pi(5) = 1$. This notation is rather awkward and provides only the most basic information about the permutation. A more compact (and more useful for our purposes) notation is known as **cycle notation**. One way to visualize how the cycle notation is constructed is by constructing a digraph from a permutation π of $[n]$. The digraph has $[n]$ as its vertex set and a directed edge from i to j if and only if $\pi(i) = j$. (Here we allow a directed edge from a vertex to itself if $\pi(i) = i$.) The digraph corresponding to the permutation π from above is shown in Figure 15.3.

---

[1] Some authors and computer algebra systems use D_n as the notation for the dihedral group of the n-gon.

Chapter 15 Pólya's Enumeration Theorem

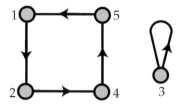

FIGURE 15.3: THE DIGRAPH CORRESPONDING TO PERMUTATION $\pi = (1245)(3)$

Since π is a permutation, every component of such a digraph is a directed cycle. We can then use these cycles to write down the permutation in a compact manner. For each cycle, we start at the vertex with smallest label and go around the cycle in the direction of the edges, writing down the vertices' labels in order. We place this sequence of integers in parentheses. For the 4-cycle in Figure 15.3, we thus obtain (1245). (If $n \geq 10$, we place spaces or commas between the integers.) The component with a single vertex is denoted simply as (3), and thus we may write $\pi = (1245)(3)$. By convention, the disjoint cycles of a permutation are listed so that their first entries are in increasing order.

Example 15.4. The permutation $\pi = (1483)(27)(56)$ has $\pi(1) = 4$, $\pi(8) = 3$, $\pi(3) = 1$, and $\pi(5) = 6$. The permutation $\pi' = (13)(2)(478)(56)$ has $\pi'(1) = 3$, $\pi'(2) = 2$, and $\pi'(8) = 4$. We say that π consists of two cycles of length 2 and one cycle of length 4. For π', we have one cycle of length 1, two cycles of length 2, and one cycle of length 3. A cycle of length k will also called a k-cycle in this chapter.

15.2.2 Multiplying permutations

Because the operation in an arbitrary group is frequently called multiplication, it is common to refer to the composition of permutations as multiplication and write $\pi_2 \pi_1$ instead of $\pi_2 \circ \pi_1$. The important thing to remember here, however, is that the operation is simply function composition. Let's see a couple of examples.

Example 15.5. Let $\pi_1 = (1234)$ and $\pi_2 = (12)(34)$. (Notice that these are the permutations r_1 and v, respectively, from D_8.) Let $\pi_3 = \pi_2 \pi_1$. To determine π_3, we start by finding $\pi_3(1) = \pi_2 \pi_1(1) = \pi_2(2) = 1$. We next find that $\pi_3(2) = \pi_2 \pi_1(2) = \pi_2(3) = 4$. Similarly, $\pi_3(3) = 3$ and $\pi_3(4) = 2$. Thus, $\pi_3 = (1)(24)(3)$, which we called n earlier.

Now let $\pi_4 = \pi_1 \pi_2$. Then $\pi_4(1) = 3$, $\pi_4(2) = 2$, $\pi_4(3) = 1$, and $\pi_4(4) = 4$. Therefore, $\pi_4 = (13)(2)(4)$, which we called p earlier. It's important to note that $\pi_1 \pi_2 \neq \pi_2 \pi_1$, which hopefully does not surprise you, since function composition is not in general commutative. To further illustrate the lack of commutativity in permutation groups, pick up a book (Not this one! You need to keep reading directions here.) so that cover is up and the spine is to the left. First, flip the book over from left to right. Then rotate it 90° clockwise. Where is the spine? Now return the book to the cover-up, spine-left position. Rotate the book 90° clockwise and then flip it over from left to right. Where is the spine this time?

It quickly gets tedious to write down where the product of two (or more) permutations sends each element. A more efficient approach would be to draw the digraph and then write down the cycle structure. With some practice, however, you can build the cycle notation as you go along, as we demonstrate in the following example.

Example 15.6. Let $\pi_1 = (123)(487)(5)(6)$ and $\pi_2 = (18765)(234)$. Let $\pi_3 = \pi_2\pi_1$. To start constructing the cycle notation for π_3, we must determine where π_3 sends 1. We find that it sends it to 3, since π_1 sends 1 to 2 and π_2 sends 2 to 3. Thus, the first cycle begins 13. Now where is 3 sent? It's sent to 8, which goes to 6, which goes to 5, which goes to 1, completing our first cycle as (13865). The first integer not in this cycle is 2, which we use to start our next cycle. We find that 2 is sent to 4, which is set to 7, which is set to 2. Thus, the second cycle is (247). Now all elements of 8 are represented in these cycles, so we know that $\pi_3 = (13865)(247)$.

We conclude this section with one more example.

Example 15.7. Let's find [(123456)][(165432)], where we've written the two permutations being multiplied inside brackets. Since we work from *right* to *left*, we find that the first permutation applied sends 1 to 6, and the second sends 6 to 1, so our first cycle is (1). Next, we find that the product sends 2 to 2. It also sends i to i for every other $i \leq 6$. Thus, the product is (1)(2)(3)(4)(5)(6), which is better known as the identity permutation. Thus, (123456) and (165432) are inverses.

In the next section, we will use standard counting techniques we've seen before in this book to prove results about groups acting on sets. We will state the results for arbitrary groups, but you may safely replace "group" by "permutation group" without losing any understanding required for the remainder of the chapter.

15.3 Burnside's Lemma

Burnside's lemma[1] relates the number of equivalence classes of the action of a group on a finite set to the number of elements of the set fixed by the elements of the group. Before stating and proving it, we need some notation and a proposition. If a group

---

[1] Again, not originally proved by Burnside. It was known to Frobenius and for the most part by Cauchy. However, it was most easily found in Burnside's book, and thus his name came to be attached.

Chapter 15 Pólya's Enumeration Theorem

G acts on a finite set C, let $\sim$ be the equivalence relation induced by this action. (As before, the action of $\pi \in G$ on C will be denoted π^*.) Denote the equivalence class containing $C \in C$ by $\langle C \rangle$. For $\pi \in G$, let $\mathrm{fix}_C(\pi) = \{C \in C : \pi^*(C) = C\}$, the set of colorings fixed by π. For $C \in C$, let $\mathrm{stab}_G(C) = \{\pi \in G : \pi(C) = C\}$ be the **stabilizer** of C in G, the permutations in G that fix C.

To illustrate these concepts before applying them, refer back to Table 15.2. Using that information, we can determine that $\mathrm{fix}_C(r_2) = \{C_1, C_{10}, C_{11}, C_{16}\}$. Determining the stabilizer of a coloring requires finding the rows of the table in which it appears. Thus, $\mathrm{stab}_{D_8}(C_7) = \{\iota, h\}$ and $\mathrm{stab}_{D_8}(C_{11}) = \{\iota, r_2, p, n\}$.

Proposition 15.8. *Let a group G act on a finite set C. Then for all $C \in C$,*

$$\sum_{C' \in \langle C \rangle} |\mathrm{stab}_G(C')| = |G|.$$

Proof. Let $\mathrm{stab}_G(C) = \{\pi_1, \ldots, \pi_k\}$ and $T(C, C') = \{\pi \in G : \pi^*(C) = C'\}$. (Note that $T(C, C) = \mathrm{stab}_G(C)$.) Take $\pi \in T(C, C')$. Then $\pi \circ \pi_i \in T(C, C')$ for $1 \leq i \leq k$. Furthermore, if $\pi \circ \pi_i = \pi \circ \pi_j$, then $\pi^{-1} \circ \pi \circ \pi_i = \pi^{-1} \circ \pi \circ \pi_j$. Thus $\pi_i = \pi_j$ and $i = j$. If $\pi' \in T(C, C')$, then $\pi^{-1} \circ \pi' \in T(C, C)$. Thus, $\pi^{-1} \circ \pi' = \pi_i$ for some i, and hence $\pi' = \pi \circ \pi_i$. Therefore $T(C, C') = \{\pi \circ \pi_1, \ldots, \pi \circ \pi_k\}$. Additionally, we observe that $T(C', C) = \{\pi^{-1} : \pi \in T(C, C')\}$. Now for all $C' \in \langle C \rangle$,

$$|\mathrm{stab}_G(C')| = |T(C', C')| = |T(C', C)| = |T(C, C')| = |T(C, C)| = |\mathrm{stab}_G(C)|.$$

Therefore,

$$\sum_{C' \in \langle C \rangle} |\mathrm{stab}_G(C')| = \sum_{C' \in \langle C \rangle} |T(C, C')|.$$

Now notice that each element of G appears in $T(C, C')$ for precisely one $C' \in \langle C \rangle$, and the proposition follows. □

With Proposition 15.8 established, we are now prepared for Burnside's lemma.

Lemma 15.9 (Burnside's Lemma). *Let a group G act on a finite set C. If N is the number of equivalence classes of C induced by this action, then*

$$N = \frac{1}{|G|} \sum_{\pi \in G} |\mathrm{fix}_C(\pi)|.$$

Before we proceed to the proof, note that the calculation in Burnside's lemma for the example of 2-coloring the vertices of a square is exactly the calculation we performed at the end of Section 15.1.

Proof. Let $X = \{(\pi, C) \in G \times \mathcal{C} \colon \pi(C) = C\}$. Notice that $\sum_{\pi \in G} |\text{fix}_\mathcal{C}(\pi)| = |X|$, since each term in the sum counts how many ordered pairs of X have π in their first coordinate. Similarly, $\sum_{C \in \mathcal{C}} |\text{stab}_G(C)| = |X|$, with each term of this sum counting how many ordered pairs of X have C as their second coordinate. Thus, $\sum_{\pi \in G} |\text{fix}_\mathcal{C}(\pi)| = \sum_{C \in \mathcal{C}} |\text{stab}_G(C)|$. Now note that the latter sum may be rewritten as

$$\sum_{\substack{\text{equivalence} \\ \text{classes } \langle C \rangle}} \left(\sum_{C' \in \langle C \rangle} |\text{stab}_G(C')| \right).$$

By Proposition 15.8, the inner sum is $|G|$. Therefore, the total sum is $N \cdot |G|$, so solving for N gives the desired equation. $\square$

Burnside's lemma helpfully validates the computations we did in the previous section. However, what if instead of a square we were working with a hexagon and instead of two colors we allowed four? Then there would be $4^6 = 4096$ different colorings and the dihedral group of the hexagon has 12 elements. Assembling the analogue of Table 15.2 in this situation would be a nightmare! This is where the genius of Pólya's approach comes into play, as we see in the next section.

15.4 Pólya's Theorem

Before getting to the full version of Pólya's formula, we must develop a generating function as promised at the beginning of the chapter. To do this, we will return to our example of Section 15.1.

15.4.1 The cycle index

Unlike the generating functions we encountered in Chapter 8, the generating functions we will develop in this chapter will have more than one variable. We begin by associating a monomial with each element of the permutation group involved. In this case, it is D_8, the dihedral group of the square. To determine the monomial associated to a permutation, we need to write the permutation in cycle notation and then determine the monomial based on the number of cycles of each length. Specifically, if π is a permutation of $[n]$ with j_k cycles of length k for $1 \le k \le n$, then the monomial associated to π is $x_1^{j_1} x_2^{j_2} \cdots x_n^{j_n}$. Note that $j_1 + 2j_2 + 3j_3 + \cdots + nj_n = n$. For example, the permutation $r_1 = (1234)$ is associated with the monomial x_4^1 since it consists of a single cycle of length 4. The permutation $r_2 = (13)(24)$ has two cycles of length 2, and thus its monomial is x_2^2. For $p = (14)(2)(3)$, we have two 1-cycles and one 2-cycle, yielding the monomial $x_1^2 x_2^1$. In Table 15.10, we show all eight permutations in D_8 along with their associated monomials.

Transformation	Monomial	Fixed colorings
$\iota = (1)(2)(3)(4)$	x_1^4	16
$r_1 = (1234)$	x_4^1	2
$r_2 = (13)(24)$	x_2^2	4
$r_3 = (1432)$	x_4^1	2
$v = (12)(34)$	x_2^2	4
$h = (14)(23)$	x_2^2	4
$p = (14)(2)(3)$	$x_1^2 x_2^1$	8
$n = (1)(24)(3)$	$x_1^2 x_2^1$	8

TABLE 15.10: MONOMIALS ARISING FROM THE DIHEDRAL GROUP OF THE SQUARE

Now let's see how the number of 2-colorings of the square fixed by a permutation can be determined from its cycle structure and associated monomial. If $\pi(i) = j$, then we know that for π to fix a coloring C, vertices i and j must be colored the same in C. Thus, the second vertex in a cycle must have the same color as the first. But then the third vertex must have the same color as the second, which is the same color as the first. In fact, all vertices appearing in a cycle of π must have the same color in C if π fixes C! Since we are coloring with the two colors white and gold, we can choose to color the points of each cycle uniformly white or gold. For example, for the permutation $v = (12)(34)$ to fix a coloring of the square, vertices 1 and 2 must be colored the same color (2 choices) and vertices 3 and 4 must be colored the same color (2 choices). Thus, there are $2 \cdot 2 = 4$ colorings fixed by v. Since there are two choices for how to uniformly color the elements of a cycle, letting $x_i = 2$ for all i in the monomial associated with π gives the number of colorings fixed by π. In Table 15.10, the "Fixed colorings" column gives the number of 2-colorings of the square fixed by each permutation. Before, we obtained this manually by considering the action of D_8 on the set of all 16 colorings. Now we only need the cycle notation and the monomials that result from it to derive this!

Recall that Burnside's Lemma states that the number of colorings fixed by the action of a group can be obtained by adding up the number fixed by each permutation and dividing by the number of permutations in the group. If we do that instead for the monomials arising from the permutations in a permutation group G in which every cycle of every permutation has at most n entries, we obtain a polynomial known as the

15.4 Pólya's Theorem

cycle index $P_G(x_1, x_2, \ldots, x_n)$. For our running example, we find

$$P_{D_8}(x_1, x_2, x_3, x_4) = \frac{1}{8}\left(x_1^4 + 2x_1^2 x_2^1 + 3x_2^2 + 2x_4^1\right).$$

To find the number of distinct 2-colorings of the square, we thus let $x_i = 2$ for all i and obtain $P_{D_8}(2, 2, 2, 2) = 6$ as before. Notice, however, that we have something more powerful than Burnside's lemma here. We may substitute *any* positive integer m for each x_i to find out how many nonequivalent m-colorings of the square exist. We no longer have to analyze how many colorings each permutation fixes. For instance, $P_{D_8}(3, 3, 3, 3) = 21$, meaning that 21 of the 81 colorings of the vertices of the square using three colors are distinct.

15.4.2 The full enumeration formula

Hopefully the power of the cycle index to count colorings that are distinct when symmetries are considered is becoming apparent. In the next section, we will provide additional examples of how it can be used. However, we still haven't seen the full power of Pólya's technique. From the cycle index alone, we can determine how many colorings of the vertices of the square are distinct. However, what if we want to know how many of them have two white vertices and two gold vertices? This is where Pólya's enumeration formula truly plays the role of a generating function.

Let's again consider the cycle index for the dihedral group D_8:

$$P_{D_8}(x_1, x_2, x_3, x_4) = \frac{1}{8}\left(x_1^4 + 2x_1^2 x_2^1 + 3x_2^2 + 2x_4^1\right).$$

Instead of substituting integers for the x_i, let's consider what happens if we substitute something that allows us to track the colors used. Since x_1 represents a cycle of length 1 in a permutation, the choice of white or gold for the vertex in such a cycle amounts to a single vertex receiving that color. What happens if we substitute $w + g$ for x_1? The first term in P_{D_8} corresponds to the identity permutation ι, which fixes all colorings of the square. Letting $x_1 = w + g$ in this term gives

$$(w + g)^4 = g^4 + 4g^3 w + 6g^2 w^2 + 4gw^3 + w^4,$$

which tells us that ι fixes one coloring with four gold vertices, four colorings with three gold vertices and one white vertex, six colorings with two gold vertices and two white vertices, four colorings with one gold vertex and three white vertices, and one coloring with four white vertices.

Let's continue establishing a pattern here by considering the variable x_2. It represents the cycles of length 2 in a permutation. Such a cycle must be colored uniformly

Chapter 15 Pólya's Enumeration Theorem

white or gold to be fixed by the permutation. Thus, choosing white or gold for the vertices in that cycle results in two white vertices or two gold vertices in the coloring. Since this happens for every cycle of length 2, we want to substitute $w^2 + g^2$ for x_2 in the cycle index. The $x_1^2 x_2^1$ terms in P_{D_8} are associated with the flips p and n. Letting $x_1 = w + g$ and $x_2 = w^2 + g^2$, we find

$$x_1^2 x_2^1 = g^4 + 2g^3 w + 2g^2 w^2 + 2gw^3 + w^4,$$

from which we are able to deduce that p and n each fix one coloring with four gold vertices, two colorings with three gold vertices and one white vertex, and so on. Comparing this with Table 15.2 shows that the generating function is right on.

By now the pattern is becoming apparent. If we substitute $w^i + g^i$ for x_i in the cycle index for each i, we then keep track of how many vertices are colored white and how many are colored gold. The simplification of the cycle index in this case is then a generating function in which the coefficient on $g^s w^t$ is the number of distinct colorings of the vertices of the square with s vertices colored gold and t vertices colored white. Doing this and simplifying gives

$$P_{D_8}(w + g, w^2 + g^2, w^3 + g^3, w^4 + g^4) = g^4 + g^3 w + 2g^2 w^2 + gw^3 + w^4.$$

From this we find one coloring with all vertices gold, one coloring with all vertices white, one coloring with three gold vertices and one white vertex, one coloring with one gold vertex and three white vertices, and two colorings with two vertices of each color.

As with the other results we've discovered in this chapter, this property of the cycle index holds up beyond the case of coloring the vertices of the square with two colors. The full version is Pólya's enumeration theorem:

Theorem 15.11 (Pólya's Enumeration Theorem). *Let S be a set with $|S| = r$ and C the set of colorings of S using the colors $c_1, \ldots, c_m$. If a permutation group G acts on S to induce an equivalence relation on C, then*

$$P_G\left(\sum_{i=1}^m c_i, \sum_{i=1}^m c_i^2, \ldots, \sum_{i=1}^m c_i^r\right)$$

is the generating function for the number of nonequivalent colorings of S in C.

If we return to coloring the vertices of the square but now allow the color blue as well, we find

$$P_{D_8}(w + g + b, w^2 + g^2 + b^2, w^3 + g^3 + b^3, w^4 + g^4 + b^4) = b^4 + b^3 g + 2b^2 g^2$$

$$+ bg^3 + g^4 + b^3 w + 2b^2 gw + 2bg^2 w + g^3 w + 2b^2 w^2 + 2bgw^2 + 2g^2 w^2$$
$$+ bw^3 + gw^3 + w^4.$$

From this generating function, we can readily determine the number of nonequivalent colorings with two blue vertices, one gold vertex, and one white vertex to be 2. Because the generating function of Pólya's Enumeration Theorem records the number of nonequivalent patterns, it is sometimes called the **pattern inventory**.

What if we were interested in making necklaces with 500 (very small) beads colored white, gold, and blue? This would be equivalent to coloring the vertices of a regular 500-gon, and the dihedral group D_{1000} would give the appropriate transformations. With a computer algebra system[1] such as *Mathematica®*, it is possible to quickly produce the pattern inventory for such a problem. In doing so, we find that there are

363602917958699368423852670795433191180233850260016230403460358325806001915838954841985082629793887833081797025344040466272877964304252714992703135653472347417085467453334179308247819807028526921872536424412922797565759360408045671032229 ≈ 3.6×10^{235}

possible necklaces. Of them,

252949184234046077349041318620101048779141729407880866280363896567824471388337043268753932294423230859058382000714795759057317766605088026968640797415175535033372572682057214340157297357996345021733060 ≈ 2.5×10^{200}

have 225 white beads, 225 gold beads, and 50 blue beads.

The remainder of this chapter will focus on applications of Pólya's Enumeration Theorem and the pattern inventory in a variety of settings.

15.5 Applications of Pólya's Enumeration Formula

This section explores a number of situations in which Pólya's enumeration formula can be used. The applications are from a variety of domains and are arranged in increasing order of complexity, beginning with an example from music theory and concluding with counting nonisomorphic graphs.

---

[1] With some more experience in group theory, it is possible to give a general formula for the cycle index of the dihedral group D_{2n}, so the computer algebra system is a nice tool, but not required.

15.5.1 Counting musical scales

Western music is generally based on a system of 12 equally-spaced **notes**. Although these notes are usually named by letters of the alphabet (with modifiers), for our purposes it will suffice to number them as $0, 1, \ldots, 11$. These notes are arranged into **octaves** so that the next pitch after 11 is again named 0 and the pitch before 0 is named 11. For this reason, we may consider the system of notes to correspond to the integers modulo 12. With these definitions, a **scale** is a subset of $\{0, 1, \ldots, 11\}$ arranged in increasing order. A **transposition of a scale** is a uniform transformation that replaces each note x of the scale by $x + a \pmod{12}$ for some constant a. Musicians consider two scales to be equivalent if one is a transposition of the other. Since a scale is a subset, no regard is paid to which note starts the scale, either. The question we investigate in this section is "How many nonequivalent scales are there consisting of precisely k notes?"

Because of the cyclic nature of the note names, we may consider arranging them in order clockwise around a circle. Selecting the notes for a scale then becomes a coloring problem if we say that selected notes are colored black and unselected notes are colored white. In Figure 15.12, we show three 5-note scales using this convention. Notice that since S_2 can be obtained from S_1 by rotating it forward seven positions, S_1 and S_2 are equivalent by the transposition of adding 7. However, S_3 is not equivalent to S_1 or S_2, as it cannot be obtained from them by rotation. (Note that S_3 could be obtained from S_1 if we allowed flips in addition to rotations. Since the only operation allowed is the transposition, which corresponds to rotation, they are inequivalent.)

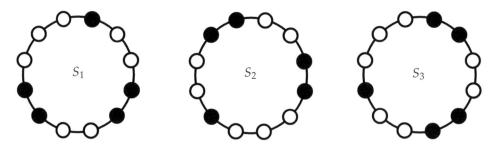

FIGURE 15.12: THREE SCALES DEPICTED BY COLORING

We have now mathematically modeled musical scales as discrete structures in a way that we can use Pólya's Enumeration Theorem. What is the group acting on our black-/white colorings of the vertices of a regular 12-gon? One permutation in the group is $\tau = (0\ 1\ 2\ 3\ 4\ 5\ 6\ 7\ 8\ 9\ 10\ 11)$, which corresponds to the transposition by one note. In fact, every element of the group can be realized as some power of τ since only rotations are allowed and τ is the smallest possible rotation. Thus, the group acting on the colorings is the **cyclic group of order** 12, denoted $C_{12} = \{\iota, \tau, \tau^2, \ldots, \tau^{11}\}$. Exercise 15.6.5 asks you to write all the elements of this group in cycle notation. The best way to do this is by multiplying τ^{i-1} by τ (i.e., compute $\tau\tau^{i-1}$) to find τ. Once you've done this, you will be able to easily verify that the cycle index is

$$P_{C_{12}}(x_1, \ldots, x_{12}) = \frac{x_1^{12}}{12} + \frac{x_2^6}{12} + \frac{x_3^4}{6} + \frac{x_4^3}{6} + \frac{x_6^2}{6} + \frac{x_{12}}{3}.$$

Since we've chosen colorings using black and white, it would make sense to substitute $x_i = b^i + w^i$ for all i in $P_{C_{12}}$ now to find the number of k-note scales. However, there is a convenient shortcut we may take to make the resulting generating function look more like those to which we grew accustomed in Chapter 8. The information about how many notes are *not* included in our scale (the number colored white) can be deduced from the number that are included. Thus, we may eliminate the use of the variable w, replacing it by 1. We now find

$$P_{C_{12}}(1 + b, 1 + b^2, \ldots, 1 + b^{12}) = b^{12} + b^{11} + 6b^{10} + 19b^9 + 43b^8$$
$$+ 66b^7 + 80b^6 + 66b^5 + 43b^4 + 19b^3 + 6b^2 + b + 1.$$

From this, we are able to deduce that the number of scales with k notes is the coefficient on b^k. Therefore, the answer to our question at the beginning of the chapter about the number of 6-note scales is 80.

15.5.2 Enumerating isomers

Benzene is a chemical compound with formula C_6H_6, meaning it consists of six carbon atoms and six hydrogen atoms. These atoms are bonded in such a way that the six carbon atoms form a hexagonal ring with alternating single and double bonds. A hydrogen atom is bonded to each carbon atom (on the outside of the ring). From benzene it is possible to form other chemical compounds that are part of a family known as **aromatic hydrocarbons**. These compounds are formed by replacing one or more of the hydrogen atoms by atoms of other elements or functional groups such as CH_3 (methyl group) or OH (hydroxyl group). Because there are six choices for which hydrogen atoms to replace, molecules with the same chemical formula but different structures can be formed in this manner. Such molecules are called **isomers**. In this subsection, we will see how Pólya's Enumeration Theorem can be used to determine the number of isomers of the aromatic hydrocarbon xylenol (also known as dimethylphenol).

Before we get into the molecular structure of xylenol, we need to discuss the permutation group that will act on a benzene ring. Much like with our example of coloring the vertices of the square, we find that there are rotations and flips at play here. In

Chapter 15 Pólya's Enumeration Theorem

fact, the group we require is the dihedral group of the hexagon, D_{12}. If we number the six carbon atoms in clockwise order as $1, 2, \ldots, 6$, then we find that the clockwise rotation by $60°$ corresponds to the permutation $r = (123456)$. The other rotations are the higher powers of r, as shown in Table 15.13. The flip across the vertical axis is the permutation $f = (16)(25)(34)$. The remaining elements of D_{12} (other than the identity ι) can all be realized as some rotation followed by this flip. The full list of permutations is shown in Table 15.13, where each permutation is accompanied by the monomial it contributes to the cycle index.

Permutation	Monomial	Permutation	Monomial
$\iota = (1)(2)(3)(4)(5)(6)$	x_1^6	$f = (16)(25)(34)$	x_2^3
$r = (123456)$	x_6^1	$fr = (15)(24)(3)(6)$	$x_1^2 x_2^2$
$r^2 = (135)(246)$	x_3^2	$fr^2 = (14)(23)(56)$	x_2^3
$r^3 = (14)(25)(36)$	x_2^3	$fr^3 = (13)(2)(46)(5)$	$x_1^2 x_2^2$
$r^4 = (153)(264)$	x_3^2	$fr^4 = (12)(36)(45)$	x_2^3
$r^5 = (165432)$	x_6^1	$fr^5 = (1)(26)(35)(4)$	$x_1^2 x_2^2$

TABLE 15.13: CYCLE REPRESENTATION OF PERMUTATIONS IN D_{12}

With the monomials associated to the permutations in D_{12} identified, we are able to write down the cycle index

$$P_{D_{12}}(x_1, \ldots, x_6) = \frac{1}{12}(x_1^6 + 2x_6^1 + 2x_3^2 + 4x_2^3 + 3x_1^2 x_2^2).$$

With the cycle index determined, we now turn our attention to using it to find the number of isomers of xylenol. This aromatic hydrocarbon has three hydrogen molecules, two methyl groups, and a hydroxyl group attached to the carbon atoms. Recalling that hydrogen atoms are the default from benzene, we can more or less ignore them when choosing the appropriate substitution for the x_i in the cycle index. If we let m denote methyl groups and h hydroxyl groups, we can then substitute $x_i = 1 + m^i + h^i$ in $P_{D_{12}}$. This substitution gives the generating function

$$1 + h + 3h^2 + 3h^3 + 3h^4 + h^5 + h^6 + m + 3hm + 6h^2m + 6h^3m$$
$$+ 3h^4m + h^5m + 3m^2 + 6hm^2 + 11h^2m^2 + 6h^3m^2 + 3h^4m^2 + 3m^3 + 6hm^3$$
$$+ 6h^2m^3 + 3h^3m^3 + 3m^4 + 3hm^4 + 3h^2m^4 + m^5 + hm^5 + m^6.$$

Since xylenol has one hydroxyl group and two methyl groups, we are looking for the coefficient on hm^2 in this generating function. The coefficient is 6, so there are six isomers of xylenol.

15.5 Applications of Pólya's Enumeration Formula

In his original paper, Pólya used his techniques to enumerate the number of isomers of the alkanes C_nH_{2n+2}. When modeled as graphs, these chemical compounds are special types of trees. Since that time, Pólya's Enumeration Theorem has been used to enumerate isomers for many different chemical compounds.

15.5.3 Counting nonisomorphic graphs

Counting the graphs with vertex set $[n]$ is not difficult. There are $C(n,2)$ possible edges, each of which can be included or excluded. Thus, there are $2^{C(n,2)}$ **labeled graphs** on n vertices. It's only a bit of extra thought to determine that if you only want to count the labeled graphs on n vertices with k edges, you simply must choose a k-element subset of the set of all $C(n,2)$ possible edges. Thus, there are

$$\binom{\binom{n}{2}}{k}$$

graphs with vertex set $[n]$ and exactly k edges.

A more difficult problem arises when we want to start counting *nonisomorphic* graphs on n vertices. (One can think of these as **unlabeled graphs** as well.) For example, in Figure 15.14, we show four different labeled graphs on four vertices. The first three graphs shown there, however, are isomorphic to each other. Thus, only two nonisomorphic graphs on four vertices are illustrated in the figure. To account for isomorphisms, we need to bring Pólya's Enumeration Theorem into play.

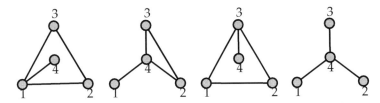

FIGURE 15.14: FOUR LABELED GRAPHS ON FOUR VERTICES

We begin by considering all $2^{C(n,2)}$ graphs with vertex set $[n]$ and choosing an appropriate permutation group to act in the situation. Since any vertex can be mapped to any other vertex, the symmetric group S_4 acts on the vertices. However, we have to be careful about how we find the cycle index here. When we were working with colorings of the vertices of the square, we realized that all the vertices appearing in the same cycle of a permutation π had to be colored the same color. Since we're concerned with edges here and not vertex colorings, what we really need for a permutation to fix a graph is that every edge be sent to an edge and every non-edge be sent to a non-edge.

Chapter 15 Pólya's Enumeration Theorem

To be specific, if $\{1, 2\}$ is an edge of some **G** and $\pi \in S_4$ fixes **G**, then $\{\pi(1), \pi(2)\}$ must also be an edge of **G**. Similarly, if vertices 3 and 4 are not adjacent in **G**, then $\pi(3)$ and $\pi(4)$ must also be nonadjacent in **G**.

To account for edges, we move from the symmetric group S_4 to its **pair group** $S_4^{(2)}$. The objects that $S_4^{(2)}$ permutes are the 2-element subsets of $\{1, 2, 3, 4\}$. For ease of notation, we will denote the 2-element subset $\{i, j\}$ by e_{ij}. To find the permutations in $S_4^{(2)}$, we consider the vertex permutations in S_4 and see how they permute the e_{ij}. The identity permutation $\iota = (1)(2)(3)(4)$ of S_4 corresponds to the identity permutation $\iota = (e_{12})(e_{13})(e_{14})(e_{23})(e_{24})(e_{34})$ of $S_4^{(2)}$. Now let's consider the permutation $(12)(3)(4)$. It fixes e_{12} since it sends 1 to 2 and 2 to 1. It also fixeds e_{34} by fixing 3 and 4. However, it interchanges e_{13} with e_{23} (3 is fixed and 1 is swapped with 2) and e_{14} with e_{24} (1 is sent to 2 and 4 is fixed). Thus, the corresponding permutation of pairs is $(e_{12})(e_{13}e_{23})(e_{14}e_{24})(e_{34})$. For another example, consider the permutation $(123)(4)$. It corresponds to the permutation $(e_{12}e_{23}e_{13})(e_{14}e_{24}e_{34})$ in $S_4^{(2)}$.

Since we're only after the cycle index of $S_4^{(2)}$, we don't need to find all 24 permutations in the pair group. However, we do need to know the types of those permutations in terms of cycle lengths so we can associate the appropriate monomials. For the three examples we've considered, the cycle structure of the permutation in the pair group doesn't depend on the original permutation in S_4 other than for *its* cycle structure. Any permutation in S_4 consisting of a 2-cycle and two 1-cycles will correspond to a permutation with two 2-cycles and two 1-cycles in $S_4^{(2)}$. A permutation in S_4 with one 3-cycle and one 1-cycle will correspond to a permutation with two 3-cycles in the pair group. By considering an example of a permutation in S_4 consisting of a single 4-cycle, we find that the corresponding permutation in the pair group has a 4-cycle and a 2-cycle. Finally, a permutation of S_4 consisting of two 2-cycles corresponds to a permutation in $S_4^{(2)}$ having two 2-cycles and two 1-cycles. (Exercise 15.6.8 asks you to verify these claims using specific permutations.)

Now that we know the cycle structure of the permutations in $S_4^{(2)}$, the only task remaining before we can find its cycle index of is to determine how many permutations have each of the possible cycle structures. For this, we again refer back to permutations of the symmetric group S_4. A permutation consisting of a single 4-cycle begins with 1 and then has 2, 3, and 4 in any of the $3! = 6$ possible orders, so there are 6 such permutations. For permutations consisting of a 1-cycle and a 3-cycle, there are 4 ways to choose the element for the 1-cycle and then 2 ways to arrange the other three as a 3-cycle. (Remember the smallest of them must be placed first, so there are then 2 ways to arrange the remaining two.) Thus, there are 8 such permutations. For a permutation consisting of two 1-cycles and a 2-cycle, there are $C(4, 2) = 6$ ways to choose the two elements for the 2-cycle. Thus, there are 6 such permutations. For a permutation to consist of two 2-cycles, there are $C(4, 2) = 6$ ways to choose two elements for the first 2-cycle. The other two are then put in the second 2-cycle. However, this counts each permutation twice, once for when the first 2-cycle is the chosen pair and once for when it is the "other two." Thus, there are 3 permutations consisting of two 2-cycles. Finally, only ι consists of four 1-cycles.

Now we're prepared to write down the cycle index of the pair group

$$P_{S_4^{(2)}}(x_1, \ldots, x_6) = \frac{1}{24} \left(x_6^1 + 9x_1^2 x_2^2 + 8x_3^2 + 6x_2 x_4 \right).$$

To use this to enumerate graphs, we can now make the substitution $x_i = 1 + x^i$ for $1 \leq i \leq 6$. This allows us to account for the two options of an edge not being present or being present. In doing so, we find

$$P_{S_4^{(2)}}(1 + x, \ldots, 1 + x^6) = 1 + x + 2x^2 + 3x^3 + 2x^4 + x^5 + x^6$$

is the generating function for the number of 4-vertex graphs with m edges, $0 \leq m \leq 6$. To find the total number of nonisomorphic graphs on four vertices, we substitute $x = 1$ into this polynomial. This allows us to conclude there are 11 nonisomorphic graphs on four vertices, a marked reduction from the 64 labeled graphs.

The techniques of this subsection can be used, given enough computing power, to find the number of nonisomorphic graphs on any number of vertices. For 30 vertices, there are

33449431630925766924943569928080028956631479935393064329967834 88721773453488058274903052159950438 $\approx 3.3 \times 10^{98}$

nonisomorphic graphs, as compared to $2^{435} \approx 8.9 \times 10^{130}$ labeled graphs on 30 vertices. The number of nonisomorphic graphs with precisely 200 edges is

31338248099707262762587724757336401854467033655017855836082677 05079969989351221982191036097960 1 $\approx 3.1 \times 10^{96}$.

The last part of the question about graph enumeration at the beginning of the chapter was about enumerating the graphs on some number of vertices in which every vertex has degree r. While this might seem like it could be approached using the techniques of this chapter, it turns out that it cannot because of the increased dependency between where vertices are mapped.

15.6 Exercises

1. Write the permutations shown below in cycle notation.

$$\pi_1 = \begin{pmatrix} 1 & 2 & 3 & 4 & 5 & 6 \\ 4 & 2 & 5 & 6 & 3 & 1 \end{pmatrix} \qquad \pi_2 = \begin{pmatrix} 1 & 2 & 3 & 4 & 5 & 6 \\ 5 & 6 & 1 & 3 & 4 & 2 \end{pmatrix}$$

$$\pi_3 = \begin{pmatrix} 1 & 2 & 3 & 4 & 5 & 6 & 7 & 8 \\ 3 & 1 & 5 & 8 & 2 & 6 & 4 & 7 \end{pmatrix} \qquad \pi_4 = \begin{pmatrix} 1 & 2 & 3 & 4 & 5 & 6 & 7 & 8 \\ 3 & 7 & 1 & 6 & 8 & 4 & 2 & 5 \end{pmatrix}$$

2. Compute $\pi_1\pi_2$, $\pi_2\pi_1$, $\pi_3\pi_4$, and $\pi_4\pi_3$ for the permutations π_i in Exercise 15.6.1.

3. Find $\text{stab}_{D_8}(C_3)$ and $\text{stab}_{D_8}(C_{16})$ for the colorings of the vertices of the square shown in Figure 15.1 by referring to Table 15.2.

4. In Figure 15.15, we show a regular pentagon with its vertices labeled. Use this labeling to complete this exercise.

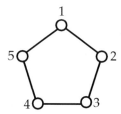

FIGURE 15.15: A PENTAGON WITH LABELED VERTICES

(a) The dihedral group of the pentagon, D_{10}, contains 10 permutations. Let $r_1 = (12345)$ be the clockwise rotation by 72° and $f_1 = (1)(25)(34)$ be the flip about the line passing through 1 and perpendicular to the opposite side. Let r_2, r_3, and r_4 be the other rotations in D_{10}. Denote the flip about the line passing through vertex i and perpendicular to the other side by f_i, $1 \leq i \leq 5$. Write all 10 elements of D_{10} in cycle notation.

(b) Suppose we are coloring the vertices of the pentagon using black and white. Draw the colorings fixed by r_1. Draw the colorings fixed by f_1.

(c) Find $\text{stab}_{D_{10}}(C)$ where C is the coloring of the vertices of the pentagon in which vertices 1, 2, and 5 are colored black and vertices 3 and 4 are colored white.

(d) Find the cycle index of D_{10}.

(e) Use the cycle index to determine the number of nonequivalent colorings of vertices of the pentagon using black and white.

(f) Making an appropriate substitution for the x_i in the cycle index, find the number of nonequivalent colorings of the vertices of the pentagon in which two vertices are colored black and three vertices are colored white. Draw these colorings.

5. Write all permutations in C_{12}, the cyclic group of order 12, in cycle notation.

6. The 12-note western scale is not the only system on which music is based. In classical Thai music, a scale with seven equally-spaced notes per octave is used. As in western music, a scale is a subset of these seven notes, and two scales are equivalent if they are transpositions of each other. Find the number of k-note scales in classical Thai music for $1 \leq k \leq 7$.

7. Xylene is an aromatic hydrocarbon having two methyl groups (and four hydrogen atoms) attached to the hexagonal carbon ring. How many isomers are there of xylene?

8. Find the permutations in $S_4^{(2)}$ corresponding to the permutations (1234) and (12)(34) in S_4. Confirm that the first consists of a 4-cycle and a 2-cycle and the second consists of two 2-cycles and two 1-cycles.

9. Draw the three nonisomorphic graphs on four vertices with 3 edges and the two nonisomorphic graphs on four vertices with 4 edges.

10.

(a) Use the method of Subsection 15.5.3 to find the cycle index of the pair group $S_5^{(2)}$ of the symmetric group on five elements.

(b) Use the cycle index from Item 15.6.10.a to determine the number of nonisomorphic graphs on five vertices. How many of them have 6 edges?

11. Tic-tac-toe is a two-player game played on a 9×9 grid. The players mark the squares of the grid with the symbols X and O. This exercise uses Pólya's enumeration theorem to investigate the number of different tic-tac-toe boards. (The analysis of *games* is more complex, since it requires attention to the order the squares are marked and stopping when one player has won the game.)

Chapter 15 Pólya's Enumeration Theorem

1	2	3
4	5	6
7	8	9

FIGURE 15.16: NUMBERED SQUARES OF A TIC-TAC-TOE BOARD

(a) Two tic-tac-toe boards are equivalent if one may be obtained from the other by rotating the board or flipping it over. (Imagine that it is drawn on a clear piece of plastic.) Since the 9×9 grid is a square, the group that acts on it in this manner is the dihedral group D_8 that we have studied in this chapter. However, as with counting nonisomorphic graphs, we have to be careful to choose the way this group is represented in terms of cycles. Here we are interested in how permutations rearrange the nine squares of the tic-tac-toe board as numbered in Figure 15.16. For example, the effect of the transformation r_1, which rotates the board 90° clockwise, can be represented as a permutation of the nine squares as $(13971)(2684)(5)$.

Write each of the eight elements of D_8 as permutations of the nine squares of a tic-tac-toe board.

(b) Find the cycle index of D_8 in terms of these permutations.

(c) Make an appropriate substitution for x_i in the cycle index to find a generating function $t(X, O)$ in which the coefficient on $X^i O^j$ is the number of nonequivalent tic-tac-toe boards having i squares filled by symbol X and j squares filled by symbol O. (Notice that some squares might be blank!)

(d) How many nonequivalent tic-tac-toe boards are there?

(e) How many nonequivalent tic-tac-toe boards have three X's and three O's?

(f) When playing tic-tac-toe, the players alternate turns, each drawing their symbol in a single unoccupied square during a turn. Assuming the first player marks her squares with X and the second marks his with O, then at each stage of the game there are either the same number of X's and O's or one more X than there are O's. Use this fact and $t(X, O)$ to determine the number of nonequivalent tic-tac-toe boards that can actually be obtained in playing a game, assuming the players continue until the board is full, regardless of whether one of them has won the game.

12. Suppose you are painting the faces of a cube and you have white, gold, and blue paint available. Two painted cubes are equivalent if you can rotate one of them so that all corresponding faces are painted the same color. Determine the number of nonequivalent ways you can paint the faces of the cube as well as the number having two faces of each color.

Hint. It may be helpful to label the faces as U ("up"), D ("down"), F ("front"), B ("back"), L ("left"), and R ("right") instead of using integers. Working with a three-dimensional model of a cube will also aid in identifying the permutations you require.

CHAPTER 16

The Many Faces of Combinatorics

16.1 On-line algorithms

Many applications of combinatorics occur in a dynamic, on-line manner. It is rare that one has all the information about the challenges a problem presents before circumstances compel that decisions be made. As examples, a decision to proceed with a major construction project must be made several years before ground is broken; investment decisions are made on the basis of today's information and may look particularly unwise when tomorrow's news is available; and deciding to exit a plane with a parachute is rarely reversible.

In this section, we present two examples intended to illustrate on-line problems in a combinatorial setting. Our first example involves graph coloring. As is customary in discussions of on-line algorithms, we consider a two-person game with the players called **Assigner** and **Builder**. The two players agree in advance on a class C of graphs, and the game is played in a series of rounds. At round 1 Builder presents a single vertex, and Assigner assigns it a color. At each subsequent rounds, Builder presents a new vertex, and provides complete information at to which of the preceding vertices are adjacent to it. In turn, Assigner must give the new vertex a color distinct from colors she has assigned previously to its neighbors.

Example 16.1. Even if Builder is constrained to build a path on 4 vertices, then Assigner can be forced to use three colors. At Round 1, Builder presents a vertex x and Assigner colors it. At Round 2, Builder presents a vertex y and declares that x and y are not adjacent.

Now Assigner has a choice. She may either give x and y the same color, or she may elect to assign a new color to y. If Assigner gives x and y different colors, then in Round 3, Builder presents a vertex z and declares that z is adjacent to both x and y. Now Assigner will be forced to use a third color on z. In Round 4, Builder will add a vertex w adjacent to y but to neither x nor z, but the damage has already been done.

On the other hand, if Assigner x and y the same color, then in Round 3, Builder presents a vertex z, with z adjacent to x but not to y. Assigner must use a second color

315

Chapter 16 The Many Faces of Combinatorics

on z, distinct from the one she gave to x and y. In Round 4, Builder presents a vertex w adjacent to z and y but not to x. Assigner must use a third color on w.

Note that a path is a tree and trees are forests. The next result shows that while forests are trivial to color off-line, there is a genuine challenge ahead when you have to work on-line. To assist us in keeping track of the colors used by Assigner, we will use the notation from Chapter 5 and write $\phi(x)$ for the color given by Assigner to vertex x.

Theorem 16.2. *Let n be a positive integer. Then there is a strategy for Builder that will enable Builder to construct a forest having at most 2^{n-1} vertices while forcing Assigner to use n colors.*

Proof. When $n = 1$, all Builder does is present a single vertex. When $n = 2$, two adjacent vertices are enough. When $n = 3$, Builder constructs a path on 4 vertices as detailed in Example 16.1. Now assume that for some $k \geq 3$, Builder has a strategy S_i for forcing Assigner to use i colors on a forest of at most 2^{i-1} vertices, for each $i = 1, 2, \ldots, k$. Here's how Builder proceeds to force $k + 1$ colors.

First, for each $i = 1, 2, \ldots, k$, Builder follows strategy S_i to build a forest F_i having at most 2^{i-1} vertices on which assigner is forced to use i colors. Furthermore, when $1 \leq i < j \leq k$, there are no edges between vertices in F_i and vertices in F_j.

Next, Builder chooses a vertex y_1 from F_1. Since Assigner uses two colors on F_2, there is a vertex y_2 from F_2 so that $\phi(y_2) \neq \phi(y_1)$. Since Assigner uses three colors on F_3, there is a vertex y_3 in F_3 so that $\{\phi(y_1), \phi(y_2), \phi(y_3)\}$ are all distinct. It follows that Builder may identify vertices $y_1, y_2, \ldots, y_k$ with $y_i \in F_i$ so that the colors $\phi(y_i)$ satisfy $\phi(y_i) \neq \phi(y_j)$ if $i \neq j$. Builder now presents a new vertex x and declares x adjacent to all vertices in $\{y_1, y_2, \ldots, y_k\}$ and to no other vertices. Clearly, the resulting graph is a forest and Assigner is forced to use a color for x distinct from the k colors she assigned previously to the vertices in $\{y_1, y_2, \ldots, y_k\}$. Also, the total number of vertices is at most $1 + [1 + 2 + 4 + 8 + \cdots + 2^{k-1}] = 2^k$. □

Discussion 16.3. Bob reads the proof and asks whether it was really necessary to treat the cases $k = 2$ and $k = 3$ separately. Wasn't it enough just to note that the case $k = 1$ holds trivially. Carlos says yes.

16.1.1 Doing Relatively Well in an On-Line Setting

Theorem 16.2 should be viewed as a negative result. It is hard to imagine a family of graphs easier to color than forests, yet in an on-line setting, graphs in this family are difficult to color. On the other hand, in certain settings, one can do reasonably well in an on-line setting, perhaps not as well as the true optimal off-line result but good enough to be useful. Here we present a particularly elegant example involving partially ordered sets.

Recall that a poset P of height h can be partitioned into h antichains—by recursively removing the set of minimal elements. But how many antichains are required in an on-line setting? Now Builder constructs a poset P one point at a time, while Assigner constructs a partition of P into antichains. At each round, Builder will present a new point x, and list those points presented earlier that are, respectively, less than x, greater than x and incomparable with x. Subsequently, Assigner will assign x to an antichain. This will be done either by adding x to an antichain already containing one or more of the points presented previously, or by assigning x to a new antichain.

Theorem 16.4. *For each $h \geq 1$, there is a on-line strategy for Assigner that will enable her to partition a poset P into at most $\binom{h+1}{2}$ antichains, provided the height of P is at most h.*

Proof. It is important to note that Assigner does not need to know the value h in advance. For example, Builder may have in mind that ultimately the value of h will be 300, but this information does not impact Assigner's strategy.

When the new point x_n enters P, Assigner computes the values r and s, where r is the largest integer for which there exists a chain C of r points in $\{x_1, x_2, \ldots, x_n\}$ having x_n as its least element. Also, s is the largest integer for which there exists a chain D of s points in $\{x_1, x_2, \ldots, x_n\}$ having x_n as its largest element. Assigner then places x in a set $A(r,s)$, claiming that any two points in this set are incomparable. To see that this claim is valid, consider the first moment where Builder has presented a new point x, Assigner places x in $A(r,s)$ and there is already a point y in $A(r,s)$ for which x and y are comparable.

When y was presented, there was at that moment in time a chain C' of r points having y as its least element. Also, there was a chain D of s points having y as its greatest element.

Now suppose that $y > x$ in P. Then we can add x to C' to form a chain of $r+1$ points having x as its least element. This would imply that x is not assigned in $A(r,s)$. Similarly, if $y < x$ in P, then we may add x to D' to form a chain of $s+1$ points having x as its greatest element. Again, this would imply that x is not assigned to $A(r,s)$.

So Assigner has indeed devised a good strategy for partitioning P into antichains, but how many antichains has she used? This is just asking how many ordered pairs (i, j) of positive integers are there subject to the restriction that $i + j - 1 \leq h$. And we learned how to solve this kind of question in Chapter 2. The answer of course is $\binom{h+1}{2}$. □

The strategy for Assigner is so simple and natural, it might be the case that a more complex strategy would yield a more efficient partitioning. Not so.

Theorem 16.5 (Szemerédi). *For every $h \geq 1$, there is a strategy S_h for builder that will enable him to build a poset P of height h so that assigner is forced to (1) use at least $\binom{h+1}{2}$ antichains in partitioning P, and (2) use at least h different antichains on the set of maximal elements.*

Proof. Strategy S_1 is just to present a single point. Now suppose that the theorem holds for some integer $h \geq 1$. We show how strategy S_{h+1} proceeds.

First Builder follows strategy S_h to form a poset P_1. Then he follows it a second time for form a poset P_2, with all points of P_1 incomparable to all points in P_2. Now we consider two cases. Suppose first that Assigner has used $h+1$ or more antichains on the set of maximal elements of $P_1 \cup P_2$. In this case, he follows strategy S_h a third time to build a poset P_3 with all points of P_3 less than all maximal elements of $P_1 \cup P_2$ and incomparable with all other points.

Clearly, the height of the resulting poset is at most $h+1$. Also, Assigner must use $h + 1 + \binom{h+1}{2} = \binom{h+2}{2}$ antichains in partitioning the poset and she has used $h+1$ on the set of maximal elements.

So it remains only to consider the case where Assigner has used a set W of h antichains on the maximal elements of P_1, and she has used exactly the same h antichains for the maximal elements of P_2. Then Builder presents a new point x and declares it to be greater than all points of P_1 and incomparable with all points of P_2. Assigner must put x in some antichain which is not in W.

Builder then follows strategy S_h a third time, but now all points of P_3 are less than x and the maximal elements of P_2. Again, Assigner has been forced to use $h+1$ different antichains on the maximal elements and $\binom{h+2}{2}$ antichains altogether. □

16.2 Extremal Set Theory

Let n be a positive integer and let $[n] = 1, 2, \ldots, n$. In this section, we consider problems having the following general form: What is the maximum size of a family of subsets of $[n]$ when the family is required to satisfy certain properties.

Here is an elementary example.

Example 16.6. The maximum size of a family $\mathcal{F}$ of subsets of $[n]$, with $A \cap B \neq \emptyset$ for all $A, B \in \mathcal{F}$, is 2^{n-1}.

For the lower bound, consider the family $\mathcal{F}$ of all subsets of $[n]$ that contain 1. Clearly this family has 2^{n-1} elements and any two sets in the family have non-empty intersection.

For the upper bound, let $\mathcal{F}$ be a family of subsets with each pair of sets in $\mathcal{F}$ having non-empty intersection. Then whenever a subset S is a member of $\mathcal{F}$, the complement S' of S cannot belong to $\mathcal{F}$. Since the entire family of all 2^n subsets of $[n]$ can be considered as 2^{n-1} complementary pairs, and at most one set from each pair can belong to $\mathcal{F}$, we conclude that $|\mathcal{F}| \leq 2^{n-1}$.

As a second example, we can revisit Sperner's Theorem from Chapter 6 and restate the result as follows.

16.2 Extremal Set Theory

Example 16.7. The maximum size of a family $\mathcal{F}$ of subsets of $[n]$ subject to the constraint that when A and B are distinct sets in $\mathcal{F}$, then neither is a subset of the other, is $\binom{n}{\lfloor n/2 \rfloor}$.

It is worth noting that in Example 16.7, there is a very small number (one or two) of extremal families, i.e., when $\mathcal{F}$ is a family of subsets of $[n]$, $|\mathcal{F}| = \binom{n}{\lfloor n/2 \rfloor}$, and no set in $\mathcal{F}$ is a proper subset of another, then either $\mathcal{F} = \{S \subseteq [n] : |S| = \lfloor n/2 \rfloor\}$ or $\mathcal{F} = \{S \subseteq [n] : |S| = \lceil n/2 \rceil\}$. And of course, when n is even, these are exactly the same family.

On the other hand, for Example 16.6, there are many extremal families, since for every complementary pair of sets, either member can be selected.

We close this brief tasting of extremal set theory with a real classic.

Theorem 16.8 (Erdős, Ko, Rado). *Let n and k be positive integers with $n \geq 2k$. Then the maximum size of a family $\mathcal{F}$ of subsets of $[n]$ subject to the restrictions that (1) $A \cap B \neq \emptyset$ for all $A, B \in \mathcal{F}$, and (2) $|A| = k$ for all $A \in \mathcal{F}$, is $\binom{n-1}{k-1}$.*

Proof. For the lower bound, consider the family $\mathcal{F}$ of all k element subset of $[n]$ that contain 1.

For the upper bound, let $\mathcal{F}$ be a family of subsets of $[n]$ satisfying the two constraints. We show that $|\mathcal{F}| \leq \binom{n-1}{k-1}$. To accomplish this, we consider a circle in the Euclidean plane with n points $p_1, p_2, \ldots, p_n$ equally spaced points around its circumference. Then there are $n!$ different ways (one for each permutation σ of $[n]$) to place the integers in $[n]$ at the points in $\{p_1, p_2, \ldots, p_n\}$ in one to one manner.

For each permutation σ of $[n]$, let $\mathcal{F}(\sigma)$ denote the subfamily of $\mathcal{F}$ consisting of all sets S from $\mathcal{F}$ whose elements occur in a consecutive block around the circle. Then let $t = \sum_\sigma |\mathcal{F}(\sigma)|$.

Our first claim is that $t \leq kn!$. To prove this, let σ be a permutation and suppose that $|\mathcal{F}(\sigma)| = s \geq 1$. Then the union of the sets from $\mathcal{F}(\sigma)$ is a set of points that form a consecutive block of points on the circle. Note that since $n \geq 2k$, this block does not encompass the entire circle. Accordingly there is a set S whose elements are the first k in a clockwise sense within this block. Since each other set in $\mathcal{F}$ represents a clockwise shift of one of more positions, it follows immediately that $|\mathcal{F}| \leq k$. Since there are $n!$ permutations, the claim follows.

We now claim that for each set $S \in \mathcal{F}$, there are exactly $nk!(n-k)!$ permutations σ for which $S \in \mathcal{F}(\sigma)$. Note that there are n positions around the circle and each can be used as the first point in a block of k consecutive positions in which the elements of S can be placed. Then there are $k!$ ways to order the elements of S and $(n-k)!$ ways to order the remaining elements. This proves our claim.

To complete the proof of the theorem, we note that we have

$$|\mathcal{F}|nk!(n-k)! \leq t \leq kn!,$$

and this implies that $|\mathcal{F}| le \binom{n-1}{k-1}$.

16.3 Markov Chains

We begin this section with a motivational example. Consider the connected graph on six vertices shown in ⟨⟨fig-markovchain⟩⟩. The first move is to choose a vertex at random and move there. Afterwards, we follow the following recursive procedures. If after i moves, you are at a vertex x and x has d neighbors, choose one of the neighbors at random, with each having probability $1/d$ and move there. We then attempt to answer questions of the following flavor:

1. For each vertex x, let $p_{x,m}$ denote the probability that you are at vertex x after m moves. Does $\lim_{m\to\infty} p_{x,m}$ exist and if so, how fast does the sequence converge to this limit?

2. How many moves must I make in order that the probability that I have walked on every edge in the graph is at least 0.999?

This example illustrates the characteristics of an important class of computational and combinatorial problems, which are collectively referred to as **Markov Chains**:

1. There is a finite set of states $S_1, S_2, \ldots, S_n$, and at time i, you are in one of these states.

2. If you are in state S_j at time i, then for each $k = 1, 2, \ldots, n$, there is a fixed probability $p(j, k)$ (which does not depend on i) that you will be in state S_k at time $i + 1$.

The $n \times n$ matrix P whose j, k entry is the probability $p(j, k)$ of moving from state S_j to state S_k is called the **transition matrix** of the Markov chain. Note that P is a **stochastic matrix**, i.e., all entries are non-negative and all row sums are 1. Conversely, each square stochastic matrix can be considered as the transition matrix of a Markov chain.

For example, here is the transition matrix for the graph in ⟨⟨fig-markovchain⟩⟩.

$$P = \begin{pmatrix} 0 & 1/4 & 1/4 & 1/4 & 1/4 & 0 \\ 1/2 & 0 & 0 & 1/2 & 0 & 0 \\ 1/3 & 0 & 0 & 1/3 & 0 & 1/3 \\ 1/3 & 1/3 & 1/3 & 0 & 0 & 0 \\ 1 & 0 & 0 & 0 & 0 & 0 \\ 0 & 0 & 1 & 0 & 0 & 0 \end{pmatrix} \qquad (16.3.1)$$

16.3 Markov Chains

A transition matrix P is **regular** if there is some integer m for which the matrix P^m has only positive entries. Here is a fundamental result from this subject, one that is easy to understand but a bit too complex to prove given our space constraints.

Theorem 16.9. *Let P be a regular $n \times n$ transition matrix. Then there is a row vector $W = (w_1, w_2, \ldots, w_n)$ of positive real numbers summing to 1 so that as m tends to infinity, each row of P^m tends to W. Furthermore, $WP = W$, and for each $i = 1, 2, \ldots, n$, the value w_i is the limiting probability of being in state S_i.*

Given the statement of Theorem 16.9, the computation of the row vector W can be carried out by eigenvalue techniques that are part of a standard undergraduate linear algebra course. For example, the transition matrix P displayed in (16.3.1) is regular since all entries of P^3 are positive. Furthermore, for this matrix, the row vector $W = (5/13, 3/13, 2/13, 2/13, 1/13, 1/13)$. However, the question involving how fast the convergence of P^m is to this limiting vector is more subtle, as is the question as to how long it takes for us to be relatively certain we have made every possible transition.

16.3.1 Absorbing Markov Chains

A state S_i in a Markov chain with transition matrix P is **absorbing** if $p_{i,i} = 1$ and $p_{i,j} = 0$ for all $j \neq i$, i.e., like the infamous Hotel California, once you are in state S_i, "you can never leave."

Example 16.10. We modify the transition matrix from (16.3.1) by making states 4 and 5 absorbing. The revised transition matrix is now:

$$P = \begin{pmatrix} 0 & 1/4 & 1/4 & 1/4 & 1/4 & 0 \\ 1/2 & 0 & 0 & 1/2 & 0 & 0 \\ 1/3 & 0 & 0 & 1/3 & 0 & 1/3 \\ 1/3 & 1/3 & 1/3 & 0 & 0 & 0 \\ 0 & 0 & 0 & 0 & 1 & 0 \\ 0 & 0 & 0 & 0 & 0 & 1 \end{pmatrix} \quad (16.3.2)$$

Now we might consider the following game. Start at one of the four vertices in $\{1, 2, 3, 4\}$ and proceed as before, making moves by choosing a neighbor at random. Vertex 4 might be considered as an "escape" point, a safe harbor that once reached is never left. On the other hand, vertex 5 might be somewhere one meets a hungry tiger and be absorbed in a way not to be detailed here.

We say the Markov chain is **absorbing** if there is at least one absorbing state and for each state S_j that is not absorbing, it is possible to reach an absorbing state—although it may take many steps to do so. Now the kinds of questions we would like to answer are:

1. If we start in non-absorbing state S_i, what is the probability of reaching absorbing state S_j (and then being absorbed in that state, a question which takes on genuine unpleasantness relative to tigers)?

2. If we are absorbed in state S_j, what is the probability that we started in non-absorbing state S_i?

3. If we start in non-absorbing state S_i, what is the expected length of time before we will be absorbed?

16.4 The Stable Matching Theorem

Now we present a light hearted optimization problem with a quite clever solution, called the *Stable Matching Theorem*. There are n eligible males $b_1, b_2, \ldots, b_n$ and n eligible females $g_1, g_2, \ldots, g_n$. (The theorem dates back many years, hence the heteronormative statement.) We will arrange n marriages, each involving one male and one female. In the process, we will try to make everyone happy—or at least we will try to keep things stable.

Each female linearly orders the males in the order of her preference, i.e., for each $i = 1, 2, \ldots, n$, there is a permutation σ_i of $[n]$ so that if g_i prefers b_j to b_k, then $\sigma_i(j) > \sigma_i(k)$. Different females may have quite different preference orders. Also, each male linearly orders the females in order of his preference, i.e., for each $i = 1, 2, \ldots, n$, there is a permutation τ_i of $[n]$ so that if b_i prefers g_j to g_k, then $\tau_i(j) > \tau(k)$.

A 1–1 matching of the n males to the n females is **stable** if there do not exist two males b and b' and two females g and g' so that

1. b is matched to g;

2. b prefers g' to g; and

3. g prefers b' to b.

The idea is that given these preferences, b and g may be mutually inclined to dissolve their relationship and initiate dalliances with other partners.

So the question is whether, regardless of their respective preferences, we can always generate a stable matching. The answer is yes and there is a quite clever argument. In fact, it is one that yields a quite efficient algorithm. At Stage 1, all males go knock on the front door of the female which is tops on their list. It may happen that some females have more than one caller while others have none. However, if a female has one or more males at her door, she reaches out and grabs the one among the group which she prefers most by the collar and tells the others, if there are any, to go away. Any male rejected at this step proceeds to the front door of the female who is second on their list. Again, a female with one or more suitors at her door chooses the best among then and sends the others away. This process continues until eventually, each female is holding onto exactly one male.

It is interesting to note that each female's prospects improve over time, i.e., once she has a suitor, things only get better. Conversely, each male's prospects deteriorate over time. Regardless, we assert that the resulting matching is stable. To see this, suppose that it is unstable and choose males b and b', females g and g' so that b is matched to g, but b prefers g' to g while g prefers b' to b. The algorithm requires that male b start at the top of his list and work his way down. Since he eventually lands on g's door step, and he prefers g' to g, it implies that once upon a time, he was actually at g''s door, and she sent him away. This means that at that exact moment, she had a male in hand that she prefers to b. Since her holdings only improve with time, it means that when the matching is finalized, female g has a mate b that she prefers to b'.

16.5 Zero–One Matrices

Matrices with all entries 0 and 1 arise in many combinatorial settings, and here we present a classic result, called the Gale-Ryser theorem. It deals with zero–one matrices with specified row and column sum strings. When M is an $m \times n$ zero–one matrix, the string $R = (r_1, r_2, \ldots, r_m)$, where $r_i = \sum_{1 \leq j \leq n} m_{i,j}$, is called the **row sum string** of M. The **column sum string** $C = (c_1, c_2, \ldots, c_n)$ is defined analogously. Conversely, let m and n be positive integers, and let $R = (r_1, r_2, \ldots, r_m)$ and $C = (c_1, c_2, \ldots, c_n)$ be strings of non-negative integers. The question is whether there exists an $m \times n$ zero–one matrix M with row sum string R and column sum string C.

To attack this problem, we pause briefly to develop some additional background material. Note that we may assume without loss of generality that there is a positive integer t so that $\sum_{i=1}^{m} r_i = \sum_{j=1}^{n} c_j = t$, else there is certainly no zero–one matrix with row sum string R and column sum string C. Furthermore, we may assume that both R and C are non-increasing strings, i.e., $r_1 \geq r_2 \geq \cdots \geq r_m$ and $c_1 \geq c_2 \geq \cdots \geq c_n$.

To see this note that whenever we exchange two rows in a zero–one matrix, the column sum string is unchanged. Accordingly after a suitable permutation of the rows, we may assume that R is non-increasing. Then the process is repeated for the columns.

Finally, it is easy to see that we may assume that all entries in R and C are positive integers, since zeroes in these strings correspond to rows of zeroes or columns of zeroes in the matrix. Accordingly, the row sum string R and the column sum string C can be viewed as partitions of the integer t, a topic we first introduced in Chapter 8.

For the balance of this section, we let t be a positive integer and we let $\mathcal{P}(t)$ denote the family of all partitions of the integer t. There is a natural partial order on $\mathcal{P}(t)$ defined by setting $V = (v_1, v_2, \ldots, v_m) \geq W = (w_1, w_2, \ldots, w_n)$ if and only if $m \leq n$

and $\sum_{1 \le i \le j} v_j \ge \sum_{1 \le i \le j} w_j$ for each $j = 1, 2, \ldots, m$, i.e., the sequence of partial sums for V is always at least as large, term by term, as the sequence of partial sums of W. For example, we show in ⟨⟨fig-partitionlattice⟩⟩ the partial order $\mathcal{P}(7)$.

FIGURE HERE

In the proof of the Gale-Ryser theorem, it will be essential to fully understand when one partition covers another. We state the following proposition for emphasis; the proof consists of just considering the details of the definition of the partial order on partitions.

Proposition 16.11. *Let $V = (v_1, v_2, \ldots, v_m)$ and $W = (w_1, w_2, \ldots, w_n)$ be partitions of an integer t. If V covers W in the poset $\mathcal{P}(t)$, then $n \le m + 1$ and there exist integers i and j with $1 \le i < j \le n$ so that the following statements hold.*

1. $v_\alpha = w_\alpha$, when $1 \le \alpha < i$.

2. $v_\beta = w_\beta$, when $j < \beta \le m$.

3. $v_i = 1 + w_i$.

4. *Either (a) $j \le m$ and $w_j = 1 + v_j$, or (b) $j = n = m + 1$ and $w_j = 1$.*

5. *If $j > i + 1$, then $w_\gamma = v_\gamma = v_i - 1$ when $i < \gamma < j$.*

To illustrate this concept, note that $(5, 4, 3)$ covers $(5, 3, 3, 1)$ in $\mathcal{P}(12)$. Also, we see $(6, 6, 4, 3, 3, 3, 1, 1, 1, 1,)$ covers $(6, 6, 3, 3, 3, 3, 2, 1, 1, 1)$ in $\mathcal{P}(29)$.

With a partition $V = (v_1, v_2, \ldots, v_m)$ from $\mathcal{P}(t)$, we associate a **dual partition** $W = (w_1, w_2, \ldots, w_n)$ defined as follows: (1) $n = v_1$ and for each $j = 1, \ldots, n$, w_j is the number of entries in V that are at least $n + 1 - j$. For example, the dual partition of $V = (8, 6, 6, 6, 5, 5, 3, 1, 1, 1)$ is $(8, 7, 7, 6, 6, 4, 1, 1)$. Of course, they are both partitions of 42, which is the secret of the universe! In what follows, we denote the dual of the partition V by V^d. Note that if $W = V^d$, then $V = W^d$, i.e., the dual of the dual is the original.

16.5.1 The Obvious Necessary Condition

Now let M be a $m \times n$ zero–one matrix with row sum string $R = (r_1, r_2, \ldots, r_m$ and column sum string $C = (c_1, c_2, \ldots, c_n)$. As noted before, we will assume that all entries in R and C are positive. Next, we modify M to form a new matrix M' as follows: For each $i = 1, 2, \ldots, t$, we push the r_i ones in row i as far to the left as possible, i.e., $m'_{i,j} = 1$ if and only if $1 \le j \le r_i$. Note that M and M' both have R for their row sum strings. However, if C' denotes the column sum string for M', then C' is a non-decreasing string, and the substring C'' of C' consisting of the positive entries is R^d, the dual partition of R. Furthermore, for each $j = 1, 2, \ldots, r_1$, we have the inequality $\sum_{1 \le i \le j} c''_i \le \sum_{1 \le i \le j} c_i$, since the operation of shift ones to the left can only increase the partial sums. It follows that $R^d \geq C$ in the poset $\mathcal{P}(t)$.

So here is the Gale-Ryser theorem.

Theorem 16.12 (Gale-Ryser). *Let R and C be partitions of a positive integer t. Then there exists a zero–one matrix with row sum string R and column sum string C if and only if $R^d \geq C$ in the poset $\mathcal{P}(t)$.*

Proof. The necessity of the condition has been established. We prove sufficiency. The proof is constructive. In the poset $\mathcal{P}(t$, let $W_0 > W_1 > \cdots > W_s$ be a chain so that (1) $W_0 = R^d$, (2) $W_s = C$ and (3) if $0 \leq p < s$, then W_p covers W_{p+1}. We start with a zero one matrix M_0 having row sum string R and column sum string W_0, as suggested in ⟨⟨fig-dualpartition⟩⟩ for the partition $(8, 4, 3, 1, 1, 1)$. If $s = 0$, we are done, so we assume that for some p with $0 \leq p < s$, we have a zero–one matrix M_p with row sum string R and column sum string W_p. Then let i and j be the integers from Proposition 16.11, which detail how W_p covers W_{p+1}. Choose a row q so that the q, i entry of M_p is 1 while the q, j entry of M is 0. Exchange these two entries to form the matrix M_{p+1}. Note that the exchange may in fact require adding a new column to the matrix. □

16.6 Arithmetic Combinatorics

In recent years, a great deal of attention has been focused on topics in arithmetic combinatorics, with a number of deep and exciting discoveries in the offing. In some sense, this area is closely aligned with Ramsey theory and number theory, but recent work shows connections with real and complex analysis, as well. Furthermore, the roots of arithmetic combinatorics go back many years. In this section, we present a brief overview of this rich and rapidly changing area.

Recall that an increasing sequence $a_1 < a_2 < a_3 < \ldots < a_t)$ of integers is called an **arithmetic progression** when there exists a positive integer d for which $a_{i+1} - a_i = d$, for all $i = 1, 2, \ldots, t-1$. The integer t is called the **length** of the arithmetic progression.

Theorem 16.13. *For pair r, t of positive integers, there exists an integer n_0, so that if $n \geq n_0$ and $\phi : \{1, 2, \ldots, n\} \to \{1, 2, \ldots, r\}$ is any function, then there exists a t-term arithmetic progression $1 \leq a_1 < a_2 < \ldots < a_t \leq n$ and an element $\alpha \in \{1, 2, \ldots, r\}$ so that $\phi(a_i) = \alpha$, for each $i = 1, 2, \ldots, t$.*

Material will be added here.

16.7 The Lovász Local Lemma

Even though humans seem to have great difficulty in providing explicit constructions for exponentially large graphs which do not have complete subgraphs or independent sets of size n, such graphs exist with great abundance. Just take one at random and you are almost certain to get one. And as a general rule, probabilistic techniques often provide a method for finding something that readily exists, but is hard to find.

Similarly, in the probabilistic proof that there exist graphs with large girth and large chromatic number (Theorem 11.7), we actually showed that almost all graphs have modest sized independence number and relatively few small cycles, provided that the edge probability is chosen appropriately. The small cycles can be destroyed without significantly changing the size of the graph.

By way of contrast, probabilistic techniques can, in certain circumstances, be used to find something which is exceedingly rare. We next present an elegant but elementary result, known as the Lovász Local Lemma, which has proved to be very, very powerful. The treatment is simplified by the following natural notation. When E is an event in a probability space, we let $\overline{E}$ denote the complement of E. Also, when $\mathcal{F} = \{E_1, E_2, \ldots, E_k\}$ we let

$$\prod_{E \in \mathcal{F}} E = \prod_{i=1}^{k} E_i = E_1 E_2 E_3 \ldots E_k$$

denote the event $E_1 \cap E_2 \cap \cdots \cap E_k$, i.e., concatenation is short hand for intersection. These notations can be mixed, so $E_1 \overline{E_2} \overline{E_3}$ represents $E_1 \cap \overline{E_2} \cap \overline{E_3}$. Now let $\mathcal{F}$ be a finite family of events, let $E \in \mathcal{F}$ and let $\mathcal{N}$ be a subfamily of $\mathcal{F} - \{E\}$. In the statement of the lemma below, we will say that E is independent of any event not in $\mathcal{N}$ when

$$P(E | \prod_{F \in \mathcal{G}} \overline{F}) = P(E)$$

provided $\mathcal{G} \cap \mathcal{N} = \emptyset$.

We first state and prove the lemma in *asymmetric* form. Later, we will give a simpler version which is called the *symmetric* version.

Lemma 16.14 (Lovász Local Lemma (Asymmetric)). *Let $\mathcal{F}$ be a finite family of events in a probability space and for each event $E \in \mathcal{F}$, let $\mathcal{N}(E)$ denote a subfamily of events from $\mathcal{F} - \{E\}$ so that E is independent of any event not in $\mathcal{N}(E)$. Suppose that for each event $E \in \mathcal{F}$, there is a real number $x(E)$ with $0 < x(E) < 1$ such that*

$$P(E) \leq x(E) \prod_{F \in \mathcal{N}(E)} (1 - x(F)).$$

16.7 The Lovász Local Lemma

Then for every non-empty subfamily $\mathcal{G} \subseteq \mathcal{F}$,

$$P(\prod_{E \in \mathcal{G}} \overline{E}) \geq \prod_{E \in \mathcal{G}} (1 - x(E)).$$

In particular, the probability that all events in $\mathcal{F}$ fail is positive.

Proof. We proceed by induction on $\mathcal{G}$. If $|\mathcal{G}| = 1$ and $\mathcal{G} = \{E\}$, we are simply asserting that $P(\overline{E}) \geq 1 - x(E)$, which is true since $P(E) \leq x(E)$. Now suppose that $|\mathcal{G}| = k \geq 2$ and that the lemma holds whenever $1 \leq |\mathcal{G}| < k$. Let $\mathcal{G} = \{E_1, E_2, \ldots, E_k\}$. Then

$$P(\prod_{i \geq 1}^{k} \overline{E_i}) = P(\overline{E_1} | \prod_{i=2}^{k} \overline{E_i}) P(\overline{E_2} | \prod_{i=3}^{k} \overline{E_i}) P(\overline{E_3} | \prod_{i=4}^{k} \overline{E_i}) \ldots$$

Now each term in the product on the right has the following form:

$$P(\overline{E} | \prod_{F \in \mathcal{F}_E} \overline{F})$$

where $|\mathcal{F}_E| < k$.

So, we done if we can show that

$$P(\overline{E} | \prod_{F \in \mathcal{F}_E} \overline{F}) \geq 1 - x(E)$$

This is equivalent to showing that

$$P(E | \prod_{F \in \mathcal{F}_E} \overline{F}) \leq x(E)$$

Suppose first that $\mathcal{F}_E \cap N(E) = \emptyset$. Then

$$P(E | \prod_{F \in \mathcal{F}_E} \overline{F}) = P(E) \leq x(E).$$

So we may assume that $\mathcal{F}_E \cap N(E) \neq \emptyset$. Let $\mathcal{F}_E = \{F_1, F_2, F_r, F_{r+1}, F_{r+2}, \ldots, F_t\}$, with $F_i \in N_E$ if and only if $r + 1 \leq i \leq t$. Then

$$P(E | \prod_{F \in \mathcal{F}_E} \overline{F}) = \frac{P(E \prod_{F \in \mathcal{F}_E \cap N(E)} \overline{F} | \prod_{F \in \mathcal{F}_E - N(E)} \overline{F})}{P(\prod_{F \in \mathcal{F}_E \cap N(E)} \overline{F})}$$

Consider first the numerator in this last expression. Note that

$$P(E \prod_{F \in \mathcal{F}_E \cap N(E)} \overline{F} | \prod_{F \in \mathcal{F}_E - N(E)} \overline{F}) \leq P(E | \prod_{F \in \mathcal{F}_E \cap N(E)} \overline{F}) \leq x(E) \prod_{F \in \mathcal{F}_E \cap N(E)} (1 - x(F))$$

Next, consider the denominator. By the inductive hypothesis, we have

$$P(\prod_F \in \mathcal{F}_E \cap \mathcal{N}(E)\overline{F} \geq \prod_{F \in \mathcal{F}_E \cap \mathcal{N}(E)} (1 - x(F)).$$

Combining these last two inequalities, we have

$$P(E | \prod_{F \in \mathcal{F}_E} \overline{F}) \leq x(E) \prod_{\mathcal{N}(E) - \mathcal{F}_E} (1 - x(F)) \leq x(E),$$

and the proof is complete. □

Now here is the symmetric version.

Lemma 16.15 (Lovász Local Lemma (Symmetric)). *Let p and d be numbers with $0 < p < 1$ and $d \geq 1$. Also, let $\mathcal{F}$ be a finite family of events in a probability space and for each event $E \in \mathcal{F}$, let $\mathcal{N}(E)$ denote the subfamily of events from $\mathcal{F} - \{E\}$ so that E is independent of any event not in $\mathcal{N}(E)$. Suppose that $P(E) \leq p$, $|\mathcal{N}(E)| \leq d$ for every event $E \in \mathcal{F}$ and that $ep(d+1) < 1$, where $e = 2.71828\ldots$ is the base for natural logarithms. Then*

$$P(\prod_{E \in \mathcal{F}} \overline{E}) \geq \prod_{E \in \mathcal{G}} (1 - x(E)),$$

i.e., the probability that all events in $\mathcal{F}$ is positive.

Proof. Set $x(E) = 1/(d+1)$ for every event $E \in \mathcal{F}$. Then

$$P(E) \leq p \leq \frac{1}{e(d+1)} \leq x(E) \prod (F \in \mathcal{N}(E)(1 - \frac{1}{d+1}). \qquad \square$$

A number of applications of the symmetric form of the Lovász Local Lemma are stated in terms of the condition that $4pd < 1$. The proof of this alternate form is just a trivial modification of the argument we have presented here.

16.8 Applying the Local Lemma

The list of applications of the Local Lemma has been growing steadily, as has the interest in how the lemma can be applied algorithmically, i.e., in a constructive setting. But here we present one of the early applications to Ramsey theory—estimating the Ramsey number $(R, 3, n)$. Recall that we have the basic inequality $R(3, n) \leq \binom{n+1}{3}$ from Theorem 11.2, and it is natural to turn to the probabilistic method to look for good lower bounds. But a few minutes thought shows that there are challenges to this approach.

16.8 Applying the Local Lemma

First, let's try a direct computation. Suppose we try a random graph on t vertices with edge probability p. So we would want no triangles, and that would say we need $t^3 p^3 = 1$, i.e., $p = 1/t$. Then we would want no independent sets of size n, which would require $n^t e^{-pn^2} = 1$, i.e., $t \ln n = pn^2$, so we can't even make t larger than n. That's not helpful.

We can do a bit better by allowing some triangles and then removing one point from each, as was done in the proof for Theorem 11.7. Along these lines, we would set $t^3 p^3 = t$, i.e., $p = t^{-2/3}$. And the calculation now yields the lower bound $R(3,n) \geq n^{6/5}/\ln^{-3/5} n$, so even the exponent of n is different from the upper bound.

So which one is right, or is the answer somewhere in between? In a classic 1961 paper, Erdős used a very clever application of the probabilistic method to show the existence of a graph from which a good lower bound could be extracted. His technique yielded the lower bound $R(3,n) \geq n^2/\ln^2 n$, so the two on the exponent of n is correct.

Here we will use the Lovász Local Lemma to obtain this same lower bound in a much more direct manner. We consider a random graph on t vertices with edge probability p. For each 3-element subset S, we have the event E_S which is true when S forms a triangle. For each n-element set T, we have the event E_T which is true when T is an independent set. In the discussion to follow, we abuse notation slightly and refer to events E_S and E_T as just S and T, respectively. Note that the probability of S is p^3 for each 3-element set S, while the probability of T is $q = (1-p)^{C(n,2)} \sim e^{-pn^2/2}$ for each n-element set T.

When we apply the Local Lemma, we will set $x = x(S)$ to be $e^2 p^3$, for each 3-element set S. And we will set $y = Y(T) = q^{1/2} \sim e^{-pn^2/4}$. It will be clear in a moment where we got those values.

Furthermore, the neighborhood of an event consists of all sets in the family which have two or more elements in common. So the neighborhood of a 3-element set S consists of $3(t-3)$ other 3-element sets and $C(t-3,n-3)+3C(t-3,n-2)$ sets of size n. Similarly, the neighborhood of an n-element set T consists of $C(n,3) + (t-n)C(n,2)$ sets of size 3 and $\sum_{i=2}^{n-1} C(n,i)C(t-n,n-i)$ other sets of size n. So the basic inequalities we need to satisfy are:

$$p^3 \leq x(1-x)^{3(t-3)}(1-y)^{C(t-3,n-3)+3C(t-3,n-2)}$$

$$q \leq y(1-x)^{C(n,3)+(t-n)C(n,2)}(1-y)^{C(t-3,n-3)+3C(t-3,n-2)}$$

Next, we assume that $n^{3/2} < t < n^2$ and then make the usual approximations, ignoring smaller order terms and multiplicative constants, to see that these inequalities can be considered in the following simplified form:

$$p^3 \leq x(1-x)^t(1-y)^{t^n}$$

329

Chapter 16 The Many Faces of Combinatorics

$$q \leq y(1-x)^{tn^2}(1-y)^{t^n}$$

A moments reflection makes it clear that we want to keep the terms involving $(1-y)$ relatively large, i.e., at least $1/e$. This will certainly be true if we keep $t^n \leq 1/y$. This is equivalent to $n \ln t \leq pn^2$, or $\ln t \leq pn$.

Similarly, we want to keep the term $(1-x)^t$ relatively large, so we keep $t \leq 1/x$, i.e., $t \leq 1/p^3$. On the other hand, we want only to keep the term $(1-x)^{tn^2} \sim e^{-xtn^2}$ at least as large as y. This is equivalent to keeping $p \leq xt$, and since $x \sim p^3$, this can be rewritten as $p^{-1} \leq t^{1/2}$.

Now we have our marching orders. We just set $\ln t = pn$ and $p^{-1} = t^{1/2}$. After substituting, we get $t = n^2/\ln^2 t$ and since $\ln t = \ln n$ (at least within the kind of approximations we are using), we get the desired result $t = n^2/\ln^2 n$.

APPENDIX A

Epilogue

Here is a progress report on our cast of characters, some five years after graduation[1].

Alice and *Bob* got married, moved to Austin, Texas, and started a high tech firm using venture capital provided by a successful Georgia Tech grad. Alice is CEO and the pattern of making quick decisions, most of which are right, continues to this day. Bob is CFO and the financial health of the firm is guaranteed. The first year though was pretty tough, but after that, their reputation got established and contracts began to walk through the door. There's even talk about an IPO in the near future. Alice and Bob don't have much time to decide whether they are happy with the way their lives are going—but we're pretty sure they are.

Carlos switched from Physics to math for graduate school and won an NSF graduate fellowship which he took at MIT. After receiving his Ph.D., he took a postdoctoral position at the American Institute for the Mathematical Sciences (AIMS). He also won an NSF Career grant. Carlos is a rapidly emerging star in the academic world. He has universities lining up to offer him tenure-track positions and he had already been invited to lecture in England, France, Germany, Hungary and Poland. He'll make a good living, not a huge salary, but the quality of life will rank with the best. He is very happy.

Dave surprised a lot of people. Somewhere along the way, he got just a bit more organized without losing that off-the-wall uniqueness that made him special. He took a job on Wall Street with a firm that just wanted really very smart people. He's making more money than any other member of the group, by far. But it comes at some cost. Long hours and lots of stress. On the occasional free Sunday (there aren't many), he wonders how much longer he can keep this up.

Xing took a job with Macrofirm in Bluemon. His group is developing new operating systems and attendant software that run on computing devices of all sizes, from smart phones through super computers. Lots of interesting challenges, for example, just in deciding how input should be done when there's no keyboard and the device screen

---

[1] Georgia Tech students do not speak of graduating. Instead, using the same phrase applied to incarceration, they talk about *getting out*.

Appendix A Epilogue

is very small. Xing is enjoying life and feels his Georgia Tech experiences were great preparation.

Yolanda used her chemistry background to go to medical school at Emory University, where she received both an M.D. and a Ph.D. Afterwards, she accepted a position at the Center for Disease Control (CDC), which is also located here in Atlanta and has a bunch of scientists with the same kind of background training. Yolanda quickly became the go to person for analyzing strange viruses which no one else was able to identify. She is part of a very important safety net which is essential to the nation's security and well-being. She is very happy with her life.

Zori didn't go down the pathway through life she once envisioned. Her first job was with a family owned company making candy bars. In that position, she helped them to make wise decisions on massive sugar purchases made on a world-wide basis. She got bored with this job, and left to accept a position with a support group for an airline company. Her group did optimization work, figuring out how best to position aircraft and crews to handle scheduling irregularities. Two years later, she moved to a position with a major chip maker where she helped optimize the movement of cutting heads in the manufacturing process, where incremental improvements could mean for hundreds of millions of dollars in savings. Zori has made lots of money, but she remains vaguely dissatisfied with life and is still looking for the right environment.

APPENDIX B

Background Material for Combinatorics

This appendix treats background material essential to the study of combinatorial mathematics. Many students will find that most—and perhaps all—of this material has been covered somewhere in their prior course work, and we expect that very few instructors will include this appendix in the syllabus. Nevertheless, students may find it convenient to consult this appendix from time to time, and we suspect that many instructors will encourage students to read this material to refresh their memories of key concepts.

B.1 Introduction

Set theory is concerned with **elements**, certain collections of elements called **sets** and a concept of **membership**. For each element x and each set X, *exactly* one of the following two statements holds:

1. x is a member of X.

2. x is *not* a member of X.

It is important to note that membership cannot be ambiguous.

When x is an element and X is a set, we write $x \in X$ when x is a member of X. Also, the statement x belongs to X means exactly the same thing as x is a member of X. Similarly, when x is not a member of X, we write $x \notin X$ and say x does not belong to X.

Certain sets will be defined explicitly by listing the elements. For example, let $X = \{a, b, d, g, m\}$. Then $b \in X$ and $h \notin X$. The order of elements in such a listing is irrelevant, so we could also write $X = \{g, d, b, m, a\}$. In other situations, sets will be defined by giving a rule for membership. As examples, let $\mathbb{N}$ denote the set of positive integers. Then let $X = \{n \in \mathbb{N} : 5 \le n \le 9\}$. Note that $6, 8 \in X$ while $4, 10, 238 \notin X$.

Appendix B Background Material for Combinatorics

Given an element x and a set X, it may at times be tedious and perhaps very difficult to determine which of the statements $x \in X$ and $x \notin X$ holds. But if we are discussing sets, it must be the case that *exactly* one is true.

Example B.1. Let X be the set consisting of the following 12 positive integers:

$$13232112332$$
$$13332112332$$
$$13231112132$$
$$13331112132$$
$$13232112112$$
$$13231112212$$
$$13331112212$$
$$13232112331$$
$$13231112131$$
$$13331112131$$
$$13331112132$$
$$13332112111$$
$$13231112131$$

Note that one number is listed twice. Which one is it? Also, does 13232112132 belong to X? Note that the apparent difficulty of answering these questions stems from (1) the size of the set X and (2) the size of the integers that belong to X. Can you think of circumstances in which it is difficult to answer whether x is a member of X even when it is known that X contains exactly one element?

Example B.2. Let P denote the set of primes. Then $35 \notin P$ since $35 = 5 \times 7$. Also, $19 \in P$. Now consider the number

$$n = 77788467064627123923601532364763319082817131766346039653933$$

Does n belong to P? Alice says yes while Bob says no. How could Alice justify her affirmative answer? How could Bob justify his negative stance? In this specific case, I know that Alice is right. Can you explain why?

B.2 Intersections and Unions

When X and Y are sets, the **intersection** of X and Y, denoted $X \cap Y$, is defined by

$$X \cap Y = \{x : x \in X, x \in Y\}$$

Note that this notation uses the convention followed by many programming languages. Namely, the "comma" in the definition means that *both* requirements for membership be satisfied. For example, if $X = \{b, c, e, g, m\}$ and $Y = \{a, c, d, h, m, n, p\}$, then $X \cap Y = \{c, m\}$.

B.2.1 The Meaning of 2-Letter Words

In the not too distant past, there was considerable discussion in the popular press on the meaning of the 2-letter word *is*. For mathematicians and computer scientists, it would have been far more significant to have a discussion of the 2-letter word *or*. The problem is that the English language uses *or* in two fundamentally different ways. Consider the following sentences:

1. A nearby restaurant has a dinner special featuring two choices for dessert: flan de casa *or* tirami-su.

2. A state university accepts all students who have graduated from in-state high schools and have SAT scores above 1000 *or* have grade point averages above 3.0.

3. A local newspaper offers customers the option of paying their for their newspaper bills on a monthly *or* semi-annual basis.

In the first and third statement, it is clear that there are two options but that *only* one of them is allowed. However, in the second statement, the interpretation is that admission will be granted to students who satisfy *at least one* of the two requirements. These interpretations are called respectively the **exclusive** and **inclusive** versions of **or**. In this class, we will assume that whenever the word "or" is used, the inclusive interpretation is intended—unless otherwise stated.

For example, when X and Y are sets, the **union** of X and Y, denoted $X \cup Y$, is defined by
$$X \cup Y = \{x : x \in x \text{ or } x \in Y\}.$$
For example, if $X = \{b, c, e, g, m\}$ and $Y = \{a, c, d, h, m, n, p\}$, then
$$X \cup Y = \{a, b, c, d, e, g, h, m, n, p\}.$$

Note that $\cap$ and $\cup$ are **commutative** and **associative** binary operations, as is the case with addition and multiplication for the set $\mathbb{N}$ of positive integers, i.e., if X, Y and Z are sets, then
$$X \cap Y = Y \cap X \quad \text{and} \quad X \cup Y = Y \cup X.$$
Also,
$$X \cap (Y \cap Z) = (X \cap Y) \cap Z \quad \text{and} \quad X \cup (Y \cup Z) = (X \cup Y) \cup Z.$$

Appendix B Background Material for Combinatorics

Also, note that each of ∩ and ∪ distributes over the other, i.e.,

$$X \cap (Y \cup Z) = (X \cap Y) \cup (X \cap Z) \quad \text{and} \quad X \cup (Y \cap Z) = (X \cup Y) \cap (X \cup Z)$$

On the other hand, in $\mathbb{N}$, multiplication distributes over addition but not vice-versa.

B.2.2 The Empty Set: Much To Do About Nothing

The **empty set**, denoted $\emptyset$ is the set for which $x \notin \emptyset$ for every element x. Note that $X \cap \emptyset = \emptyset$ and $X \cup \emptyset = X$, for every set X.

The empty set is unique in the sense that if $x \notin X$ for every element x, then $X = \emptyset$.

B.2.3 The First So Many Positive Integers

In this text, we will use the symbols $\mathbb{N}, \mathbb{Z}, \mathbb{Q}$ and $\mathbb{R}$ to denote respectively the set of positive integers, the set of all integers (positive, negative and zero), the set of rational numbers (fractions) and the set of real numbers (rationals and irrationals). On occasion, we will discuss the set $\mathbb{N}_0$ of **non-negative integers**. When n is a positive integer, we will use the abbreviation $[n]$ for the set $\{1, 2, \ldots, n\}$ of the first n positive integers. For example, $[5] = \{1, 2, 3, 4, 5\}$. For reasons that may not be clear at the moment but hopefully will be transparent at the right moment, we use the notation **n** to denote the n-element set $\{0, 1, 2, \ldots, n-1\}$. Of course, **n** is just the set of the first n non-negative integers. For example, **5** $= \{0, 1, 2, 3, 4\}$.

B.2.4 Subsets, Proper Subsets and Equal Sets

When X and Y are sets, we say X is a **subset** of Y and write $X \subseteq Y$ when $x \in Y$ for every $x \in X$. When X is a subset of Y and there exists at least one element $y \in Y$ with $y \notin X$, we say X is a **proper subset** of Y and write $X \subsetneq Y$. For example, the P of primes is a proper subset of the set $\mathbb{N}$ of positive integers.

Surprisingly often, we will encounter a situation where sets X and Y have different rules for membership yet both are in fact the same set. For example, let $X = \{0, 2\}$ and $Y = \{z \in \mathbb{Z} : z + z = z \times z\}$. Then $X = Y$. For this reason, it is useful to have a test when sets are equal. If X and Y are sets, then

$$X = Y \quad \text{if and only if} \quad X \subseteq Y \text{ and } Y \subseteq X.$$

B.3 Cartesian Products

When X and Y are sets, the **cartesian product** of X and Y, denoted $X \times Y$, is defined by
$$X \times Y = \{(x, y) : x \in X \text{ and } y \in Y\}$$
For example, if $X = \{a, b\}$ and $Y = [3]$, then
$$X \times Y = \{(a, 1), (b, 1), (a, 2), (b, 2), (a, 3), (b, 3)\}.$$
Elements of $X \times Y$ are called **ordered pairs**. When $p = (x, y)$ is an ordered pair, the element x is referred to as the **first coordinate** of p while y is the **second coordinate** of p. Note that if either X or Y is the empty set, then $X \times Y = \emptyset$.

Example B.3. Let $X = \{\emptyset, (1, 0), \{\emptyset\}\}$ and $Y = \{(\emptyset, 0)\}$. Is $((1, 0), \emptyset)$ a member of $X \times Y$?

Cartesian products can be defined for more than two factors. When $n \geq 2$ is a positive integer and $X_1, X_2, \ldots, X_n$ are non-empty sets, their **cartesian product** is defined by
$$X_1 \times X_2 \times \cdots \times X_n = \{(x_1, x_2, \ldots, x_n) : x_i \in X_i \text{ for } i = 1, 2, \ldots, n\}$$

B.4 Binary Relations and Functions

A subset $R \subseteq X \times Y$ is called a **binary relation** on $X \times Y$, and a binary relation R on $X \times Y$ is called a **function from X to Y** when for every $x \in X$, there is *exactly one* element $y \in Y$ for which $(x, y) \in R$.

Many authors prefer to write the condition for being a function in two parts:

1. For every $x \in X$, there is *some* element $y \in Y$ for which $(x, y) \in R$.

2. For every $x \in X$, there is *at most one* element $y \in Y$ for which $(x, y) \in R$.

The second condition is often stated in the following alternative form: If $x \in X$, $y_1, y_2 \in Y$ and $(x, y_1), (x, y_2) \in R$, then $y_1 = y_2$.

Example B.4. For example, let $X = [4]$ and $Y = [5]$. Then let
$$\begin{aligned} R_1 &= \{(2, 1), (4, 2), (1, 1), (3, 1)\} \\ R_2 &= \{(4, 2), (1, 5), (3, 2)\} \\ R_3 &= \{(3, 2), (1, 4), (2, 2), (1, 1), (4, 5)\} \end{aligned}$$

Of these relations, only R_1 is a function from X to Y.

Appendix B Background Material for Combinatorics

In many settings (like calculus), it is customary to use letters like f, g and h to denote functions. So let f be a function from a set X to a set Y. In view of the defining properties of functions, for each $x \in X$, there is a unique element $y \in Y$ with $(x, y) \in f$. And in this case, the convention is to write $y = f(x)$. For example, if $f = R_1$ is the function in Example B.4, then $2 = f(4)$ and $f(3) = 1$.

The shorthand notation $f : X \to Y$ is used to indicate that f is a function from the set X to the set Y.

In calculus, we study functions defined by algebraic rules. For example, consider the function f whose rule is $f(x) = 5x^3 - 8x + 7$. This short hand notation means that $X = Y = \mathbb{R}$ and that

$$f = \{(x, 5x^3 - 8x + 7) : x \in \mathbb{R}\}$$

In combinatorics, we sometimes study functions defined algebraically, just like in calculus, but we will frequently describe functions by other kinds of rules. For example, let $f : \mathbb{N} \to \mathbb{N}$ be defined by $f(n) = |n/2|$ if n is even and $f(n) = 3|n| + 1$ when n is odd.

A function $f : X \to Y$ is called an **injection** from X to Y when for every $y \in Y$, there is *at most* one element $x \in X$ with $y = f(x)$.

When the meaning of X and Y is clear, we just say f is an **injection**. An injection is also called a 1–1 function (read this as "one to one") and this is sometimes denoted as $f : X \xrightarrow{1-1} Y$.

A function $f : X \to Y$ is called a **surjection** from X to Y when for every $y \in Y$, there is *at least* one $x \in X$ with $y = f(x)$.

Again, when the meaning of X and Y is clear, we just say f is a **surjection**. A surjection is also called an **onto function** and this is sometimes denoted as $f : X \xrightarrow{onto} Y$.

A function f from X to Y which is both an injection and a surjection is called a **bijection**. Alternatively, a bijection is referred to as a 1–1, onto function, and this is sometimes denoted as $f : X \xrightarrow[onto]{1-1} Y$. A bijection is also called a **1–1-correspondence**.

Example B.5. Let $X = Y = \mathbb{R}$. Then let f, g and h be the functions defined by

1. $f(x) = 3x - 7$.

2. $g(x) = 3(x - 2)(x + 5)(x - 7)$.

3. $h(x) = 6x^2 - 5x + 13$.

Then f is a bijection; g is a surjection but not an injection (*Why?*); and h is neither an injection nor a surjection (*Why?*).

Proposition B.6. *Let X and Y be sets. Then there is a bijection from X to Y if and only if there is a bijection from Y to X.*

B.5 Finite Sets

A set X is said to be **finite** when either (1) $X = \emptyset$; or (2) there exists positive integer n and a bijection $f : [n] \xrightarrow[\text{onto}]{1-1} X$. When X is not finite, it is called **infinite**. For example, $\{a, \emptyset, (3,2), \mathbb{N}\}$ is a finite set as is $\mathbb{N} \times \emptyset$. On the other hand, $\mathbb{N} \times \{\emptyset\}$ is infinite. Of course, $[n]$ and **n** are finite sets for every $n \in \mathbb{N}$.

Proposition B.7. *If X be a non-empty finite set, then there is a unique positive integer n for which there is a bijection $f : [n] \xrightarrow[\text{onto}]{1-1} X$.*

In some cases, it may take some effort to determine whether a set is finite or infinite. Here is a truly classic result.

Proposition B.8. *The set P of primes is infinite.*

Proof. Suppose that the set P of primes is finite. It is non-empty since $2 \in P$. Let n be the unique positive integer for which there exists a bijection $f : [n] \to P$. Then let

$$p = 1 + f(1) \times f(2) \times f(3) \times \cdots \times f(n)$$

Then p is not divisible by any of the primes in P but is larger than any element of P. Thus, either p is prime or there is a prime that does not belong to P. The contradiction completes the proof. $\square$

Here's a famous example of a set where no one knows if the set is finite or not.

Conjecture B.9. *It is conjectured that the following set is infinite:*

$$T = \{n \in \mathbb{N} : n \text{ and } n + 2 \text{ are both primes}\}.$$

This conjecture is known as the *Twin Primes Conjecture*. Guaranteed A + + for any student who can settle it!

Proposition B.10. *Let X and Y be finite sets. If there exists an injection $f : X \xrightarrow{1-1} Y$ and an injection $g : Y \xrightarrow{1-1} X$, then there exists a bijection $h : X \xrightarrow[\text{onto}]{1-1} Y$.*

When X is a finite non-empty set, the **cardinality** of X, denoted $|X|$ is the unique positive integer n for which there is a bijection $f : [n] \xrightarrow[\text{onto}]{1-1} X$. Intuitively, $|X|$ is the number of elements in X. For example,

$$|\{(6,2), (8, (4, \emptyset)), \{3, \{5\}\}\}| = 3.$$

By convention, the cardinality of the empty set is taken to be zero, and we write $|\emptyset| = 0$.

Appendix B Background Material for Combinatorics

Proposition B.11. *If X and Y are finite non-empty sets, then $|X \times Y| = |X| \times |Y|$.*

We note that the statement in Proposition B.11 is an example of "operator overloading", a technique featured in several programming languages. Specifically, the times sign $\times$ is used twice but has different meanings. As part of $X \times Y$, it denotes the cartesian product, while as part of $|X| \times |Y|$, it means ordinary multiplication of positive integers. Programming languages can keep track of the data types of variables and apply the correct interpretation of an operator like $\times$ depending on the variables to which it is applied.

We also have the following general form of Proposition B.11:

$$|X_1 \times X_2 \times \cdots \times X_n| = |X_1| \times |X_2| \times \cdots \times |X_n|$$

Theorem B.12.

1. *There is a bijection between any two of the following infinite sets $\mathbb{N}$, $\mathbb{Z}$ and $\mathbb{Q}$.*

2. *There is an injection from $\mathbb{Q}$ to $\mathbb{R}$.*

3. *There is no surjection from $\mathbb{Q}$ to $\mathbb{R}$.*

B.6 Notation from Set Theory and Logic

In set theory, it is common to deal with statements involving one or more elements from the universe as variables. Here are some examples:

1. For $n \in \mathbb{N}$, $n^2 - 6n + 8 = 0$.

2. For $A \subseteq [100]$, $\{2, 8, 25, 58, 99\} \subseteq A$.

3. For $n \in \mathbb{Z}$, $|n|$ is even.

4. For $x \in \mathbb{R}$, $1 + 1 = 2$.

5. For $m, n \in \mathbb{N}$, $m(m+1) + 2n$ is even.

6. For $n \in \mathbb{N}$, $2n + 1$ is even.

7. For $n \in \mathbb{N}$ and $x \in \mathbb{R}$, $n + x$ is irrational.

These statements may be true for some values of the variables and false for others. The fourth and fifth statements are true for *all* values of the variables, while the sixth is false for all values.

Implications are frequently abbreviated using with a double arrow $\implies$; the quantifier $\forall$ means "for all" (or "for every"); and the quantifier $\exists$ means "there exists" (or "there is"). Some writers use ∧ and ∨ for logical "and" and "or", respectively. For example,

$$\forall A, B \subseteq [4] \quad ((1, 2 \in A) \wedge (|B| \geq 3)) \implies ((A \subseteq B) \vee (\exists n \in A \cup B, n^2 = 16))$$

The double arrow $\iff$ is used to denote logical equivalence of statements (also "if and only if"). For example

$$\forall A \subseteq [7] \quad A \cap \{1, 3, 6\} \neq \emptyset \iff A \not\subseteq \{2, 4, 5, 7\}$$

We will use these notational shortcuts *except* for the use of ∧ and ∨, as we will use these two symbols in another context: binary operators in lattices.

B.7 Formal Development of Number Systems

Up to this point, we have been discussing number systems in an entirely informal manner, assuming everyone knew all that needed to be known. Now let's pause and put things on a more firm foundation. So for the time being, do a memory dump and forget everything you have ever learned about numbers and arithmetic. The set of natural numbers has just been delivered on our door step in a big box with a warning label saying *Assembly Required.* We open the box and find a single piece of paper on which the following "instructions" are printed. These defining properties of the natural numbers are known as the **Peano Postulates**.

i There is a non-empty set of elements called **natural numbers**. There is natural number called **zero** which is denoted 0. The set of all natural numbers is denoted $\mathbb{N}_0$

ii There is a one-to-one function $s : \mathbb{N}_0 \xrightarrow{1\text{-}1} \mathbb{N}_0$ called the **successor function**. For each $n \in \mathbb{N}_0$, $s(n)$ is called the **successor** of n.

iii There is no natural number n for which $0 = s(n)$.

iv Let $\mathbb{M} \subseteq \mathbb{N}_0$. Then $\mathbb{M} = \mathbb{N}_0$ if and only if

  a $0 \in \mathbb{M}$; and

  b $\forall k \in \mathbb{N}_0 \quad (k \in \mathbb{M}) \implies (s(k) \in \mathbb{M})$.

Property Item iv in the list of Peano Postulates is called the **Principle of Mathematical Induction**, or just the **Principle of Induction**. As a first application of the Principle of Induction, we prove the following basic property of the natural numbers.

Proposition B.13. *Let n be a natural number with $n \neq 0$. Then there is a natural number m so that $n = s(m)$.*

Proof. Let $S = \{n \in \mathbb{N}_0 : \exists m \in \mathbb{N}_0, n = s(m)\}$. Then set $M = \{0\} \cup S$. We show that $M = \mathbb{N}_0$. First, note that $0 \in M$. Next, we will show that for all $k \in \mathbb{N}_0$, if $k \in M$, then $s(k) \in M$. However, this is trivial since for all $k \in \mathbb{N}_0$, we have $s(k) \in S \subseteq M$. We conclude that $M = \mathbb{N}_0$. □

B.7.1 Addition as a Binary Operation

A **binary operation** $*$ on set X is just a function $* : X \times X \to X$. So the image of the ordered pair (x, y) would normally be denoted $*((x, y))$. However, this is usually abbreviated as $*(x, y)$ or even more compactly as $x * y$. With this convention, we now define a binary operation $+$ on the set $\mathbb{N}_0$ of natural numbers. This operation is defined as follows for every natural number $n \in \mathbb{N}_0$:

i $n + 0 = n$.

ii For all $k \in \mathbb{N}_0$, $n + s(k) = s(n + k)$.

We pause to make it clear why the preceding two statements define $+$. Let n be an arbitrary natural number. Then let M denote the set of all natural numbers m for which $n + m$ is defined. Note that $0 \in M$ by part (i). Also note that for all $k \in \mathbb{N}_0$, $s(k) \in M$ whenever $k \in M$ by part (ii). This shows that $M = \mathbb{N}_0$. Since n was arbitrary, this allows us to conclude that $n + m$ is defined for all $n, m \in \mathbb{N}_0$.

We read $n + m$ as n **plus** m. The operation $+$ is also called **addition**.

Among the natural numbers, the successor of zero plays a very important role, so important that it deserves its own special symbol. Here we follow tradition and call the natural number $s(0)$ **one** and denote it by 1. Note that for every natural number n, we have $n + 1 = n + s(0) = s(n)$. In particular, $0 + 1 = 1$.

With this notation, the Principle of Induction can be restated in the following form.

Principle B.14 (Principle of Induction). *Let $M \subseteq \mathbb{N}_0$. Then $M = \mathbb{N}_0$ if and only if*

a $0 \in M$; and

b $\forall k \in \mathbb{N}_0 \quad (k \in M) \implies (k + 1 \in M)$.

Theorem B.15 (Associative Property of Addition). $m + (n + p) = (m + n) + p$, *for all $m, n, p \in \mathbb{N}_0$.*

Proof. Let $m, n \in \mathbb{N}_0$. Then let M denote the set of all natural numbers p for which $m + (n + p) = (m + n) + p$. We show that $M = \mathbb{N}_0$.

B.7 Formal Development of Number Systems

Note that
$$m + (n + 0) = m + n = (m + n) + 0$$
which shows that $0 \in M$.

Now assume that $k \in M$, i.e., $m + (n + k) = (m + n) + k$. Then
$$m + [n + (k+1)] = m + [(n+k)+1] = [m+(n+k)]+1 = [(m+n)+k]+1 = (m+n)+(k+1).$$

Notice here that the first, second, and fourth equalities follow from the second part of the definition of addition while the third uses our inductive assumption that $m + (n + k) = (m + n) + k$. This shows that $k + 1 \in M$. Therefore, $M = \mathbb{N}_0$. Since m and n were arbitrary elements of $\mathbb{N}_0$, the theorem follows. □

In proofs to follow, we will trim out some of the wording and leave only the essential mathematical steps intact. In particular, we will (i) omit reference to the set M, and (ii) drop the phrase "For all $k \in \mathbb{N}_0$" For example, to define addition, we will just write (i) $n + 0 = n$, and (ii) $n + (k + 1) = (n + k) + 1$.

Lemma B.16. $m + (n + 1) = (m + 1) + n$, for all $m, n \in \mathbb{N}_0$.

Proof. Fix $m \in \mathbb{N}_0$. Then
$$m + (0 + 1) = m + 1 = (m + 0) + 1.$$
Now assume that $m + (k + 1) = (m + 1) + k$. Then
$$m + [(k+1) + 1] = [m + (k+1)] + 1 = [(m+1) + k] + 1 = (m+1) + (k+1). \quad \square$$

We next prove the commutative property, a task that takes two steps. First, we prove the following special case.

Lemma B.17. $n + 0 = 0 + n = n$, for all $n \in \mathbb{N}_0$.

Proof. The statement is trivially true when $n = 0$. Now suppose that $k + 0 = 0 + k = k$ for some $k \in \mathbb{N}_0$. Then
$$(k+1) + 0 = k + 1 = (0 + k) + 1 = 0 + (k+1). \quad \square$$

Theorem B.18 (Commutative Law of Addition). $m + n = n + m$ for all $m, n \in \mathbb{N}_0$.

Proof. Let $m \in \mathbb{N}_0$. Then $m + 0 = 0 + m$ from the preceding lemma. Assume $m + k = k + m$. Then
$$m + (k+1) = (m+k) + 1 = (k+m) + 1 = k + (m+1) = (k+1) + m. \quad \square$$

Lemma B.19. *If $m, n \in \mathbb{N}_0$ and $m + n = 0$, then $m = n = 0$.*

Proof. Suppose that either of m and n is not zero. Since addition is commutative, we may assume without loss of generality that $n \neq 0$. Then there exists a natural number p so that $n = s(p)$. This implies that $m + n = m + s(p) = s(m + p) = 0$, which is impossible since 0 is not the successor of any natural number. □

Theorem B.20 (Cancellation Law of Addition). *If $m, n, p \in \mathbb{N}_0$ and $m + p = n + p$, then $m = n$.*

Proof. Let $m, n \in \mathbb{N}_0$. Suppose that $m + 0 = n + 0$. Then $m = n$. Now suppose that $m = n$ whenever $m + k = n + k$. If $m + (k + 1) = n + (k + 1)$, then

$$s(m + k) = (m + k) + 1 = m + (k + 1) = n + (k + 1) = (n + k) + 1 = s(n + k).$$

Since s is an injection, this implies $m + k = n + k$. Thus $m = n$. □

B.8 Multiplication as a Binary Operation

We define a binary operation ×, called **multiplication**, on the set of natural numbers. When m and n are natural numbers, $m \times n$ is also called the **product** of m and n, and it sometimes denoted $m * n$ and even more compactly as mn. We will use this last convention in the material to follow. Let $n \in \mathbb{N}_0$. We define

  i $n0 = 0$, and

  ii $n(k + 1) = nk + n$.

Note that $10 = 0$ and $01 = 00 + 0 = 0$. Also, note that $11 = 10 + 1 = 0 + 1 = 1$. More generally, from (ii) and Lemma B.19, we conclude that if $m, n \neq 0$, then $mn \neq 0$.

Theorem B.21 (Left Distributive Law). $m(n + p) = mn + mp$, *for all* $m, n, p \in \mathbb{N}_0$.

Proof. Let $m, n \in \mathbb{N}_0$. Then

$$m(n + 0) = mn = mn + 0 = mn + m0.$$

Now assume $m(n + k) = mn + mk$. Then

$$m[n + (k + 1)] = m[(n + k) + 1] = m(n + k) + m$$
$$= (mn + mk) + m = mn + (mk + m) = mn + m(k + 1). \quad \square$$

Theorem B.22 (Right Distributive Law). $(m + n)p = mp + np$, *for all* $m, n, p \in \mathbb{N}_0$.

Proof. Let $m, n \in \mathbb{N}_0$. Then
$$(m+n)0 = 0 = 0 + 0 = m0 + n0.$$
Now assume $(m+n)k = mk + nk$. Then
$$(m+n)(k+1) = (m+n)k + (m+n) = (mk + nk) + (m+n)$$
$$= (mk + m) + (nk + n) = m(k+1) + n(k+1). \qquad \square$$

Theorem B.23 (Associative Law of Multiplication). $m(np) = (mn)p$, for all $m, n, p \in \mathbb{N}_0$.

Proof. Let $m, n \in \mathbb{N}_0$. Then
$$m(n0) = m0 = 0 = (mn)0.$$
Now assume that $m(nk) = (mn)k$. Then
$$m[n(k+1)] = m(nk+n) = m(nk) + mn = (mn)k + mn = (mn)(k+1). \qquad \square$$

The commutative law requires some preliminary work.

Lemma B.24. $n0 = 0n = 0$, for all $n \in \mathbb{N}_0$.

Proof. The lemma holds trivially when $n = 0$. Assume $k0 = 0k = 0$. Then
$$(k+1)0 = 0 = 0 + 0 = 0k + 0 = 0(k+1). \qquad \square$$

Lemma B.25. $n1 = 1n = n$, for every $n \in \mathbb{N}_0$.

Proof. $01 = 00 + 0 = 0 = 10$. Assume $k1 = 1k = k$. Then
$$(k+1)1 = k1 + 11 = 1k + 1 = 1(k+1). \qquad \square$$

Theorem B.26 (Commutative Law of Multiplication). $mn = nm$, for all $m, n \in \mathbb{N}_0$.

Proof. Let $m \in \mathbb{N}_0$. Then $m0 = 0m$. Assume $mk = km$. Then
$$m(k+1) = mk + m = km + m = km + 1m = (k+1)m. \qquad \square$$

B.9 Exponentiation

We now define a binary operation called **exponentiation** which is defined only on those ordered pairs (m, n) of natural numbers where not both are zero. The notation for exponentiation is non-standard. In books, it is written m^n while the notations $m\text{**}n$, $m \wedge n$ and $\exp(m, n)$ are used in-line. We will use the m^n notation for the most part.

When $m = 0$, we set $0^n = 0$ for all $n \in \mathbb{N}_0$ with $n \neq 0$. Now let $m \neq 0$. We define m^n by (i) $m^0 = 1$ and (ii) $m^{k+1} = mm^k$.

Theorem B.27. *For all $m, n, p \in \mathbb{N}_0$ with $m \neq 0$, $m^{n+p} = m^n \, m^p$.*

Proof. Let $m, n \in \mathbb{N}_0$ with $m \neq 0$. Then $m^{n+0} = m^n = m^n \, 1 = m^n \, m^0$. Now suppose that $m^{n+k} = m^n \, m^k$. Then

$$m^{n+(k+1)} = m^{(n+k)+1} = m \, m^{n+k} = m(m^n \, m^k) = m^n(m \, m^k) = m^n \, m^{k+1}.$$ □

Theorem B.28. *For all $m, n, p \in \mathbb{N}_0$ with $m \neq 0$, $(m^n)^p = m^{np}$.*

Proof. Let $m, n \in \mathbb{N}_0$ with $m \neq 0$. Then $(m^n)^0 = 1 = m^0 = m^{n0}$. Now suppose that $(m^n)^k = m^{nk}$. Then

$$(m^n)^{k+1} = m^n(m^n)^k = m^n(m^{nk}) = m^{n+nk} = m^{n(k+1)}.$$ □

B.10 Partial Orders and Total Orders

A binary relation R on a set X is just a subset of the cartesian product $X \times X$. In discussions of binary relations, the notation $(x, y) \in R$ is sometimes written as xRy.

A binary relation R is:

  i **reflexive** if $(x, x) \in R$ for all $x \in X$.

 ii **antisymmetric** if $x = y$ whenever $(x, y) \in R$ and $(y, x) \in R$, for all $x, y \in X$.

iii **transitive** if $(x, y) \in R$ and $(y, z) \in R$ imply $(x, z) \in R$, for all $x, y, z \in X$.

A binary relation R on a set X is called a **partial order** on X when it is reflexive, antisymmetric, and transitive. Traditionally, symbols like $\leq$ and $\subseteq$ are used to denote partial orders. As an example, recall that if X is a family of sets, we write $A \subseteq B$ when A is a subset of B.

When using the ordered pair notation for binary relations, to indicate that a pair (x, y) is not in the relation, we simply write $(x, y) \notin R$. When using the alternate notation, this is usually denoted by using the negation symbol from logic and writing $\neg(xRy)$. Most of the special symbols used to denote partial orders come with negative versions, e.g., $x \not\leq y$, $x \not\subseteq y$.

A partial order is called a **total order** on X when for all $x, y \in X$, $(x, y) \in R$ or $(y, x) \in R$. For example, if

$$X = \{\emptyset, \{\emptyset\}, \{\emptyset, \{\emptyset\}\}\}$$

then $\subseteq$ is a total order on X.

When $\leq$ is a partial order on a set X, we write $x < y$ when $x \leq y$ and $x \neq y$.

B.11 A Total Order on Natural Numbers

Let $m, n \in \mathbb{N}_0$. Define a binary relation $\leq$ on $\mathbb{N}_0$ by setting $m \leq n$ if and only if there exists a natural number p so that $m + p = n$.

Proposition B.29. $\leq$ *is a total order on* $\mathbb{N}_0$.

Proof. $\leq$ is reflexive since $n + 0 = n$ and therefore $n \leq n$, for all $n \in \mathbb{N}_0$. Next, we show that $\leq$ is antisymmetric. Let $m, n \in \mathbb{N}_0$ and suppose that $m \leq n$ and $n \leq m$. Then there exist natural numbers p and q so that $m + p = n$ and $n + q = m$. It follows that

$$m + (p + q) = (m + p) + q = n + q = m = m + 0$$

Therefore $p + q = 0$, which implies that $p = q = 0$. Thus $m + p = m + 0 = m = n$.

Next, we show that $\leq$ is transitive. Suppose that $m, n, p \in \mathbb{N}_0$, $m \leq n$ and $n \leq p$. Then there exist natural numbers q and r so that $m + q = n$ and $n + r = p$. Then

$$m + (q + r) = (m + q) + r = n + r = p.$$

Thus $m \leq p$, and we have now shown that $\leq$ is a partial order on $\mathbb{N}_0$.

Finally, we show that $\leq$ is a total order. To accomplish this, we choose an arbitrary element $m \in \mathbb{N}_0$ and show that for every $n \in \mathbb{N}_0$, either $m \leq n$ or $n \leq m$. We do this by induction on n. Suppose first that $n = 0$. Since $0 + m = m$, we conclude that $0 \leq m$. Now suppose that for some $k \in \mathbb{N}_0$, we have $m \leq k$. Then there is a natural number p so that $m + p = k$. Then $m + (p + 1) = (m + p) + 1 = k + 1$, so $m \leq k + 1$.

On the other hand, suppose that for some $k \in \mathbb{N}_0$, we have $k \leq m$. If $k = m$, then $m \leq k$ and $m \leq k + 1$ as above. Now suppose that $k \leq m$ and $k \neq m$. Since $k \leq m$, there exists a natural number p so that $k + p = m$. Since $k \neq m$, we know $p \neq 0$. Therefore, there is a natural number q so that $p = q + 1$. Then $m = k + p = k + (q + 1) = (k + 1) + q$ which shows that $k + 1 \leq m$. □

Note that if $m, n \in \mathbb{N}_0$, then $m < n$ if and only if there exists a natural number $p \neq 0$ so that $m + p = n$.

Theorem B.30 (Monotonic Law for Addition). *Let* $m, n, p \in \mathbb{N}_0$. *If* $m \leq n$, *then* $m + p \leq n + p$. *Furthermore, if* $m < n$, *then* $m + p < n + p$.

Proof. It suffices to prove that if $m, n \in \mathbb{N}_0$ with $m < n$, then $m + p < n + p$ for every $p \in \mathbb{N}_0$. Let $q \neq 0$ be the natural number so that $m + q = n$. Now let $p \in \mathbb{N}_0$. Then $(m + p) + q = (m + q) + p = n + p$, so $m + p < n + p$. □

Lemma B.31. *If* $m, n \in \mathbb{N}_0$, $m \neq 0$ *and* $n \neq 0$, *then* $mn \neq 0$.

Appendix B Background Material for Combinatorics

Proof. Assume to the contrary, that $m, n \in \mathbb{N}_0$, $m \neq 0$, $n \neq 0$ and $mn = 0$. Let $n = s(p)$. Then $0 = mn = ms(p) + m$ which requires $m = 0$. This is a contradiction. □

Theorem B.32 (Monotonic Law for Multiplication). *Let $m, n, p \in \mathbb{N}_0$. If $m \leq n$, then $mp \leq np$. Furthermore, if $m < n$ and $p \neq 0$, then $mp < np$.*

Proof. Only the last statement requires proof. Let $m, n \in \mathbb{N}_0$ with $m < n$. Then $m + q = n$ for some $q \neq 0$. Then $np = (m + q)p = mp + pq$. Since $pq \neq 0$, we conclude $mp < np$. □

Corollary B.33 (Cancellation Law of Multiplication). *If $m, n, p \in \mathbb{N}_0$, $mp = np$, and $p \neq 0$, then $m = n$.*

Proof. If $m < n$, then $mp < np$, and if $n < m$, then $np < mp$. We conclude that $m = n$. □

B.12 Notation for Natural Numbers

In some sense, we already have a workable notation for natural numbers. In fact, we really didn't need a special symbol for $s(0)$. The natural number 0 and the sucessor function s are enough. For example, the positive integer associated with the number of fingers (including the thumb) on one hand is $s(s(s(s(s(0)))))$, our net worth is 0, and the age of Professor Trotter's son in years when this section was first written was

$$s(s(s(s(s(s(s(s(s(s(s(s(s(s(s(s(s(s(0)))))))))))))))))).$$

Admittedly, this is not very practical, especially if some day we win the lottery or want to discuss the federal deficit. So it is natural (ugh!) to consider alternative notations.

Here is one such scheme. First, let's decide on a natural $b > s(0)$ as **base**. We will then develop a notation which is called the **base b notation**. We already have a special symbol for zero, namely 0, but we need additional symbols for each natural number n with $0 < n < b$. These symbols are called **digits**. For example, the positive integer $b = s(s(s(s(s(s(s(s(0))))))))$ is called **eight**, and it makes a popular choice as a base. Here are the symbols (digits) customarily chosen for this base: $1 = s(0)$, $2 = s(1)$; $3 = s(2)$; $4 = s(3)$; $5 = s(4)$; $6 = s(5)$; and $7 = s(6)$. Technically speaking, it is not necessary to have a separate symbol for b, but it might be handy regardless. In this case, most people prefer the symbol 8. We like this symbol, unless and until it gets lazy and lays down sideways.

So the first 8 natural numbers are then 0, 1, 2, 3, 4, 5, 6 and 7. To continue with our representation, we want to use the following basic theorem.

B.12 Notation for Natural Numbers

Theorem B.34. *Let $n, d \in \mathbb{N}_0$ with $d > 0$. Then there exist unique natural numbers q and r so that $n = qd + r$ and $0 \le r < d$.*

Proof. Let $d \in \mathbb{N}_0$ with $d > 0$. We first show that for each $n \in \mathbb{N}_0$, there exists $q, r \in \mathbb{N}_0$ so that $n = qd + r$ and $0 \le r < d$. If $n = 0$, we can take $q = 0$ and $r = 0$. Now suppose that $k = qd + r$ and $0 \le r < m$ for some $k \in \mathbb{N}_0$.

Note that $r < d$ implies $r + 1 \le d$. If $r + 1 < d$, then $k + 1 = qd + (r + 1)$. On the other hand, if $r + 1 = d$, then $k + 1 = (q + 1)d + 0$.

Now that existence has been settled, we note that the uniqueness of q and r follow immediately from the cancellation properties. □

Now suppose that for some $k \in \mathbb{N}_0$, with $k \ge 7$, we have defined a base eight notation for the representation of k, for all n with $0 \le n \le k$, and that in each case, this representation consists of a string of digits, written left to right, and selected from $\{0, 1, 2, 3, 4, 5, 6, 7\}$. Write $k + 1 = qb + r$ where $0 \le r < b$. Note that $q \le k$, so that we already have a representation for q. To obtain a representation of $k + 1$, we simply append r at the (right) end.

For example, consider the age of Professor Trotter's son. It is then written as 22. And to emphasize the base eight notation, most people would say 22, base 8 and write $(22)_8$.

Among the more popular bases are base 2, where only the digits 0 and 1 are used, and base sixteen, where sixteen is the popular word for $(20)_8$. Here the digit symbols are

$$0, 1, 2, 3, 4, 5, 6, 7, 8, 9, A, B, C, D, E, F$$

Another popular choice, in fact the one in most widespread use in banks, shopping centers and movie theatres, is base **ten**. Ten is the natural number A, base sixteen. Also, ten is $(12)_8$. Most folks use the digits $0, 1, 2, 3, 4, 5, 6, 7, 8, 9$ for base ten notation. And when no other designation is made, then it is assumed that the natural number is written base ten. So of course, Professor Trotter's son is 18 and is a freshman at Georgia Tech. Which explains why his hair is as white as it is.

For any base $b > 1$, caution must be exercised when discussing multiplication, since writing the product $m \times n$ in the abbreviated form mn causes us some grief. For example, if $b = 8$, then writing the product 372×4885 as 3724885 is ambiguous. For this reason, when using base b notation, the product symbol $\times$ (or some variation of $\times$) is always used.

B.12.1 Alternate Versions of Induction

Many authors prefer to start the development of number systems with the set of **positive integers** and defer the introduction of the concept of zero. In this setting, you have a non-empty set $\mathbb{N}$, a one-to-one **successor** function $s : \mathbb{N} \xrightarrow{1\text{-}1} \mathbb{N}$ and a positive

Appendix B Background Material for Combinatorics

integer called **one** and denoted 1 that is not the successor of any positive integer. The Principle of Induction then becomes: If $M \subseteq \mathbb{N}$, then $M = \mathbb{N}$ if and only if

a $1 \in M$; and

b $\forall k \in \mathbb{N}_0 \ (k \in M) \Longrightarrow (s(k) \in M)$.

More generally, to show that a set M contains all integers greater than or equal to an integer n, it is sufficient to show that (i) $n \in M$, and (ii) For all $k \in \mathbb{Z}$, $(k \in M \Longrightarrow (k+1 \in M)$.

Here is another version of induction, one that is particularly useful in combinatorial arguments.

Theorem B.35. *Let $M \subseteq \mathbb{N}$. If $M \neq \mathbb{N}$, then there is a unique least positive integer n that does not belong to M.*

B.13 Equivalence Relations

A binary relation R is **symmetric** if $(x, y) \in R$ implies $(y, x) \in R$ for all $x, y \in X$.

A binary relation R on a set X is called an **equivalence relation** when it is reflexive, symmetric, and transitive. Typically, symbols like, $=$, $\cong$, $\equiv$ and $\sim$ are used to denote equivalence relations. An equivalence relation, say $\cong$, defines a partition on the set X by setting

$$\langle x \rangle = \{y \in X : x \cong y\}.$$

Note that if $x, y \in X$ and $\langle x \rangle \cap \langle y \rangle \neq \emptyset$, then $\langle x \rangle = \langle y \rangle$. The sets in this partition are called **equivalence classes**.

When using the ordered pair notation for binary relations, to indicate that a pair (x, y) is not in the relation, we simply write $(x, y) \notin R$. When using the alternate notation, this is usually denoted by using the negation symbol from logic and writing $\neg(xRy)$. Many of the special symbols used to denote equivalence relations come with negative versions: $x \neq y$, $x \not\cong y$, $x \not\sim y$, etc.

B.14 The Integers as Equivalence Classes of Ordered Pairs

Define a binary relation $\cong$ on the set $Z = \mathbb{N}_0 \times \mathbb{N}_0$ by

$$(a, b) \cong (c, d) \quad \text{iff} \quad a + d = b + c.$$

Lemma B.36. $\cong$ *is reflexive.*

Proof. Let $(a,b) \in Z$. Then $a + b = b + a$, so $(a,b) \cong (b,a)$. □

Lemma B.37. $\cong$ *is symmetric.*

Proof. Let $(a,b), (c,d) \in Z$ and suppose that $(a,b) \cong (c,d)$. Then $a + d = b + c$, so that $c + b = d + a$. Thus $(c,d) \cong (a,b)$. □

Lemma B.38. $\cong$ *is transitive.*

Proof. Let $(a,b), (c,d), (e,f) \in Z$. Suppose that

$$(a,b) \cong (c,d) \quad \text{and} \quad (c,d) \cong (e,f).$$

Then $a + d = b + c$ and $c + f = d + e$. Therefore,

$$(a + d) + (c + f) = (b + c) + (d + e).$$

It follows that

$$(a + f) + (c + d) = (b + e) + (c + d).$$

Thus $a + f = b + e$ so that $(a,b) \cong (e,f)$. □

Now that we know that $\cong$ is an equivalence relation on Z, we know that $\cong$ partitions Z into equivalence classes. For an element $(a,b) \in Z$, we denote the equivalence class of (a,b) by $\langle (a,b) \rangle$.

Let $\mathbb{Z}$ denote the set of all equivalence classes of Z determined by the equivalence relation $\cong$. The elements of $\mathbb{Z}$ are called **integers**.

B.15 Properties of the Integers

For the remainder of this chapter, most statements will be given without proof. Students are encouraged to fill in the details.

We define a binary operation $+$ on $\mathbb{Z}$ by the following rule:

$$\langle (a,b) \rangle + \langle (c,d) \rangle = \langle (a + c, b + d) \rangle.$$

Note that the definition of addition is made in terms of representatives of the class, so we must pause to make sure that $+$ is **well defined**, i.e., independent of the particular representatives.

Lemma B.39. *If* $\langle (a,b) \rangle = \langle (c,d) \rangle$ *and* $\langle (e,f) \rangle = \langle (g,h) \rangle$, *then* $\langle (a,b) \rangle + \langle (e,f) \rangle = \langle (c,d) \rangle + \langle (g,h) \rangle$.

Appendix B Background Material for Combinatorics

Proof. Since $(a,b) \cong (c,d)$, we know $a + d = b + c$. Since $(e, f) \cong (g, h)$, we know $e + h = f + g$. It follows that $(a + d) + (e + h) = (b + c) + (f + g)$. Thus $(a + e) + (d + h) = (b + f) + (c + g)$, which implies that $\langle(a, b)\rangle + \langle(e, f)\rangle = \langle(c, d)\rangle + \langle(g, h)\rangle$. □

In what follows, we use a single symbol, like x, y or z to denote an integer, but remember that each integer is in fact an entire equivalence class whose elements are ordered pairs of natural numbers.

Theorem B.40. *For all $x, y, z \in \mathbb{Z}$,*

1. $x + y = y + x$;

2. $x + (y + z) = (x + y) + z$; *and*

3. $x + y = x + z$ *implies* $y = z$.

Next, we define a second binary operation called **multiplication**, and denoted $x \times y$, $x * y$ or just xy. When $x = \langle(a,b)\rangle$ and $y = \langle(c,d)\rangle$, we define:

$$xy = \langle(a,b)\rangle\langle(c,d)\rangle = \langle(ac + bd, ad + bc)\rangle.$$

Theorem B.41. *Multiplication is well defined. Furthermore,*

1. $xy = yx$, *for every* $x, y \in \mathbb{Z}$.

2. $x(yz) = (xy)z$, *for every* $x, y, z \in \mathbb{Z}$.

3. $x(y + z) = xy + xz$, *for every* $x, y, z \in \mathbb{Z}$.

The integer $\langle(0,0)\rangle$ has a number of special properties. Note that for all $x \in \mathbb{Z}$, $x + \langle(0,0)\rangle = x$ and $x\langle(0,0)\rangle = \langle(0,0)\rangle$. So most folks call $\langle(0,0)\rangle$ **zero** and denote it by 0. This is a terrible abuse of notation, since we have already used the word zero and the symbol 0 to denote a particular natural number.

But mathematicians, computer scientists and even real people do this all the time. We use the same word and even the same phrase in many different settings expecting that the listener will make the correct interpretation. For example, how many different meanings do you know for *You're so bad?*

If $x = \langle(a,b)\rangle$ is an integer and $y = \langle(b,a)\rangle$, then $x + y = \langle(a + b, a + b)\rangle = 0$. The integer y is then called the **additive inverse** of x and is denoted $-x$. The additive inverse of x is also called **minus** x. The basic property is that $x + (-x) = 0$, for every $x \in \mathbb{Z}$.

We can now define a new binary operation, called **subtraction** and denoted $-$, on $\mathbb{Z}$ by setting $x - y = x + (-y)$. In general, subtraction is neither commutative nor associative. However, we do have the following basic properties.

Theorem B.42. For all $x, y, z \in \mathbb{Z}$,

1. $x(-y) = -xy$;
2. $x(y - z) = xy - xz$; and
3. $-(x - y) = y - x$.

Next, we define a total order on $\mathbb{Z}$ by setting $x \leq y$ in $\mathbb{Z}$ when $x = \langle(a, b)\rangle$, $y = \langle(c, d)\rangle$ and $a + d \leq b + c$ in $\mathbb{N}_0$.

Theorem B.43 (Monotonic Law for Addition). Let $x, y, z \in \mathbb{Z}$. If $x \leq y$, then $x+z \leq y+z$. Furthermore, if $x < y$, then $x + z < y + z$.

For multiplication, the situation is more complicated.

Theorem B.44 (Monotonic Law for Multiplication). Let $x, y, z \in \mathbb{Z}$. If $x < y$, then

1. $xz < yz$, if $z > 0$,
2. $xz = yz = 0$, if $z = 0$, and
3. $xz > yz$, if $z < 0$.

Now consider the function $f : \mathbb{N}_0 \longrightarrow \mathbb{Z}$ defined by $f(n) = \langle(n, 0)\rangle$. It is easy to show that f is an injection. Furthermore, it respects addition and multiplication, i.e., $f(n + m) = f(n) + f(m)$ and $f(nm) = f(n)f(m)$. Also, note that if $x \in \mathbb{Z}$, then $x > 0$ if and only if $x = f(n)$ for some $n \in \mathbb{N}_0$. So, it is customary to abuse notation slightly and say that $\mathbb{N}_0$ is a "subset" of $\mathbb{Z}$. Similarly, we can either consider the set $\mathbb{N}$ of positive integers as the set of natural numbers that are successors, or as the set of integers that are greater than 0.

When n is a positive integer and 0 is the zero in $\mathbb{Z}$, we define $0^n = 0$. When $x \in \mathbb{Z}$, $x \neq 0$ and $n \in \mathbb{N}_0$, we define x^n inductively by (i) $x^0 = 1$ and $x^{k+1} = xx^k$.

Theorem B.45. If $x \in \mathbb{Z}$, $x \neq 0$, and $m, n \in \mathbb{N}_0$, then $x^m x^n = x^{m+m}$ and $(x^m)^n = x^{mn}$.

B.16 Obtaining the Rationals from the Integers

We consider the set $\mathbb{Q}$ of all ordered pairs in $\mathbb{Z} \times \mathbb{Z}$ of the form (x, y) with $y \neq 0$. Elements of $\mathbb{Q}$ are called **rational numbers**, or **fractions**. Define an equivalence relation, denoted =, on $\mathbb{Q}$ by setting $(x, y) = (z, w)$ if and only if $xw = yz$. Here we should point out that the symbol = can be used (and often is) to denote an equivalence relation. It is not constrained to mean "identically the same."

Appendix B Background Material for Combinatorics

When $q = (x, y)$ is a fraction, x is called the **numerator** and y is called the **denominator** of q. Remember that the denominator of a fraction is never zero.

Addition of fractions is defined by

$$(a, b) + (c, d) = (ad + bc, bd),$$

while multiplication is defined by

$$(a, b)(c, d) = (ac, bd).$$

As was the case with integers, it is important to pause and prove that both operations are well defined.

Theorem B.46. *Let $x, y, z, w \in \mathbb{Q}$. If $x = y$ and $z = w$, then $x + z = y + w$ and $xz = yw$.*

Addition and multiplication are both associative and commutative. Also, we have the distributive property.

Theorem B.47. *Let $x, y, z \in \mathbb{Q}$. Then*

1. *$x + y = y + x$ and $xy = yx$.*
2. *$x + (y + z) = (x + y) + z$ and $x(yz) = (xy)z$.*
3. *$x(y + z) = xy + xz$.*

The additive inverse of a fraction (a, b) is just $(-a, b)$. Using this, we define subtraction for fractions: $(a, b) - (c, d) = (a, b) + (-c, d)$.

When (a, b) is a fraction, and $a \neq 0$, the fraction (b, a) is the **reciprocal** of (a, b). The reciprocal is also called the **multiplicative inverse**, and the reciprocal of x is denoted x^{-1}. When $y \neq 0$, we can then define **division** by setting $x/y = xy^{-1}$, i.e., $(a, b)/(c, d) = (ad, bc)$. Of course, division by zero is not defined, a fact that you probably already knew!

As was the case for both $\mathbb{N}_0$ and $\mathbb{Z}$, when n is a positive integer, and 0 is the zero in $\mathbb{Q}$, we define $0^n = 0$. When $x = (a, b)$ is a fraction with $x \neq 0$ and n is a non-negative integer, we define x^n inductively by (i) $x^0 = 1$ and (ii) $x^{n+1} = xx^n$.

Theorem B.48. *If $x \in \mathbb{Q}$, $x \neq 0$, and $m, n \in \mathbb{Z}$, then $x^m x^n = x^{m+n}$ and $(x^m)^n = x^{mn}$.*

Many folks prefer an alternate notation for fractions in which the numerator is written directly over the denominator with a horizontal line between them, so $(2, 5)$ can also be written as $\frac{2}{5}$.

Via the map $g(x) = (x, 1) = \frac{x}{1}$, we again say that the integers are a "subset" of the rationals. As before, note that $g(x + y) = g(x) + g(y)$, $g(x - y) = g(x) - g(y)$ and $g(xy) = g(x)g(y)$.

B.17 Obtaining the Reals from the Rationals

In the third grade, you were probably told that $5 = \frac{5}{1}$, but by now you are realizing that this is not exactly true. Similarly, if you had told your teacher that $\frac{3}{4}$ and $\frac{6}{8}$ weren't really the same and were only "equal" in the broader sense of an equivalence relation defined on a subset of the cartesian product of the integers, you probably would have been sent to the Principal's office.

Try to imagine the trouble you would have gotten into had you insisted that the real meaning of $\frac{1}{2}$ was

$$\frac{1}{2} = \langle(\langle(s(s(0)), s(0))\rangle, \langle(s(s(0)), 0)\rangle)\rangle$$

We can also define a total order on $\mathbb{Q}$. To do this, we assume that $(a, b), (c, d) \in \mathbb{Q}$ have $b, d > 0$. (If $b < 0$, for example, we would replace it by $(a', b') = (-a, -b)$, which is in the same equivalence class as (a, b) and has $b' > 0$.) Then we set $(a, b) \leq (c, d)$ in $\mathbb{Q}$ if $ad \leq bc$ in $\mathbb{Z}$.

B.16.1 Integer Exponents

When n is a positive integer and 0 is the zero in $\mathbb{Q}$, we define $0^n = 0$. When $x \in \mathbb{Q}$, $x \neq 0$ and $n \in \mathbb{N}_0$, we define x^n inductively by (i) $x^0 = 1$ and $x^{k+1} = xx^k$. When $n \in \mathbb{Z}$ and $n < 0$, we set $x^n = 1/x^{-n}$.

Theorem B.49. *If $x \in \mathbb{Q}$, $x \neq 0$, and $m, n \in \mathbb{Z}$, then $x^m x^n = x^{m+n}$ and $(x^m)^n = x^{mn}$.*

B.17 Obtaining the Reals from the Rationals

A full discussion of this would take us far away from a discrete math class, but let's at least provide the basic definitions. A subset $S \subset \mathbb{Q}$ of the rationals is called a **cut** (also, a **Dedekind cut**), if it satisfies the following properties:

1. $\emptyset \neq S \neq \mathbb{Q}$, i.e, S is a proper non-empty subset of $\mathbb{Q}$.

2. $x \in S$ and $y < x$ in $\mathbb{Q}$ implies $y \in S$, for all $x, y \in \mathbb{Q}$.

3. For every $x \in S$, there exists $y \in S$ with $x < y$, i.e., S has no greatest element.

Cuts are also called **real numbers**, so a real number is a particular kind of set of rational numbers. For every rational number q, the set $\bar{q} = \{p \in \mathbb{Q} : p < q\}$ is a cut. Such cuts are called **rational cuts**. Inside the reals, the rational cuts behave just like the rational numbers and via the map $h(q) = \bar{q}$, we abuse notation again (we are getting used to this) and say that the rational numbers are a subset of the real numbers.

But there are cuts which are not rational. Here is one: $\{p \in \mathbb{Q} : p \leq 0\} \cup \{p \in \mathbb{Q} : p^2 < 2\}$. The fact that this cut is not rational depends on the familiar proof that there is no rational q for which $q^2 = 2$.

Appendix B Background Material for Combinatorics

The operation of addition on cuts is defined in the natural way. If S and T are cuts, set $S + T = \{s + t : s \in S, t \in T\}$. Order on cuts is defined in terms of inclusion, i.e., $S < T$ if and only if $S \subsetneq T$. A cut is **positive** if it is greater than $\bar{0}$. When S and T are positive cuts, the product ST is defined by

$$ST = \bar{0} \cup \{st : s \in S, t \in T, s \geq 0, t \geq 0\}.$$

One can easily show that there is a real number r so that $r^2 = \bar{2}$. You may be surprised, but perhaps not, to learn that this real number is denoted $\sqrt{2}$.

There are many other wonders to this story, but enough for one day.

B.18 Obtaining the Complex Numbers from the Reals

By now, the following discussion should be transparent. The **complex number system** $\mathbb{C}$ is just the cartesian product $\mathbb{R} \times \mathbb{R}$ with

1. $(a, b) = (c, d)$ in $\mathbb{C}$ if and only if $a = c$ and $b = d$ in $\mathbb{R}$.

2. $(a, b) + (c, d) = (a + c, b + d)$.

3. $(a, b)(c, d) = (ac - bd, ad + bc)$.

Now the complex numbers of the form $(a, 0)$ behave just like real numbers, so is natural to say that the complex number system **contains** the real number system. Also, note that $(0, 1)^2 = (0, 1)(0, 1) = (-1, 0)$, i.e., the complex number $(0, 1)$ has the property that its square is the complex number behaving like the real number -1. So it is convenient to use a special symbol like i for this very special complex number and note that $i^2 = -1$.

With this beginning, it is straightforward to develop all the familiar properties of the complex number system.

B.18.1 Decimal Representation of Real Numbers

Every real number has a decimal expansion—although the number of digits after the decimal point may be infinite. A rational number $q = m/m$ from $\mathbb{Q}$ has an expansion in which a certain block of digits repeats indefinitely. For example,

$$\frac{2859}{35} = 81.685714285714285714285714285714285714285714285714\ldots$$

In this case, the block 857142 of size 6 is repeated forever.

B.18 Obtaining the Complex Numbers from the Reals

Certain rational numbers have **terminating decimal expansions**. For example, we know that 385/8 = 48.125. If we chose to do so, we could write this instead as an infinite decimal by appending trailing 0's, as a repeating block of size 1:

$$\frac{385}{8} = 48.125000000000000000000000000000000\ldots$$

On the other hand, we can also write the decimal expansion of 385/8 as

$$\frac{385}{8} = 48.124999999999999999999999999999999\ldots$$

Here, we intend that the digit 9, a block of size 1, be repeated forever. Apart from this anomaly, the decimal expansion of real numbers is unique.

On the other hand, irrational numbers have non-repeating decimal expansions in which there is no block of repeating digits that repeats forever.

You know that $\sqrt{2}$ is irrational. Here is the first part of its decimal expansion:

$$\sqrt{2} = 1.41421356237309504880168872420969807856967187537694807317667973\ldots$$

An irrational number is said to be **algebraic** if it is the root of polynomial with integer coefficients; else it is said to be **transcendental**. For example, $\sqrt{2}$ is algebraic since it is the root of the polynomial $x^2 - 2$.

Two other famous examples of irrational numbers are π and e. Here are their decimal expansions:

$$\pi = 3.14159265358979323846264338327950288419716939937510582097494459\ldots$$

and

$$e = 2.71828182845904523536028747135266249775724709369995957496696 76277\ldots$$

Both π and e are transcendental.

Example B.50. Amanda and Bilal, both students at a nearby university, have been studying rational numbers that have large blocks of repeating digits in their decimal expansions. Amanda reports that she has found two positive integers m and n with $n < 500$ for which the decimal expansion of the rational number m/n has a block of 1961 digits which repeats indefinitely. Not to be outdone, Bilal brags that he has found such a pair s and t of positive integers with $t < 300$ for which the decimal expansion of s/t has a block of 7643 digits which repeats indefinitely. Bilal should be (politely) told to do his arithmetic more carefully, as there is no such pair of positive integers (*Why?*). On the other hand, Amanda may in fact be correct—although, if she has done her work with more attention to detail, she would have reported that the decimal expansion of m/n has a smaller block of repeating digits (*Why?*).

Proposition B.51. *There is no surjection from $\mathbb{N}$ to the set $X = \{x \in \mathbb{R} : 0 < x < 1\}$.*

Proof. Let f be a function from $\mathbb{N}$ to X. For each $n \in \mathbb{N}$, consider the decimal expansion(s) of the real number $f(n)$. Then choose a positive integer a_n so that (1) $a_n \leq 8$, and (2) a_n is not the nth digit after the decimal point in any decimal expansion of $f(n)$. Then the real number x whose decimal expansion is $x = .a_1 a_2 a_3 a_4 a_5 \ldots$ is an element of X which is distinct from $f(n)$, for every $n \in \mathbb{N}$. This shows that f is not a surjection. □

B.19 The Zermelo-Fraenkel Axioms of Set Theory

In the first part of this appendix, we put number systems on a firm foundation, but in the process, we used an intuitive understanding of sets. Not surprisingly, this approach is fraught with danger. As was first discovered more than 100 years ago, there are major conceptual hurdles in formulating consistent systems of axioms for set theory. And it is very easy to make statements that sound "obvious" but are not.

Here is one very famous example. Let X and Y be sets and consider the following two statements:

1. There exists an injection $f : X \to Y$.

2. There exists a surjection $g : Y \to X$.

If X and Y are finite sets, these statements are equivalent, and it is perhaps natural to surmise that the same is true when X and Y are infinite. But that is not the case.

Here is the system of axioms popularly known as ZFC, which is an abbreviation for Zermelo-Fraenkel plus the Axiom of Choice. In this system, the notion of **set** and the membership operator $\in$ are undefined. However, if A and B are sets, then exactly one of the following statements is true: (i) $A \in B$ is *true*; (ii) $A \in B$ is *false*. When $A \in B$ is false, we write $A \notin B$. Also, there is an equivalence relation $=$ defined on sets.

Axiom B.52 (Zermelo-Fraenkel Axioms with Axiom of Choice).

Axiom of extensionality *Two sets are equal if and only if they have the same elements.*

Axiom of empty set *There is a set $\emptyset$ with no elements.*

Axiom of pairing *If x and y are sets, then there exists a set containing x and y as its only elements, which we denote by $\{x, y\}$. Note: If $x = y$, then we write only $\{x\}$.*

Axiom of union *For any set x, there is a set y such that the elements of y are precisely the elements of the elements of x.*

Axiom of infinity *There exists a set x such that $\emptyset \in x$ and whenever $y \in x$, so is $\{y, \{y\}\}$.*

B.19 The Zermelo-Fraenkel Axioms of Set Theory

Axiom of power set *Every set has a power set. That is, for any set x, there exists a set y, such that the elements of y are precisely the subsets of x.*

Axiom of regularity *Every non-empty set x contains some element y such that x and y are disjoint sets.*

Axiom of separation (or subset axiom) *Given any set and any proposition $P(x)$, there is a subset of the original set containing precisely those elements x for which $P(x)$ holds.*

Axiom of replacement *Given any set and any mapping, formally defined as a proposition $P(x, y)$ where $P(x, y_1)$ and $P(x, y_2)$ implies $y_1 = y_2$, there is a set containing precisely the images of the original set's elements.*

Axiom of choice *Given any set of mutually exclusive non-empty sets, there exists at least one set that contains exactly one element in common with each of the non-empty sets.*

A good source of additional (free) information on set theory is the collection of Wikipedia articles. Do a web search and look up the following topics and people:

1. Zermelo-Fraenkel set theory.

2. Axiom of Choice.

3. Peano postulates.

4. Georg Cantor, Augustus De Morgan, George Boole, Bertrand Russell and Kurt Gödel.

APPENDIX C

List of Notation

Symbol	Description	Page
$n!$	n factorial	20
$P(m,n)$	number of permutations	20
$\binom{n}{k}$	binomial coefficient	21
$C(n,k)$	binomial coefficient (inline)	21
$\binom{n}{k_1,k_2,k_3,\ldots,k_r}$	multinomial coefficient	30
$\mathcal{P}$	polynomial time problems	65
$\mathcal{NP}$	nondeterministic polynomial time problems	66
$\deg_{\mathbf{G}}(v)$	degree of vertex v in graph $\mathbf{G}$	70
$\mathbf{K}_n$	complete graph on n vertices	70
$\mathbf{I}_n$	independent graph on n vertices	70
$\mathbf{P}_n$	path with n vertices	71
$\mathbf{C}_n$	path with n vertices	71
$\chi(\mathbf{G})$	chromatic number of a graph $\mathbf{G}$	81
$\omega(\mathbf{G})$	clique number of $\mathbf{G}$	84
$x \| y$	x and y are incomparable	118
height($\mathbf{P}$)	height of poset $\mathbf{P}$	119
width($\mathbf{P}$)	width of poset $\mathbf{P}$	119
$D(x), D(S), D[x], D[S]$	down set	123
$U(x), U(S), U[x], U[S]$	up set	123
$\mathbf{n}$	chain with n points	128
$\mathbf{P}+\mathbf{Q}$	disjoint sum of posets	128
$\phi(n)$	Euler ϕ function	149
$\binom{p}{k}$	generalized binomial coefficient	166
$Af(n)$	advancement operator applied to $f(n)$	188

(Continued on next page)

Appendix C List of Notation

Symbol	Description	Page
$P(A\|B)$	probability of A given B	216
$C(X, k)$	family of all k-element subsets of X	229
$R(m, n)$	Ramsey number	230
$\langle C \rangle$	equivalence class of C	298
$\text{stab}_G(C)$	stabilizer of C under action of G	298
$\overline{E}$	complement of event E	326
$x \in X$	x is a member of the set X	333
$x \notin X$	x is not a member of the set X	333
$X \cap Y$	intersection of X and Y	334
$X \cup Y$	union of X and Y	335
$\emptyset$	empty set	336
$\mathbb{N}$	set of positive integers	336
$\mathbb{Z}$	set of integers	336
$\mathbb{Q}$	set of rational numbers	336
$\mathbb{R}$	set of real numbers	336
$\mathbb{N}_0$	set of non-negative integers	336
$[n]$	$\{1, 2, \ldots, n\}$	336
$X \subseteq Y$	X is a subset of Y	336
$X \subsetneq Y$	X is a proper subset of Y	336
$X \times Y$	cartesian product of X and Y	337
$f: X \to Y$	f is a function from X to Y	338
$f: X \xrightarrow{1\text{-}1} Y$	f is an injection from X to Y	338
$f: X \xrightarrow{\text{onto}} Y$	f is a surjection from X to Y	338
$f: X \xrightarrow[\text{onto}]{1\text{-}1} Y$	f is a bijection from X to Y	338
$|X|$	cardinality of set X	339

Index

absorbing
    Markov chain, 321
    state of a Markov chain, 321
addition
    formal definition of, 342
adjacent vertices, 69
algorithm
    on-line, 315
    polynomial time, 65
alphabet, 17
antichain, 118
antisymmetric, 346
arithmetic progression, 325
array, 17
automorphism
    of poset, 119

basis step, 51
Bernoulli trials, 217
big Oh notation, 63
bijection, 338
binomial coefficient, 21, 25, 29
    formula for, 21
    generalized, 166
    recursive formula for, 42
binomial theorem, 29
    Newton's, 166
bit string, *see* string, binary

capacity
    of a cut, 261
    of an edge, 259
cardinality, 339

cartesian product, 337
Catalan number, 27
Cayley's formula, 97
certificate, 62
chain, 118
chain partition, 284
characters, 17
chromatic number, 81, 234
circuit, 76
clique, 84
    maximum size, 84
clique number, 84
Collatz sequence, 9
coloring
    proper, 81
combination, 21
    number of
        formula for, 21
comparable, 118
complex number
    formal definition of, 356
component, 73
    of poset, 120
connected
    poset, 120
conservation law, 259, 260
cover, 115
cut, 261
    Dedekind, 355
cycle, 71
    directed, 245
cycle index, 301

363

INDEX

degree of a vertex, 70
denominator, 354
derangement, 147
digraph, 245
Dijkstra's algorithm, 246
Dilworth's theorem, 122
    dual of, 122
dimension, 139
distance, 245
divides, 45
division theorem, 45
divisor, 45
    common, 45
    greatest common, 45
down set, 123
drawing of a graph, 89
    planar, 89
dual, 120

edge, 69
    directed, 245
    multiple, 75
element, 333
embedding, 119
equivalence classes, 350
Erdős-Ko-Rado Theorem, 319
Euclidean algorithm, 45
Euler ϕ function, 149
Euler's formula, 90
eulerian
    circuit, 76
    trail, 105
event, 215
    dependent, 217
    independent, 217
expectation, 218
expected value, *see* expectation

face, 89
factorial
    definition, 20
    recursive definition, 41
Fibonacci
    numbers, 183
    sequence, 184, 187
flow, 259
    value of, 260
Ford-Fulkerson labeling algorithm, 268
forest, 73, 316
full house (poker hand), 216
function, 337
    injective, 59
    one-to-one, 59, 338
    onto, 338

Gale-Ryser theorem, 325
generating function, 157
    and solving recurrences, 202
    exponential, 170
    ordinary, 170
girth, 108, 234
graph, 69
    2-colorable, 82
    acyclic, 73
    bipartite, 83
        matching in, 280
    comparability, 118, 120
    complete, 70
    connected, 73
    cover, 115, 121
    directed, *see* digraph
    disconnected, 73
    eulerian, 75
    hamiltonian, 79
    incomparability, 118
    independent, 70
    intersection, 87
    interval, 87
    labeled, 307

oriented, 259
perfect, 88
planar, 89
regular, 291
shortest path in, 246
simple, 75
unlabeled, 307
greatest common divisor, *see* divisor, greatest common
Euler ϕ function, 149
ground set, 114
group, 294
permutation, 294
symmetric, 295

hamiltonian
cycle, 79
Hasse diagrams, 116
hat check problem, 147
height, 119
homeomorphic, 92

incident to, 69
incomparable, 118
independent
event, 217
random variables, 222
induction
principle of mathematical, 49, 341
strong, 54
inductive hypothesis, 51
inductive step, 51
injection, 59, 338
input size, 63
integers
formal definition of, 351
positive, 349
intersection, 334
interval order, 128
interval representation, 128
distinguishing, 128
inverse, 294
isomorphism
of graphs, 72
of posets, 119

Kruskal's algorithm, 243

labeling algorithm, 268
lattice, 126
subset, 126
lattice path, 26
counting, 27
number not crossing $y = x$, 27
leaf, 74
length, 245
of arithmetic progression, 325
of path or cycle, 71
letter, 17
linear diophantine equation, 46
linear extension, 125, 134
little oh notation, 64
loop, 75
Lovász Local Lemma, 329
asymmetric, 326
symmetric, 328

Markov chain, 320
matching, 280
maximum, 280
stable, 322
matrix
stochastic, 320
transition, 320
zero–one, 323, 325
maximal
antichain, 123
chain, 123
points of a poset, 123
maximum
antichain, 123

INDEX

chain, 123
mean, *see* expectation
membership, 333
merge sort, 47
minimal
    point of a poset, 122
minimum weight spanning tree
    Kruskal's algorithm for, 243
minimum weight spanning trees
    Prim's algorithm for, 244
multigraph, 75
multinomial coefficient, 30
multinomial theorem, 30
multiplicative inverse, 354

natural numbers, 341
neighbor, 70
neighborhood (of a vertex), 70
network, 259
network flow, 259
nondeterministic polynomial time, 66
notes
    musical, 304
numerator, 354

octave, 304
operation
    binary, 342
operations, 62
operator
    advancement, 188
    linear, 199
or
    exclusive, 335
    inclusive, 335
order
    linear, 116
    partial, 114, 346
    total, 116, 346
        on natural numbers, 347

ordered pairs, 337
outcomes, 215

partially ordered set, 114
partition
    antichain, 317
    dual, 324
    of an integer, 168, 325
path, 71
    augmenting, 264
    directed, 245
pattern inventory, 303
Peano postulates, 341
permutation, 20, 294
    cycle notation for, 295
    function, 143
pigeon hole principle, 59
    generalized, 85
planar drawing, *see* drawing of a
        graph, planar
poset, 114
potential, 268
Prim's algorithm, 244
principle of inclusion-exclusion, 145
probability, 215
    conditional, 216
    measure, 215
    space, 214
proof
    combinatorial, 22
Prüfer code, 97
pseudo-alphabetic order, 267

Ramsey number, 230, 328
    small, 231
    symmetric, 232
Ramsey's theorem, 230, 233, 234
random variable, 218
rational numbers, 353
real number

formal definition of, 355
reciprocal, 354
recurrence equation
    constant coefficients, 187, 189
    general solution, 191
    homogeneous, 187
    linear, 186
    nonconstant coefficients, 187
    nonlinear, 205
    particular solution, 195
recurrence equations
    nonhomogeneous, 194
recursive definition, 42
reflexive, 346
regular
    transition matrix, 321
relation
    binary, 337, 346
    equivalence, 134, 350
    symmetric, 134

sampling
    without replacement, 215
scale, 304
sequence, 17
series
    finite geometric, 158
        sum of, 159
    infinite geometric, 157
        sum of, 157, 158
set, 333
    empty, 336
    finite, 339
    infinite, 339
    Zermelo-Fraenkel axioms, 358
Sigma-notation
    definition of, 41
sink, 259
sorting, 47
source, 259

stabilizer, 298
standard deviation, 222
statement
    open, 49
statements
    meaning of, 40
string, 17
    binary, 18, 22, 185
    column sum, 323
    row sum, 323
    ternary, 18, 171, 185, 191
subdivision
    elementary, 92
subgraph, 70
    induced, 70
    spanning, 70
subposet, 118
subset, 336
    proper, 336
successor, 341
Sudoku puzzle, 15
surjection, 338
symmetric, 350

threshold probability, 236, 237
transitive, 346
transposition
    of a scale, 304
tree, 73
    binary, 205
    ordered, 205
    rooted, 205
    spanning, 73, 239
    unlabeled, 205
trees
    labeled, 96

union, 335
up set, 123

variance, 222

INDEX

vertex, 69

walk, 71

weight, 239
well ordered property, 40
word, 17

This book was authored in PreTeXt. For the LaTeX version, TeX Gyre Pagella was used as the body font with newpxmath used to select the font for mathematical symbols. The LaTeX document class is scrbook from the KOMA-Script package. The HTML version uses the mathbook-4.css color scheme.

Made in the USA
Middletown, DE
10 August 2019